The Isolated Hepatocyte

**Use in Toxicology and
Xenobiotic Biotransformations**

CELL BIOLOGY: A Series of Monographs

EDITORS

D. E. BUETOW
*Department of Physiology
and Biophysics
University of Illinois
Urbana, Illinois*

I. L. CAMERON
*Department of Cellular and
Structural Biology
The University of Texas
Health Science Center at San Antonio
San Antonio, Texas*

G. M. PADILLA
*Department of Physiology
Duke University Medical Center
Durham, North Carolina*

A. M. ZIMMERMAN
*Department of Zoology
University of Toronto
Toronto, Ontario, Canada*

Recently published volumes

Gary L. Whitson (editor). NUCLEAR-CYTOPLASMIC INTERACTIONS IN THE CELL CYCLE, 1980

Danton H. O'Day and Paul A. Horgen (editors). SEXUAL INTERACTIONS IN EUKARYOTIC MICROBES, 1981

Ivan L. Cameron and Thomas B. Pool (editors). THE TRANSFORMED CELL, 1981

Arthur M. Zimmerman and Arthur Forer (editors). MITOSIS/CYTOKINESIS, 1981

Ian R. Brown (editor). MOLECULAR APPROACHES TO NEUROBIOLOGY, 1982

Henry C. Aldrich and John W. Daniel (editors). CELL BIOLOGY OF *PHYSARUM* AND *DIDYMIUM*. Volume I: Organisms, Nucleus, and Cell Cycle, 1982; Volume II: Differentiation, Metabolism, and Methodology, 1982

John A. Heddle (editor). MUTAGENICITY: New Horizons in Genetic Toxicology, 1982

Potu N. Rao, Robert T. Johnson, and Karl Sperling (editors). PREMATURE CHROMOSOME CONDENSA-TION: Application in Basic, Clinical, and Mutation Research, 1982

George M. Padilla and Kenneth S. McCarty, Sr. (editors). GENETIC EXPRESSION IN THE CELL CYCLE, 1982

David S. McDevitt (editor). CELL BIOLOGY OF THE EYE, 1982

P. Michael Conn (editor). CELLULAR REGULATION OF SECRETION AND RELEASE, 1982

Govindjee (editor). PHOTOSYNTHESIS, Volume I: Energy Conversion by Plants and Bacteria, 1982; Volume II: Development, Carbon Metabolism, and Plant Productivity, 1982

John Morrow. EUKARYOTIC CELL GENETICS, 1983

John F. Hartmann (editor). MECHANISM AND CONTROL OF ANIMAL FERTILIZATION, 1983

Gary S. Stein and Janet L. Stein (editors). RECOMBINANT DNA AND CELL PROLIFERATION, 1984

Prasad S. Sunkara (editor). NOVEL APPROACHES TO CANCER CHEMOTHERAPY, 1984

Burr G. Atkinson and David B. Walden (editors). CHANGES IN EUKARYOTIC GENE EXPRESSION IN RESPONSE TO ENVIRONMENTAL STRESS, 1985

Reginald M. Gorczynski (editor). RECEPTORS IN CELLULAR RECOGNITION AND DEVELOPMENTAL PROCESSES, 1986

Govindjee, Jan Amesz, and David Charles Fork (editors). LIGHT EMISSION BY PLANTS AND BACTERIA, 1986

Peter B. Moens (editor). MEIOSIS, 1987

Robert A. Schlegel, Margaret S. Halleck, and Potu N. Rao (editors). MOLECULAR REGULATION OF NUCLEAR EVENTS IN MITOSIS AND MEIOSIS, 1987

Monique C. Braude and Arthur M. Zimmerman (editors). GENETIC AND PERINATAL EFFECTS OF ABUSED SUBSTANCES, 1987

E. J. Rauckman and George M. Padilla (editors). THE ISOLATED HEPATOCYTE: USE IN TOXICOLOGY AND XENOBIOTIC BIOTRANSFORMATIONS, 1987

The Isolated Hepatocyte
Use in Toxicology and Xenobiotic Biotransformations

Edited by

E. J. Rauckman
National Institute of Environmental Health Sciences
Research Triangle Park, North Carolina

George M. Padilla
Department of Physiology
Duke University Medical Center
Durham, North Carolina

1987

ACADEMIC PRESS, INC.
Harcourt Brace Jovanovich, Publishers
Orlando   San Diego   New York   Austin
Boston   London   Sydney   Tokyo   Toronto

ACADEMIC PRESS, INC.
Orlando, Florida 32887

United Kingdom Edition published by
ACADEMIC PRESS INC. (LONDON) LTD.
24–28 Oval Road, London NW1 7DX

Library of Congress Cataloging in Publication Data

The Isolated hepatocyte.

(Cell biology)
Includes index.
1. Liver cells.   2. Toxicity testing.   3. Carcinogenicity testing.   4. Xenobiotics—Metabolism.
5. Biotransformation (Metabolism)   I. Rauckman, E. J.
II. Padilla, George M.   III. Series.   [DNLM:
1. Biotransformation.   2. Cytological Technics.
3. Liver—drug effects.   4. Liver—physiology.
5. Toxicology—methods.   WI 702 I85]
RA1199.4.L57I86   1987   615.9'07   86-32064
ISBN 0–12–582870–5 (alk. paper)

PRINTED IN THE UNITED STATES OF AMERICA

87 88 89 90   9 8 7 6 5 4 3 2 1

Contents

4 Cytochrome *P*-450-Dependent Monooxygenase Systems in Mouse Hepatocytes

Kenneth W. Renton

5 Control of Hepatocyte Proliferation *in Vitro*

Noreen C. Luetteke and George K. Michalopoulos

6 Cytotoxicity Measures: Choices and Methods

Charles A. Tyson and Carol E. Green

7 Metals, Hepatocytes, and Toxicology

Curtis D. Klaassen and Neill H. Stacey

8 The Metabolism and Toxicity of Xenobiotics in a Primary Culture System of Postnatal Rat Hepatocytes

Daniel Acosta, David B. Mitchell, Elsie M. B. Sorensen, and James V. Bruckner

9 The Analysis of Carcinogen-Induced Pleiotropic Drug Resistance in Rat Hepatocytes

Brian I. Carr

10 Measurement of Chemically Induced DNA Repair in Hepatocytes *in Vivo* and *in Vitro* as an Indicator of Carcinogenic Potential

Byron E. Butterworth

11 Genotoxicity Studies with Human Hepatocytes

Stephen C. Strom, David K. Monteith, K. Manoharan,
and Alan Novotny

Preface

In vitro studies using isolated mammalian cells are rapidly becoming an important alternative to whole-animal experimentation. This is a development which merits our serious consideration from a practical and ethical point of view. Moreover, the use of isolated mammalian cells is recognized as a unique experimental approach in which exactly defined experimental conditions are readily set and maintained by an investigator, thus leading to sophisticated biomechanistic studies. With the development of cell dispersion techniques, applicable to a wide variety of tissues and organs, studies requiring multiple types of experimental conditions can be initiated with populations of cells obtained from a single animal. More significantly, cells isolated from human tissues as well as lower animals now permit a direct evaluation of toxicologic studies which were previously based solely on nonhuman, animal experimentation. For this purpose, the isolated hepatocyte, which is rapidly becoming the most useful *in vitro* cellular system for the toxicologist, was chosen as the cellular prototype for this treatise.

This work links research on the isolated hepatocyte to the disciplines of cell culture, toxicology, metabolism, and molecular biology. It is an instructional and reference volume for the design and implementation of current, state-of-the-art procedures. Critical reviews of the literature and evaluation of experimental protocols have been provided to prevent the use of inappropriate techniques by those less familiar with this area of biomedical experimentation. Where necessary, extensive technical detail has been included to provide investigators with a central resource and ready access to methodologies using isolated hepatocytes in the field of toxicology.

E. J. Rauckman
George M. Padilla

The Isolated Hepatocyte

Use in Toxicology and
Xenobiotic Biotransformations

1

Regulation of Liver Growth: Historical Perspectives and Future Directions

NANCY L. R. BUCHER

Department of Pathology
Boston University School of Medicine
Boston, Massachusetts 02118

I. INTRODUCTION

The popularity of the liver as a model for growth regulation is based on its normal state of growth arrest, coupled with a latent capacity for remarkably rapid but controlled proliferation that is readily inducible and quantifiable.

In the rat, the species most widely studied, the adult liver exhibits a high degree of polyploidy, about 70–80% of the parenchymal cells being tetraploid and 1–2% octaploid. Regardless of the ploidy, the cells, as in other species, must replicate their DNA before dividing (Bucher and Malt, 1971). Liver growth occurs in response to a relative liver deficit, which may arise physiologically, as in the case of pregnancy, or through tissue loss by surgery, physical damage, or chemical injury from drugs or toxins. Liver growth is also activated by extreme metabolic and hormonal imbalances. The most dramatic display of proliferative activity follows partial hepatectomy, carried out in the rat in standard fashion by surgical ablation of the two main liver lobes, which comprise 68% of the whole organ with remarkable consistency (Higgins and Anderson, 1931; Russell and Bucher, 1983a).

A brief survey of many years of whole-animal experimentation carried out in an effort to define the mechanisms that control liver growth will serve to characterize features of the growth process for which regulatory mechanisms are now

1

being sought in isolated hepatocytes cultured *in vitro*. Following partial hepatectomy there is a prereplicative lag period during which the hepatocytes, arrested in
the so called G_0 phase, reengage in the cell cycle and progress through the G_1
phase. DNA synthesis, marking the S phase, begins after 14–16 hours, rising
steeply to a peak at 22–24 hr, then subsiding. This course is reflected in a similar
peak of mitosis occurring 6–8 hr later. The regenerative process continues at a
diminishing rate until the original complement of liver tissue is restored. The
multiplicative rate at its peak is comparable for a brief period to that of embryonic cells. The tissue is mostly restored during the first 3 days, and in young
animals the process is completed in about a week (Bucher and Malt, 1971;
Leffert *et al.*, 1979; Bucher and McGowan, 1985).

The primary questions are the following: What mechanisms regulate this remarkable outburst of cell proliferation in an organ that is normally quiescent?
How are the body's needs for increased liver functions (needs which may vary
qualitatively as well as quantitatively) communicated and translated into new
liver production? Conversely, how is the growth process terminated when the
requisite hepatic complement is restored?

II. WHOLE-ANIMAL STUDIES

A. Characterization of the Growth Process
Following Partial Hepatectomy

Determination of DNA replication by [^{3}H]thymidine incorporation is the
means most generally employed to quantify liver growth and serves usefully to
characterize salient features of the growth process. Autoradiographs show that
following partial hepatectomy the parenchymal cells are the first to become
labeled and undergo mitosis in response to a liver loss, the other types of cells
lagging by many hours (Grisham, 1962). Hepatocyte activation *in vivo* is therefore not secondary to proliferative activity on the part of these other cells, and
indeed the converse seems likely.

Excision of lesser amounts of liver than the usual two-thirds causes little or no
difference in the length of the prereplicative lag, but whereas large liver losses
elicit high rates of DNA synthesis, small losses are repaired slowly over a similar
time period despite the liver's large growth potential (Bucher and Swaffield,
1964). This proportional adjustment of the growth rate to the size of the liver
deficit is also reflected by the continually diminishing proliferative rate following
partial hepatectomy as repair of the liver loss progresses toward completion.

B. Growth Signals

The signals by which this adjustment of growth is regulated are blood borne, as evidenced by experiments with liver grafts and transplants (Sigel *et al.*, 1963; Leong *et al.*, 1964) and exchange of blood between rats. For example, when blood was cross-circulated between a normal and a partially hepatectomized rat by means of arteriovenous shunting via polyethylene cannulas, DNA synthesis and mitosis in the liver of the intact partner increased in proportion to the amount of liver removed from the hepatectomized partner. (Moolten and Bucher, 1967; Lieberman, 1969; Sakai, 1970) Continuation of the cross-circulation for 12–14 hours was necessary to stimulate the expected rise in DNA synthesis at 22 hr; shorter periods of blood exchange caused no stimulation. Although a 6-hr exchange was by itself ineffective, if the exchange was resumed after a several hour interruption and continued so that the total elapsed time was 14 hours, a stimultion resulted (Bucher *et al.*, 1969). It is of interest in this regard that parallel behavior is observable in cell cultures. For example, interaction between a mitogenic agent in the medium and the surface receptors of the cells must take place for a number of hours before a significant number of cells are committed to enter S phase. This is a considerably longer time than is required to achieve maximal binding of the growth factor to its receptor (approximately 1 hr) (Carpenter *et al.*, 1982). It is noteworthy that in quiescent cell cultures activation by growth factors also occurs stepwise, first leading to a stage of "competence," and then to actual "progression" through the cell cycle (Stiles *et al.*, 1979).

Within the animal the signals are specific for liver. In partially hepatectomized rats proliferation is prominent only in this organ although small perturbations in other tissues have sometimes been reported.

The metabolic state of the liver cells at the time the signals are received is important. Comparison of *ad libitum*-fed and "meal-fed" rats, which are given access to food for only 4 hr (2–6 AM) daily, has shown that although the animals accustomed to the "meal-fed" regimen gained weight at the normal rate, they exhibited a 2- to 3-fold higher peak rate of DNA synthesis, presumed to be largely due to improved synchrony in mitotic cycling compared to animals fed *ad libitum*. When the final meal was entirely omitted in the meal-fed animals, the peak rate of DNA synthesis was reduced by half (Bucher *et al.*, 1978a). Moreover, when normal intact rats, previously conditioned by 3 days on a protein-free diet, were given an amino acid meal (casein hydrolyzate), a peak of hepatic DNA synthesis appeared that was nearly two-thirds as high as the peak following partial hepatectomy in rats fed *ad libitum*, a highly significant rise. (Leduc, 1949; Short *et al.*, 1974). On the other hand, in animals subjected to a period of fasting, an identical amino acid meal elicited no such response (Bucher *et al.*, 1978a).

C. Hormones and Growth Factors

The question arises whether nutritional influences on liver growth are due to direct effects of the nutrients themselves upon hepatocyte metabolism or are secondary to diet induced-alterations in the hormonal environment, or both. Pancreatic hormones in particular are well known to be highly responsive to nutrient availability. Following partial hepatectomy, determination of insulin and glucagon concentrations in the portal venous blood showed that glucagon rose substantially, whereas insulin fell somewhat. Glucagon increased to a similar extent after either protein depletion or fasting when followed by amino-acid feeding, although, as noted, in the fasted animals DNA synthesis was not stimulated (Bucher et al., 1978a). Therefore, the elevated concentration of glucagon received by the liver following partial hepatectomy was not in itself sufficient to induce liver growth. Although insulin levels were generally lower than normal under conditions in which the rate of liver regeneration was high, insulin was nevertheless essential. Depression of the already low insulin concentrations present in partially hepatectomized animals by administration of anti-insulin serum significantly reduced DNA synthesis (Bucher et al., 1978b). That lower than normal insulin concentrations could support even the highest rates of liver growth (Bucher et al., 1978a) was consonant with an earlier observation that livers regenerated at a substantial rate in rats made severely diabetic with alloxan (Younger et al., 1966).

Further probing into the possible influence upon liver growth of glucagon and insulin, and of enteric hormones as well, led to an examination of hepatic DNA synthesis in splanchnic eviscerated, partially hepatectomized rats, i.e., animals from which the entire gastrointestinal tract, pancreas, spleen, and main lobes of the liver were extirpated. In these animals, a highly significant rise in DNA synthesis occurred compared to nonhepatectomized-eviscerated controls, although it was delayed and greatly diminished compared to partially hepatectomized but otherwise normal animals. This suggested that hepatic growth-initiating substances need not all derive from portal splanchnic organs. The diminished DNA synthetic rate in the eviscerated animals was restored to normal by simultaneous infusion of insulin and glucagon, but not by either hormone alone. It should be emphasized, however, that in this instance the synergistic interactions of insulin and glucagon were demonstrated in animals that were under duress; the treatment did not substantially alter the rate of DNA synthesis in normal animals, either intact or partially hepatectomized (Bucher and Swaffield, 1973, 1975). Along similar lines, in a collaborative study with investigators working with the A59 mouse hepatitis virus, we showed that insulin and glucagon, together but not separately, significantly increased survival of mice with fulminant hepatitis due to this virus which was fatal in all untreated animals under the same conditions (Farivar et al., 1976).

Taken together, these several studies implicate certain hormones, stressful conditions, and the nutritional state of the animal as important growth-regulating influences for liver cells.

Epidermal growth factor (EGF), a polypeptide first isolated from the salivary gland of the male mouse and similar in molecular weight to insulin, is the mouse equivalent of the human hormone urogastrone. It is potently mitogenic for a wide variety of cells. When infused continuously into the portal vein or peritoneal cavity of normal rats, EGF slightly, but significantly, stimulated hepatic DNA synthesis and mitosis. This action was greatly augmented by insulin. In these normal animals the combination of EGF and insulin caused the liver cells to proliferate even though the full complement of hepatocytes was already present (Bucher et al., 1978b).

Glucagon also enhanced the mitogenic action of EGF, but only in pharmacologic doses, as was also the case in the experiments with eviscerated rats and A59 virus-infected mice mentioned previously. On the other hand, in combination with insulin and EGF, glucagon appeared to have an inhibitory influence. (Bucher et al., 1978c).

In experiments in which the combination of insulin and EGF were infused intraperitonally, the stimulus appeared to be specific for the liver since no rise in DNA synthesis was observed in a number of other organs (Bucher et al., 1978c). This finding, however, may be of doubtful significance because of the infusion route employed. As EGF is very rapidly cleared from the portal venous blood by the liver, the amounts reaching the systemic circulation, and hence other tissues, may have been very small (St. Hilaire et al., 1983).

The experiments just described, involving insulin, glucagon, and EGF, underscore the importance of synergistic interactions among these hormonal agents in the regulation of liver growth.

The pituitary hormone vasopressin is a small peptide that, like glucagon, influences hepatocyte carbohydrate and lipid metabolism and can modulate liver cell proliferation as well. The Brattleboro rat, a mutant derived from the Long–Evans strain, completely lacks this hormone. Following partial hepatectomy in these animals, both the rate of DNA synthesis and the accumulation of DNA in the growing liver remnant (indicative of the increase in liver cell number) were found to be impaired (Russell and Bucher, 1983a). The impairment could be at least partially overcome by continuous infusion of vasopressin in doses that approximately corrected the diabetes insipidus, which is the hallmark of these animals.

Both vasopressin and glucagon modulate hepatocyte DNA synthesis, and both also stimulate glycogenolysis, inhibit glycogen synthesis, and promote gluconeogenesis (Hems and Whitton, 1973). In the liver, glucagon action is mediated through the intracellular messenger cyclic AMP, whereas vasopressin functions through the messengers generated by breakdown of phosphatidylinositols, mobilization of intracellular calcium, and activation of protein kinase C (Creba et al., 1983).*

Various actions of the catecholamines resemble those of glucagon and vasopressin, signal transduction in the liver being by way of the calcium and phosphatidylinositol mechanism for the α_1-adrenergic receptors and through cyclic AMP for the β receptors. It has been suggested that the catecholamines may be modulators of hepatocyte proliferation in vivo (Hasegawa and Koga, 1977; Morley, 1981), and this is supported by recent studies in cultures (Friedman et al., 1981; Michalopoulos et al., 1982; Cruise et al., 1985).

A number of additional agents can be mentioned that may influence liver growth, including growth hormone, somatomedin C, triiodothyronine, parathryoid hormone, calcitonin, prostaglandins, inhibitors or ''chalones,'' and several partially characterized substances from serum.

The preceding brief overview is intended to provide historical perspective. I have described experiments drawn largely from our own work to illustrate salient features of hepatic growth regulation that have emerged from years of wholeanimal research. The studies point to synergistic interactions among hormonal agents and nutrients as the probable growth signals. The multiplicity of agents seems confusing, in view of the direct hormone/target-organ feedback mechanisms evident in various other systems. That the liver should exhibit greater complexity may not be surprising, however, when the array of its functions is considered in relation to the limited activities of other organs. Multiple, variable signals may be necessary for inducing responses to so wide a range of functional demands.

III. HEPATOCYTE CULTURES

Modes of action of hormones and growth factors at the molecular level are difficult to explore in whole-animal experiments where it is often uncertain whether administered substances influence liver growth by direct action on the hepatocytes or through indirect ancillary mechanisms: nutrients alter hormone levels, hormones alter transport and metabolism of nutrients, and hormones influence actions of other hormones in numerous ways. Isolated hepatocytes in culture permit dissection and probing into these interactions in ways impossible in vivo. The availability of readily workable techniques for isolation and culture of adult hepatocytes and the demonstration that these cells in synthetic media in the absence of serum can be stimulated to proliferate by purified hormones and growth factors that are also effective in vivo (Hasegawa et al., 1982; McGowan and Bucher, 1983a,b) make this an attractive approach for in depth investigation of the molecular mechanisms governing liver growth.

*A more recent report shows that glucagon operates in the liver through both the cyclic AMP and the phosphatidylinositol systems (Wakilan, M. J. O., et al., 1986, Nature 323, 68–71).

A. Methods

The technique for isolation of hepatocytes in high yield and virtually free of other cell types by perfusion of the liver with collagenase is now widely used (Berry and Friend, 1969; Seglen, 1976). The procedure has been described in detail and appears in many publications, each laboratory having adopted its own modifications. Basically, it consists of perfusion of the liver *in situ* in the anesthetized animal (usually an adult rat), first with calcium-free buffer, then with calcium-containing buffer to which collagenase has been added. The liver is dissected free during or after the perfusion, and its capsule gently peeled back to permit release of the hepatocytes by gentle agitation into buffer or culture medium. The residue, consisting mainly of connective tissue and biliary and vascular structures, is discarded. The hepatocyte suspension is strained through nylon mesh, and the cells are then washed by one or more cycles of sedimentation by gravity or very low-speed centrifugation, followed by resuspension in culture media. After determination of cell number and viability by trypan blue dye exclusion, aliquots of the suspension are cultured as monolayers. Modifications of the procedure, introduced to accommodate requirements of individual investigators, appear to be nearly as numerous as the laboratories engaged in this research. Besides numerous variations in the cell isolation and washing procedures, an assortment of media with and without different sera are employed, and specific substances are added to preserve or promote special functions. For example, to preserve cytochrome P-450, important in the function of the oxidative drug-metabolizing enzymes, the composition of the culture medium is altered in various ways (Allen *et al.,* 1981; Nelson and Acosta, 1982); to promote longevity, hepatocytes are cultured on a substratum of "biomatrix" prepared from homogenates of whole-rat liver (Reid *et al.,* 1980) or are cocultured with liver-derived diploid epithelial cells that are not hepatocytes (Clement *et al.,* 1984) or, most recently, in medium containing 2% dimethyl sulfoxide (DMSO) (Isom *et al.,* 1985).

In my own laboratory, which focuses on growth regulation, modifications have been introduced that promote DNA replication and mitosis. Currently we perfuse the liver with Krebs-Ringer bicarbonate buffer containing glucose (5.5 mM) and HEPES (20 mM), first without calcium, then with calcium (5 mM) and collagenase (0.05%). The hepatocytes are released from the perfused liver into modified Waymouth's MAB 87/3 medium. Modifications are the omission of glutamine, thymidine, and insulin; reduction of glucose to 5 mM; addition of hydrocortisone (2 μg/ml) and gentamycin (60 μg/ml); and replacement of arginine with ornithine. During cell isolation, 5% bovine serum, insulin (200 mU/ml), and glucose (27 mM) are included. Cells are plated at a density of 35,000 per cm^2 in 35-mm collagen-coated Lux dishes in 1.5 ml of medium and allowed 60 min for attachment. The medium is then replaced with modified

Waymouth's medium without serum, insulin, and extra glucose, and this medium is renewed daily, so that after the first medium change the cells are exposed only to purified chemicals and hormones. The serum, which is present only during the period of cell isolation and plating, is intended primarily to counteract proteolytic enzymes. We have shown that although the serum treatment appears to improve the general condition of the cells, its complete omission does not alter the experimental outcome in terms of DNA synthesis (McGowan *et al.*, 1981). DNA synthesis as determined by incorporation of [^{3}H]thymidine is usually maximal at 2–3 days. The present technique is detailed in several publications. We are continually exploring further modifications (McGowan *et al.*, 1981; McGowan and Bucher, 1983b, 1985). Variables in the procedure that may affect hepatocyte performance are thoughtfully considered in a review by McGowan (1985) which deals with the liver perfusion procedure, composition of media, choice of substrata, cell plating density, and supplemental nutrients, hormones, and growth factors.

B. Effects of Hormones, Growth Factors, and Nutrients on Hepatocyte Cultures

To study growth regulation *in vitro* it was first necessary to establish that cultures of adult rat hepatocytes, prepared as described and maintained in an artifical medium containing only purified chemicals, were responsive to hormones and growth factors known to influence hepatocyte proliferation *in vivo*. We employed short-term cultures so that the behavior of the cells *in vivo* would be reflected *in vitro* insofar as possible under the rigorous conditions of an artificial environment. It has been found that, upon placement in culture, hepatocytes respond poorly to growth factors and certain liver-specific functions decline, associated with a rapid loss of liver-specific mRNAs (Clayton and Darnell, 1983). Recovery occurs somewhat variably during the first day or two depending on experimental conditions and the functions being studied (Bissell, 1980), but within a few days thereafter a process described as "dedifferentiation" or "retrodifferentiation" is seen, manifested by reversion towards a fetal phenotype (Leffert *et al.*, 1978; Sirica *et al.*, 1979; Guguen-Guillouzo and Guillouzo, 1983). We have, therefore, until now confined our studies to short-term cultures. Restricted retrodifferentiation, however, occurs normally during hepatic regeneration *in vivo* (Uriel, 1976), and the manifestation of phenotypic reversion observed in culture probably reflects transition toward a proliferative state (Leffert *et al.*, 1978; Sirica *et al.*, 1979). It has been proposed that the ability to respond to specific hormones that rapidly modulate metabolic activity or promote their growth and development may be the best indicator of the preservation of differentiated functions (Wolffe and Tata, 1984).

Under the conditions that we employ, DNA replication, determined both analytically and autoradiographically following incorporation of [^{3}H]thymidine, is low but consistently higher than occurs within the liver *in vivo*. During the process of isolation and placement in culture, the hepatocytes are subjected to stressful conditions combined with hormone and nutrient down and up shifts, procedures which can activate hepatocyte proliferation *in vivo* as mentioned earlier. The cells are subjected to drastically unphysiological conditions including mechanical trauma, hypoxia, temperature shifts, and perturbation of the cell surface by proteases, so it may not seem surprising that, when freshly isolated, the hepatocytes may display a rounded shape and preferential synthesis of "heat-shock" proteins which gradually disappear as the cells flatten and form aggregates over a period of several days (Wolffe and Tata, 1984; Wolffe *et al.*, 1984). These studies on heat-shock proteins were carried out on *Xenopus* hepatocytes at 20°C, in which the time course of recovery is probably slower than in the rat where cell flattening and spreading begins within a few hours after the plating, and where responsiveness to growth-promoting hormones (insulin and EGF) is only impaired for 1 day or less (McGowan *et al.*, 1981). It remains to be determined whether the low level of DNA synthesis, involving relatively few hepatocytes seen under "baseline" conditions, is related to the "stressful" conditions mentioned or to other influences. We mention these aspects of the culture system here, as the initial status of the cells merits consideration when toxicity studies are to be undertaken.

At the time we began our studies, most investigators agreed that despite the latent capacity of G_0-arrested adult hepatic parenchymal cells to proliferate *in vivo*, they survived *in vitro* for only limited periods and exhibited little evidence of proliferative activity (Laishes and Williams, 1976; Sirica *et al.*, 1979; Ichihara *et al.*, 1980). Reports to the contrary were thought to be based upon outgrowth of a relatively undifferentiated type of liver epithelial cell of uncertain origin that can actively multiply and demonstrate limited hepatocytic functions in culture (Clement *et al.*, 1984; Tsao *et al.*, 1984). Friedman and his associates soon showed, however, that adult hepatocytes in cultures containing serum could be stimulated to synthesize DNA by EGF, whose action was augmented by insulin and glucagon, consonant with the *in vivo* studies described earlier. Although these growth-promoting substances interacted synergistically to cause a considerable increase in [^{3}H]thymidine incorporation compared to control cultures, mitotic activity was minimal (Richman *et al.*, 1976; Friedman *et al.*, 1981). Subsequent studies with insulin and EGF under more rigorous, serum-free conditions were in agreement and are illustrated here by the open bars in Fig. 1 (McGowan *et al.*, 1981, 1984).

Whether or not EGF, which is a wide spectrum mitogen, has a role in liver regeneration is uncertain; it may be unlikely because in partially hepatectomized animals only the liver is induced to proliferate. On the other hand, the possibility

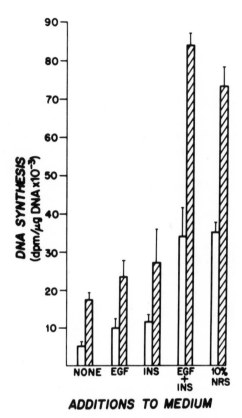

Fig. 1. Stimulation of DNA synthesis in hepatocyte cultures by hormones, growth factors and pyruvate. EGF concentration was 10 ng/ml, insulin 20 mU /ml, dialyzed normal rat serum (NRS) 10%, and pyruvate (20 mM). [³H]Thymidine was added at 48 hr and cells were harvested at 72 hr after plating. Open bars, no pyruvate; hatched bars, plus pyruvate. Bars are mean values for triplicate dishes; vertical lines show standard deviation. [Reproduced from McGowan *et al.* (1984), with permission.]

that EGF may be involved is supported by its avid uptake from the bloodstream by the liver (St. Hilaire *et al.*, 1983), by its stimulatory action upon hepatocytes both *in vivo* and *in vitro,* and possibly also by the down-regulation of EGF receptors and the parallel decrease in EGF-dependent EGF receptor tyrosine phosphorylation on plasma membranes isolated from regenerating livers (Earp and O'Keefe, 1981; Rubin *et al.*, 1982).

Efforts to maximize the growth response to EGF and insulin in serum-free hepatocyte cultures led to the finding that high (20–60 mM) concentrations of pyruvate caused striking enhancement of growth factor-stimulated DNA synthesis, regardless of whether the factors were defined (insulin and EGF) or

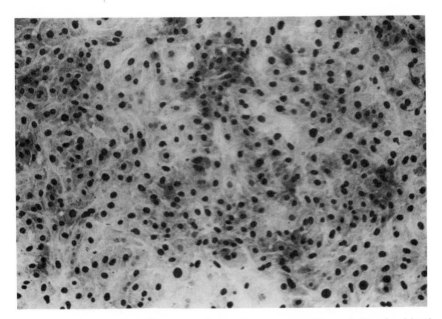

Fig. 2.   Replicate cultures of hepatocytes isolated from the same rat liver as in Figs. 3 and 4 and cultured in modified Waymouth's medium supplemented with insulin (20 mU/ml), EGF (100 ng/ml), and sodium pyruvate (20 mM). Culture labeled with [³H]thymidine (1 μCi/ml) from 24 to 72 hr after cell plating and fixed at 72 hr to show number of replicating cells. Autoradiograph, giemsa stain (×125).

undefined (normal rat serum) (Fig. 1, hatched bars) (McGowan and Bucher, 1981). Moreover, mitotic figures appeared in abundance, whether or not serum was present (Figs. 2, 3, and 4) (McGowan and Bucher, 1983a,b). Koga and associates (Hasegawa *et al.*, 1980; Hasegawa and Koga, 1981) obtained similar results, noting enhancement of DNA synthesis by high concentration of pyruvate in medium supplemented with insulin and glucagon (with fibrinogen digest, without serum or EGF). Moreover, they induced mitosis under similar conditions by reducing the usual 1 mM calcium level to 0.1 mM (Hasegawa *et al.*, 1982). These observations illustrate the considerable influence on hepatocyte behavior of variations in nutrient and electrolyte concentrations.

Lactate equaled pyruvate in potency, indicating no preferential utilization of reduced over oxidized form of substrate. These two substances regularly surpassed a number of other intermediates of carbohydrate metabolism which were also maximally effective in the 20–50 mM concentration range, except for two (α-ketobutyrate and oxaloacetate), which were among the least potent and functioned best at somewhat lesser concentrations (McGowan and Bucher, 1983b). As in general these substances do not readily enter cells (Mapes and Harris, 1975), the supraphysiologic concentrations found to be optimal may be neces-

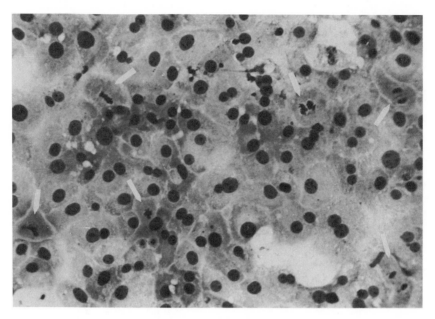

Fig. 3. Replicate culture of hepatocytes isolated from the same rat liver and cultured under the same conditions as in Fig. 2, except without [^{3}H]thymidine. Culture fixed at 72 hr and stained with giemsa to show mitotic activity (white arrows) ($\times$250).

sary to obtain adequate levels inside the cell or an intracellular compartment. In any case, the observations underscore the importance of the metabolic state of the cells, as previously implied in the *in vivo* experiments.

In further correlation with whole-animal studies, vasopressin was found substantially to augment the stimulation afforded by insulin and EGF or by normal rat serum, especially when these growth factors were present at low (near threshold) concentrations (Russell and Bucher, 1983b). The possible implication is that, under physiological conditions within the animal, small shifts in balance among synergistically interacting growth signals could greatly amplify a subliminal proliferative stimulus.

Glucagon, some of whose actions resemble those of vasopressin on hepatocyte carbohydrate metabolism and whose somewhat inconsistent effects in whole-animal experiments were noted earlier, has also yielded divergent results *in vitro* (Friedman *et al.*, 1981; Tomita *et al.*, 1981; Bucher *et al.*, 1983; Bronstad *et al.*, 1983). Although insulin and EGF were consistently stimulatory, glucagon sometimes enhanced this action, sometimes had little effect, and was sometimes inhibitory. Its actions were reproduced by substances that elevated intracellular concentrations of cyclic AMP (Hasegawa *et al.*, 1980; McGowan *et al.*, 1981; Friedman *et al.*, 1981). Although the stimulatory or inhibitory effects might be

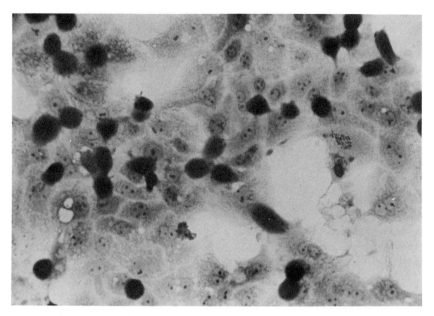

Fig. 4.   Replicate culture of hepatocytes isolated from the same rat liver and cultured as in Fig. 3. Colcemid ($10^{-5} M$) present in medium for the final 18 hr. Fixed at 72 hr. Darkly stained, rounded cells are blocked in mitosis by the colcemid, indicative of the extent of mitotic activity. Giemsa stain ($\times 250$).

partly due to variations in cell density and glucagon concentration (Bronstad *et al.*, 1983), our present impression is that glucagon may influence hepatocyte proliferation positively or negatively depending upon the metabolic state of the cells.

Catecholamines have effects upon the liver resembling those of vasopressin and glucagon and may also be modulators of liver regeneration *in vivo* (Hasegawa and Koga, 1977; Morley, 1981). In hepatocyte cultures, DNA synthesis was enhanced by norepinephrine acting through the α_1-adrenergic receptors (Friedman *et al.*, 1981; Cruise *et al.*, 1985). With time in culture, the number of α_1-type receptors, which is high *in vivo,* declined while the β type increased in reciprocal fashion (Nakamura *et al.*, 1983, 1984c).

The foregoing observations point to possible involvement of a number of substances in regulation of liver growth. They cast little light, however, upon a major facet of the problem, which is that of the specificity of the physiological growth stimulus for hepatocytes, since they implicate substances that are mitogenic for a wide variety of cells. The likely importance of synergy has been noted, and reports of interactions among growth-promoting agents that can influence binding of growth promoters to their receptors may provide clues as to possible mechanisms for restricting the stimulus to hepatocytes. It has been found, for example, that EGF binding to its own receptor can be modulated by other growth factors; platelet derived growth factor (PDGF) induced a transient down-regulation of the EGF receptor, at least in fibroblasts (Fox *et al.*, 1982), whereas vasopressin inhibited EGF binding by lowering the affinity of the receptor for its ligand (Rozengurt *et al.*, 1981). Moreover, the sensitivity of various fibroblast cell lines to EGF stimulation was related to the intracellular concentration of cyclic AMP (Olashaw *et al.*, 1984). In hepatocytes, glucocorticoids promoted a rise in EGF binding to its cell surface receptors largely through increasing the number of receptors, but when insulin and the glucocortocoid were added simultaneously, insulin inhibited the increase (Lin *et al.*, 1984). Information of this kind is accumulating rapidly. What it suggests are possible means through which specificity could be conferred by key combinations of factors interacting in these or other ways mutually to reinforce growth signals. Mechanisms in addition to those mentioned undoubtedly may participate. Conceivably more than one set of signals could set the growth process in motion, depending upon the metabolic state of the cells.

Alternatively, there are undefined growth promoting substances in serum that may determine the specificity and/or amplitude of the growth signals. Extensively dialyzed normal rat serum, which in the concentrations employed would contain negligible amounts of insulin or EGF, exhibited a potency similar to these two substance in stimulating hepatocyte DNA synthesis, indicating that larger molecules are also stimulatory. Rat serum exceeded sera from five other species (including fetal bovine serum) by 2-fold or more (Strain *et al.*, 1982). At least half of this activity resided in a polypeptide fraction derived from the platelets which differed in a number of respects from the well-characterized platelet-derived growth factor (PDGF) from human platelets (Russell *et al.*, 1984a,b; Paul and Piasecki, 1984). Other polypeptide growth factors that stimulate cultured rat hepatocytes have also been at least partially purified from rat serum and appear to be increased during hepatic regeneration (Michalopoulos *et al.*, 1982, 1984; Nakamura *et al.*, 1984a; Thaler and Michalopoulos, 1985). There are divergencies among the findings of the several laboratories studying these serum factors, probably because each one collects, stores, and assays the serum fractions under quite different experimental conditions. This underscores the importance of attention to technical details and experimental variables.

An additional aspect of hepatic growth regulation that has been less extensively studied involves growth inhibitors (Iype and McMahon, 1984; Nakamura *et al.*, 1984b). Whether these will eventually emerge as having an important role as counterbalances to growth promoters or in some other fashion is speculative at the present time.

IV. FUTURE DIRECTIONS

Certain substances that modulate proliferation in cultured hepatocytes appear to act similarly upon hepatocytes in the whole animal. Whether these or other substances will eventually prove to be true physiological regulators with hepatocyte specificity will not be ascertained with finality until the molecular mechanisms by which they implement and modulate the proliferative process are resolved. This is due to deficiencies inherent in the assays; in whole-animal experiments, growth regulatory substances cannot be tested singly nor their direct effects upon hepatocytes readily evaluated, making positive identification difficult. On the other hand, in cell cultures putative growth regulators may be merely fulfilling needs peculiar to the artifical conditions employed and may entirely lack physiological relevance.

Future directions require a coming to grips with these issues. This is evident from the tentative nature of much that has been presented in this brief overview. While many aspects of the studies so far are valuable, solid information at the molecular level is the ultimate key to an understanding of hepatic growth regulation.

ACKNOWLEDGMENT

These studies were supported for a number of years by grants from the American Cancer Society, the Damon Runyon Memorial Fund for Cancer Research, and later by continuing grants from the National Institutes of Health: CA-02146, CA-39099, AM-19435, and AM-33347.

REFERENCES

Allen, C. M., Hockin, L. J., and Paine, A. J. (1981). The control of glutathione and cytochrome P-450 concentrations. *Biochem. Pharmacol.* **30,** 2739–2742.

Berry, M. N., and Friend, D. S. (1969). High yield preparation of isolated rat liver parenchymal cells. A biochemical and fine structural study. *J. Cell Biol.* **43,** 506–520.

Bissell, D. M. (1980). Phenotypic stability of adult rat hepatocytes in primary monolayer cultures. *Ann. N.Y. Acad. Sci.* **349,** 85–98.

Bronstad, G. O., Sand, T. E., and Christoffersen, T. (1983). Bidirectional concentration-dependent effects of glucagon and dibutyryl cyclic-AMP on DNA synthesis in cultured adult rat hepatocytes. *Biochim. Biophys. Acta* **763,** 58–63.

Bucher, N. L. R., and McGowan, J. A. (1985). Regulatory mechanisms in hepatic regeneration. *In* "Liver and Biliary Disease: A Pathophysiological Approach" (R. Wright, K. M. G. G. Alberti, S. Karran, and H. Millward-Sadler, eds.), pp. 251–265. Saunders, London.

Bucher, N. L. R., and Malt, R. A. (1971). "Regeneration of Liver and Kidney." Little, Brown, Boston, Massachusetts.

Bucher, N. L. R., and Swaffield, M. N. (1964). The rate of incorporation of labeled thymidine into the deoxyribonucleic acid of regenerating rat liver in relation to the amount of liver excised. *Cancer Res.* **24,** 1611–1625.

Bucher, N. L. R., and Swaffield, M. N. (1973). Regeneration of liver in rats in the absence of portal splanchnic organs and a portal blood supply. *Cancer Res.* **33,** 3189–3194.

Bucher, N. L. R., and Swaffield, M. N. (1975). Regulation of hepatic regeneration in rats by synergistic action of insulin and glucagon. *Proc. Natl. Acad. Sci. U.S.A.* **72,** 1157–1160.

Bucher, N. L. R., Schrock, T. R., and Moolten, F. L. (1969). An experimental view of hepatic regeneration. *Johns Hopkins Med. J.* **125,** 150–257.

Bucher, N. L. R., McGowan, J. A., and Patel, U. (1978a). Hormonal regulation of liver growth. *ICN–UCLA Symp. Mol. Cell. Biol.* **12,** 661–670.

Bucher, N. L. R., Patel, U., and Cohen, S. (1978b). Hormonal factors concerned with liver regeneration. *Hepatotrophic Factors, Ciba Found. Symp.* No. 55, 95–107.

Bucher, N. L. R., Patel, U., and Cohen, S. (1978c). Hormonal factors and liver growth. *Adv. Enzyme Regul.* **16,** 205–213.

Bucher, N. L. R., McGowan, J. A., and Russell, W. E. (1983). Control of liver regeneration: Present status. *In* "Nerve, Organ and Tissue Regeneration: Research Perspectives" (F. J. Seil, ed.), pp. 455–469. Academic Press, New York.

Carpenter, G., Stotscheck, C. M., and Soderquist, A. M. (1982). Epidermal growth factor. *Ann,. N.Y. Acad. Sci.* **397,** 11–17.

Clayton, D. F., and Darnell, J. E., Jr. (1983). Changes in liver-specific compared to common gene transcription during primary culture of mouse hepatocytes. *Mol. Cell. Biol.* **3,** 1552–1561.

Clement, B., Guguen-Guillouzo, C., Campion, J.-P., Glasie, D., Bourel, M., and Guillouzo, A. (1984). *Hepatology* **4,** 373–380.

Creba, J. A., Downes, C. P., Hawkins, P. T., Brewster, G., Michell, R. M., and Kirk, C. J. (1983). Rapid breakdown of phosphatidylinositol 4-phosphate and phosphatidylinositol 4,5-biphosphate in rat hepatocyte stimulated by vasopressin and other Ca^{2+}-mobilizing hormones. *Biochem. J.* **212,** 733–747.

Cruise, J. L., Houck, K. A., and Michalopoulos, G. K. (1985). Induction of DNA synthesis in cultured rat hepatocytes through stimulation of α_1 adrenoreceptor by norepinephrine. *Science* **227,** 749–751.

Earp, H. S., and O'Keefe, E. J. O. (1981). Epidermal growth factor receptor number decreases during rat liver regeneration. *J. Clin. Invest.* **67,** 1580–1583.

Farivar, M., Wands, J. D., Isselbacher, K. J. and Bucher, N. L. R. (1976). Effects of insulin and glucagon on fulminant murine hepatitis. *N. Engl. J. Med.* **295,** 1517–1519.

Fox, C. F., Linsley, P. S., and Wrann, M. (1982). Receptor remodeling and regulation in the action of epidermal growth factor. *Fed. Proc., Fed. Am. Soc. Exp. Biol.* **41,** 2988–2995.

Friedman, D., Claus, T., Pilkis, S., and Pine, G. (1981). Hormonal regulation of DNA synthesis in primary cultures of adult rat hepatocytes; action of glucagon. *Exp. Cell Res.* **135,** 283–290.

Grisham, J. W. (1962). Morphologic study of deoxyribonucleic acid synthesis and cell proliferation in regenerating rat liver: Autoradiography with thymidine-H³. *Cancer Res.* **22,** 842–849.

Guguen-Guillouzo, C., and Guillouzo, A. (1983). Modulation of functional activation in cultured rat hepatocytes. *Mol. Cell Biochem.* **53/54,** 34–56.

Hasegawa, K., and Koga, M. (1977). Induction of liver cell proliferation in intact rats by amines and glucagon. *Life Sci.* **21,** 1723–1728.

Hasegawa, K., and Koga, M. (1981). A high concentration of pyruvate is essential for survival and DNA synthesis in primary cultures of adult rat hepatocytes in a serum-free medium. *Biomed. Res.* **2,** 217–221.

Hasegawa, K., Namai, K., and Koga, M. (1980). Induction of DNA synthesis in adult rat hepatocytes cultured in a serum free medium. *Biochem. Biophys. Res. Commun.* **95**, 243–249.

Hasegawa, K., Watanabe, K., and Koga, M. (1982). Induction of mitosis in primary cultures of adult rat hepatocytes under serum-free conditions. *Biochem. Biophys. Res. Commun.* **104**, 259–265.

Hems, D. A., and Witton, P. D. (1973). Stimulation by vasopressin of glycogen breakdown, and gluconeogenesis in the perfused rat liver. *Biochem. J.* **136**, 705–709.

Higgins, G. M., and Anderson, R. M. (1931). Experimental pathology of the liver. I. Restoration of the liver of the white rat following partial surgical removal. *Arch. Pathol.* **12**, 186–202.

Ichihara, A., Nakamura, T., Tamaka, K., Tomita, Y., Aoyama, K., Kato, S., and Shinno, H. (1980). Biochemical functions of adult rat hepatocytes in primary culture. *Ann. N.Y. Acad. Sci.* **349**, 77–84.

Isom, H. C., Secott, T., Georgoff, I., Woodworth, C., and Mummaw, J. (1985). Maintenance of differentiated rat hepatocytes in primary culture. *Proc. Natl. Acad. Sci. U.S.A.* **82**, 3252–3256.

Iype, P. T., and McMahon, J. B. (1984). Hepatic proliferation inhibitor. *Mol. Cell. Biochem.* **59**, 57–80.

Laishes, B. A., and Williams, G. M. (1976). Conditions affecting primary cell cultures of functional adult rat hepatocytes. *In Vitro* **12**, 821–832.

Leduc, E. H. (1949). Mitotic activity in the liver of the mouse during inactivation followed by refeeding different levels of protein. *Am. J. Anat.* **84**, 397–430.

Leffert, H., Moran, T., Sell, S., Kelley, H., Ibsen, K., Mueller, M., and Arias, I. (1978). Growth state-dependent phenotypes of adult hepatocytes in primary monolayer culture. *Proc. Natl. Acad. Sci. U.S.A.* **75**, 1834–1838.

Leffert, H. L., Koch, K. S., Moran, T., and Rubalcava, B. (1979). Hormonal control of rat liver regeneration. *Gastroenterology* **76**, 1470–1482.

Leong, G. F., Grisham, J. W., Hole, B. V., and Albright, M. L. (1964). Effect of partial hepatectomy on DNA synthesis and mitosis in heterotopic partial autografts of rat liver. *Cancer Res.* **24**, 1496–1501.

Lieberman, I. (1969). *In* "Biochemistry of Cell Division" (R. Baserga, Ed.), pp. 119–137. Thomas, Springfield, Illinois.

Lin, P., Blaisdell, J., O'Keefe, E., and Earp, M. S. (1984). Insulin inhibits the glucocorticoid-mediated increase in hepatocyte EGF binding. *J. Cell. Physiol.* **119**, 267–272.

McGowan, J. A. (1986). Hepatocyte proliferation in culture. *In* "Isolated and Cultured Hepatocytes" (A. Guillouzo and C. Guguen-Guillouzo, eds.), pp. 1–12. Libbey Eurotext INSERM.

McGowan, J. A., and Bucher, N. L. R. (1981). Enhancement of DNA synthesis in primary cultures of adult rat liver hepatocytes by pyruvate. *In Vitro* **17**, 256.

McGowan, J. A., and Bucher, N. L. R. (1983a). Hepatotrophic activity of pyruvate. *In* "Isolation, Characterization and Use of Hepatocytes" (R. A. Harris and N. W. Cornell, eds.), pp. 165–170. Elsevier, New York.

McGowan, J. A., and Bucher, N. L. R. (1983b). Pyruvate promotion of DNA synthesis in serum-free primary cultures of adult rat hepatocytes. *In Vitro* **19**, 159–166.

McGowan, J. A., and Bucher, N. L. R. (1985). The isolation of adult rat liver parenchymal cells. *J. Tissue Cult. Methods* **9**, 49–52.

McGowan, J. A., Strain, A. J., and Bucher, N. L. R. (1981). DNA synthesis in primary cultures of adult rat hepatocytes in a defined medium: effects of epidermal growth factor, insulin, glucagon, and cyclic-AMP. *J. Cell. Physiol.* **108**, 353–363.

McGowan, J. A., Russell, W. E., and Bucher, N. L. R. (1984). Hepatocyte DNA replication: Effect of nutrients and intermediary metabolites. *Fed. Proc., Fed. Am. Soc. Exp. Biol.* **43**, 131–133.

Mapes, J. P., and Harris, R. A. (1975). On the oxidation of succinate by parenchymal cells isolated from rat liver. *FEBS Lett.* **51**, 80–83.

Michalopoulos, G., Cianciulli, H. D., Novotny, A. R., Kligerman, A. D., Strom, S. C., and Jirtle, R. L. (1982). Liver regeneration studies with rat hepatocytes in primary cultures. *Cancer Res.* **42**, 4673–4682.

Michalopoulos, G., Houck, V. A., Dolan, M. G., and Luetteke, N. C. (1984). Control of hepatocyte replication by two serum factors. *Cancer Res.* **44**, 4414–4419.

Moolten, F. L., and Bucher, N. L. R. (1967). Regeneration of rat liver: transfer of "humoral" agent by cross circulation. *Science* **158**, 272–274, 1967.

Morley, C. (1981). Adrenergic agents as possible regulators of liver regeneration. *Int. J. Biochem.* **13**, 969–973.

Nakamura, T., Tomomura, A., Noda, C., Shimagi, M., and Ichihara, A. (1983). Acquisition of a β-adrenergic response by adult rat hepatocytes during primary culture. *J. Biol. Chem.* **258**, 9283–9289.

Nakamura, T., Nawa, K., and Ichihara, A. (1984a). Partial purification and characterization of hepatocyte growth factors from serum of hepatectomized rats. *Biochem. Biophys. Res. Commun.* **122**, 1450–1457.

Nakamura, T., Nakyama, K., and Ichihara, A. (1984b). Reciprocal modulation of growth and liver functions of mature rat hepatocytes in primary culture by an extract of hepatic plasma membranes. *J. Biol. Chem.* **259**, 8056–8058.

Nakamura, T., Tomomura, A., Kato, S., Noda, C., and Ichihara, A. (1984c). Reciprocal expression of α₁ and β-adrenergic receptors, but constant expression of glucagon receptor by rat hepatocytes during development and primary culture. *J. Biochem. (Tokyo)* **96**, 127–136.

Nelson, K. F., and Acosta, D. (1982). Development of a primary culture system of postnatal rat hepatocytes which retain high levels of cytochrome P-450 activity. *In Vitro* **18**, 303.

Olashaw, N. E., Leaf, E. B., O'Keefe, E. J., and Pledger, W. J. (1984). Differential sensitivity of fibroblasts to epidermal growth factors is related to cyclic AMP concentration. *J. Cell. Physiol.* **118**, 291–297.

Paul, D., and Piasecki, A. (1984). Rat platelets contain growth factor(s) distinct from PDGF which stimulate DNA synthesis in primary adult rat hepatocyte cultures. *Exp. Cell Res.* **154**, 95–100.

Reid, L. M., Gatmaitan, Z., Arias, I., Ponce, P., and Rojkind, M. (1980). Long-term cultures of normal rat hepatocytes on liver biomatrix. *Ann. N.Y. Acad. Sci.* **349**, 70–76.

Richman, R. A., Claus, J. A., Pilkis, S. S., and Friedman, D. L. (1976). Hormonal stimulation of DNA synthesis in primary cultures of adult rat hepatocytes. *Proc. Natl. Acad. Sci. U.S.A.* **73**, 3589–3593.

Rozengurt, E., Brown, K. D., and Pellican, P. (1981). Vasopressin inhibition of epidermal growth factor binding to cultured mouse cells. *J. Biol. Chem.* **256**, 716–722.

Rubin, R. A., O'Keefe, E. J., and Earp, H. S. (1982). Alterations of epidermal growth factor-dependent phosphorylation during rat liver regeneration. *Proc. Natl. Acad. Sci. U.S.A.* **79**, 776–780.

Russell, W. E., and Bucher, N. L. R. (1983a). Vasopressin modulates liver regeneration in the Brattleboro Rat. *Am. J. Physiol.* **245**, G321–G324.

Russell, W. E., and Bucher, N. L. R. (1983b). Vasopressin as a regulator of liver growth. *In* "Isolation, Characterization and Use of Hepatocytes" (R. A. Harris and N. W. Cornell, eds.), pp. 171–176. Elsevier, New York.

Russell, W. E., McGowan, J. A., and Bucher, N. L. R. (1984a). Partial characterization of a hepatocyte growth factor from rat platelets. *J. Cell. Physiol.* **119**, 183–192.

Russell, W. E., McGowan, J. A., and Bucher, N. L. R. (1984b). Biological properties of a hepatocyte growth factor from rat platelets. *J. Cell. Physiol.* **119**, 193–197.

St. Hilaire, R. J., Hradek, G. T., and Jones, A. L. (1983). Hepatic regeneration and biliary secretion of epidermal growth factor: Evidence for a high capacity uptake system. *Proc. Natl. Acad. Sci. U.S.A.* **80,** 3797–3801.

Sakai, A. (1970). Humoral factor triggering DNA synthesis after partial hepatectomy in the rat. *Nature (London)* **228,** 1186–1187.

Seglen, P. O. (1976). Preparation of isolated rat liver cells. *Methods Cell Biol.* **13,** 29–83.

Short, J., Armstrong, N. B., Kolitsky, M. A., Mitchell, R. A., Zemel, R., and Lieberman, I. (1974). Amino acids and the control of nuclear DNA replication in liver. *Cold Spring Harbor Conf. Cell Proliferation* **1,** 37–48.

Sigel, B., Acevedo, F. J., and Dunn, M. R. (1963). Effect of partial hepatectomy an autotransplanted liver tissue. *Surg., Gynecol. Obstet.* **117,** 29–36.

Sirica, A. E., Richards, W., Tsukada, Y., Sattler, C. A., and Pitot, H. C. (1979). Fetal phenotypic expression by adult rat hepatocytes on collagen gel/nylon meshes. *Proc. Natl. Acad. Sci. U.S.A.* **76,** 283–287.

Stiles, C. D., Capone, G. T., Scher, C. D., Antoniades, H. N., VanWyk, J. J., and Pledger, W. J. (1979). Dual control of cell growth by somatomedins and platelet-derived growth factor. *Proc. Natl. Acad. Sci. U.S.A.* **76,** 1279–1283.

Strain, A. J., McGowan, J. A., and Bucher, N. L. R. (1982). Stimulation of DNA synthesis in primary cultures of adult rat hepatocytes by rat platelet-associated substance(s). *In Vitro* **18,** 106–116.

Thaler, F. J., and Michalopoulos, G. K. (1985). Hepatopoietin A: Partial characterization and trypsin activation of a hepatocyte growth factor. *Cancer Res.* **45,** 2545–2549.

Tomita, Y., Nakamura, T., and Ichihara, A. (1981). Control of DNA synthesis and ornithine decarboxylase activity by hormones and amino acids in primary cultures of adult rat hepatocytes. *Exp. Cell Res.* **135,** 363–371.

Tsao, M. S., Smith, J. D., Nelson, K. G., and Grisham, J. W. (1984). A diploid epithelial cell line from normal adult rat liver with phenotypic properties of "oval" cells. *Exp. Cell Res.* **154,** 38–52.

Uriel, J. (1976). Cancer, retrodifferentiation, and the myth of Faust. *Cancer Res.* **36,** 4269–4275.

Wolffe, A. P., and Tata, J. R. (1984). Primary culture, cellular stress and differentiated function. *FEBS Lett.* **176,** 8–15.

Wolffe, A. P., Glover, J. F., and Tata, J. R. (1984). Culture shock. Synthesis of heat-shock-like proteins in fresh primary cell cultures. *Exp. Cell Res.* **154,** 581–590.

Younger, L. P., King, J., and Steiner, D. F. (1966). Hepatic proliferative response to insulin in severe alloxan diabetes. *Cancer Res.* **26,** 1408–1414.

2

Regulation of Sulfation and Glucuronidation of Xenobiotics in Periportal and Pericentral Regions of the Liver Lobule

JAMES G. CONWAY,*,[1] FREDRICK C. KAUFFMAN,[†] AND RONALD G. THURMAN*

*Department of Pharmacology and Curriculum in Toxicology
University of North Carolina
Chapel Hill, North Carolina 27514
†Department of Pharmacology and Experimental Therapeutics
University of Maryland
Baltimore, Maryland 21201

I. INTRODUCTION

Considerable pressure has mounted recently to minimize the use of animals in testing in the drug and cosmetic industries. This has led to increased interest in the development of *in vitro* tests to evaluate toxicity. We suspect that some *in vitro* tests may be inappropriate and will not meet rigid validation criteria because important physiological changes have occurred in the preparations used for *in vitro* assays. These changes, however, may yield important information of relevance physiologically. For example, information on hepatic processes can be obtained with cultured hepatocytes, isolated hepatocytes, liver slices, and perfused livers *in vitro*. Comparison of results obtained with these various techniques has and will lead to important discoveries related to cell, organ, and whole-animal physiology and metabolism of xenobiotics.

[1]Present address: Chemical Industry Institute of Toxicology, Research Triangle Park, North Carolina 27709.

THE ISOLATED
HEPATOCYTE

Cultured hepatocytes are convenient and efficient models which yield reproducible data; however, following several passages in culture, many enzyme systems revert to the fetal state and levels of cytochrome P-450, a key enzyme in toxification and detoxification reactions, declines. Primary cultures of hepatocytes are also convenient and efficient; however, they do not make bile and lack the natural architecture of the liver lobule. Evaluation of drugs or chemicals which may be cholestatic or which affect specific zones of the liver lobule are difficult with hepatocytes.

The perfused liver, on the other hand, maintains the natural architecture of the liver lobule. It is a nearly physiologic preparation which makes bile and has an intact microcirculation. Although data cannot be collected as efficiently with liver perfusion as with isolated hepatocytes, it is considerably more efficient than studies *in vivo* (e.g., complete dose–response curves can be obtained from one liver). Information obtained from perfused livers may be easier to validate, since the preparation closely resembles the liver *in vivo*. The excessive use of experimental animals and the difficulty in interpretation of data from isolated hepatocyte studies, where intraorgan complexities are not operative, make the perfused liver an attractive model. In addition, the availability of the perfused liver as a model truly bridges the gap between intact animals and hepatocytes.

Sulfation and glucuronidation involve the incorporation of inorganic sulfate and glucuronic acid into substrates with phenolic or alcoholic hydroxyl groups. Formation of these water-soluble conjugates, in most cases, inactivates potentially toxic xenobiotics and facilitates their elimination in the bile and urine. Conjugation may also lead to the formation of unstable, reactive compounds, particularly the sulfate conjugates of N-hydroxyarylamines (1–6). Understanding the rate-limiting components of sulfation and glucuronidation is fundamental to evaluating the mechanism of toxicity of numerous hepatotoxins. This chapter will review factors involved in the regulation of sulfation and glucuronidation in intact cells, with particular emphasis on the heterogenity of conjugation in periportal and pericentral regions of the liver lobule.

A. Regulation of Sulfation in Intact Hepatocytes

Enzyme concentration, cofactor supply, and substrate availability influence rates of sulfation in intact cells (Fig. 1). Inorganic sulfate taken up actively from the extracellular fluid or generated intracellularly via degradation of cysteine and methionine must be activated to form 3′-phosphoadenosine- 5′-phosphosulfate (PAPS). PAPS is produced by a two-reaction sequence requiring ATP which is catalyzed by ATP-sulfurylase and APS kinase. The sulfate group of PAPS can be donated to a large variety of substrates by sulfotransferases localized in the cytosol.

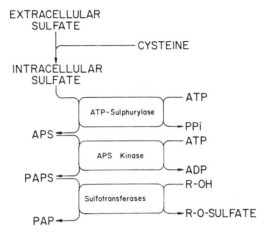

Fig. 1. Scheme depicting pathways involved in sulfation. APS, Adenosine 5'-phosphate; PAPS, 3-phosphoadenosine 5'-phosphosulfate; PAP, adenosine 3',5'-diphosphate; ATP, adenosine triphosphate; R-OH, substrate with phenolic or alcoholic hydroxyl group; PPi, inorganic pyrophosphatate.

Considerable progress has been made in the separation and *in vitro* characterization of sulfotransferases. Jakoby and co-workers have used isoelectric focusing and DEAE–cellulose column chromatography to characterize seven sulfotransferase isoenzymes (7). Four enzymes are catalytically active with simple phenols but inactive with primary and secondary alcohols, bile acids, and arylamines (7,8). The other three enzymes are active with alcohols, including endogenous hydroxysteroids, but inactive toward phenols and bile acids (7,8). In addition, Chen *et al.* (9) and Green and Singer (10) have isolated and characterized sulfotransferases from rat liver specific for bile acids and estradiol. Predicting the activity of sulfotransferases *in vivo* from assays of purified enzymes is complicated by the loss of activity during purification and limitations imposed by the ionic composition and concentration of cofactors and inhibitors in intact cells. Identification of rate-limiting factors for sulfation under physiological conditions has been accomplished in studies with intact cells.

The concentration of inorganic sulfate in serum (about 0.9 mM in rats) (11–13) is maintained primarily by the degradation of L-cysteine. This amino acid is derived from catabolism of ingested proteins, but may also be generated from methionine via the cystathionine pathway (14). Krijgsheld and Mulder (12) used a low-protein diet to decrease the concentration of inorganic sulfate in serum to about 0.15 mM. Concentrations of inorganic sulfate in serum were then increased by stepwise intravenous infusions of inorganic sulfate during steady-state rates of excretion of harmol sulfate in the urine. The half-maximal concentration of inorganic sulfate in serum for sulfation in the intact animal was calculated to be 0.4 mM (12). Half-maximal concentrations of extracellular inorganic sulfate for sulfation of naphthol (15) and harmol (16,17) in isolated hepatocytes and perfused livers ranged from 0.2 to 0.5 mM. Taken together, the above data suggest that normal concentrations of inorganic sulfate (0.9 mM) in serum are not limiting for sulfation in intact hepatocytes.

Recent studies using isolated hepatocytes (15) and perfused livers (18) indicate that the uptake of inorganic sulfate across the plasma membrane is very rapid. In the presence of 1 mM extracellular inorganic sulfate, rates of uptake of inorganic sulfate into sulfate-depleted hepatocytes were at least 60 μmol/g liver/hr (15). Under steady-state conditions of sulfate exchange in perfused liver, the transport of sulfate was about 600 μmol/g liver/hr (18). Thus, the rate of transport of sulfate across the plasma membrane greatly exceeds maximal rates of sulfation, which range from 2 to 6 μmol/g/hr (15,19–21). In the presence of physiological concentrations of inorganic sulfate and excess substrate, rates of PAPS synthesis and/or sulfotransferase activity likely limit rates of sulfation.

Under special conditions, high rates of sulfation *in vivo* can decrease the concentration of inorganic sulfate in blood and thus limit the supply of inorganic sulfate for sulfation. For example, Krijgsheld and Mulder (12) showed that continuous intravenous infusion of harmol *in vivo* decreased inorganic sulfate concentrations in rat serum to about 0.3 mM and, subsequently, rates of harmol sulfate excretion. The decrease in concentrations of inorganic sulfate and rates of sulfation of harmol were reversed by intravenous infusion of sodium sulfate (12). This mechanism for inhibition of sulfation may have clinical relevance since a large dose of acetaminophen *in vivo* also lowered the concentration of inorganic sulfate in serum to about 0.3 mM (11–13). Interestingly, concentrations of PAPS in liver paralleled changes in inorganic sulfate concentrations in serum (13). Administration of sodium sulfate and N-acetylcysteine eliminated the acetamino-phen-induced decrease in serum sulfate concentrations (11). Moreover, sodium sulfate (22,23) or N-acetylcysteine (23) increased sulfation of acetaminophen administered as a large bolus dose (22) or as continuous intravenous infusion (23). Schwarz (15) studied the effect of extracellular inorganic sulfate on intra-cellular sulfate concentrations and rates of sulfation of 1-naphthol and found that sulfation of 1-naphthol was half maximal with 0.5 mM inorganic sulfate in the intracellular and extracellular space. Thus, limitation of PAPS synthesis by intracellular concentrations of sulfate limits sulfation when normal concentra-tions of inorganic sulfate in serum fall below 0.5 mM, such as after large doses of drugs like harmol (12) and acetaminophen (11–13). Regulation of inorganic sulfate concentrations in serum via catabolism of cysteine and methionine war-rants further investigation.

The concentration of the cofactor PAPS in rat liver is about 30 μM (24). Thus, the concentration of PAPS in the cell may be an important rate determinant for sulfation, since the apparent K_m of sulfotransferases for PAPS ranges from 2 to 30

μM (7,8,25,26). Because the steady-state concentrations of PAPS are quite low, the turnover of PAPS in the liver must be quite rapid during sulfation (19,27). Steady-state rates of sulfation of p-nitrophenol (19), acetaminophen (28), and harmol (21) requiring the synthesis of 3–6 μmol/g/hr PAPS have been observed in perfused livers. Under these conditions, the pool of PAPS in liver must be resynthesized every 4 sec. Therefore, decreases in rates of PAPS synthesis quickly decrease rates of sulfation (19). For example, decreases in PAPS synthesis by limited ATP supply are important during hypoxia since inhibition of sulfation at low oxygen tensions corresponded closely with decreases in ATP/ADP ratios and the reduction of cytochrome oxidase (29). Genetic comparisons indicate that PAPS synthesis limits sulfation in the brachymorphic mouse (30). Brachymorphic mice excrete less p-nitrophenyl sulfate in urine than control mice after administration of p-nitrophenol *in vivo* and liver cystosol from this mouse has decreased ability to synthesize PAPS from added ATP and inorganic sulfate (30).

It is not clear whether the rate of PAPS synthesis under physiological conditions directly limits sulfotransferase activity. Some sulfotransferase isoenzymes (7,31) can transfer the active sulfate group of p-nitrophenyl sulfate to adenosine 3',5'-diphosphate (PAP), thus artifically generating PAPS (32). Schwartz (15,33) added p-nitrophenyl sulfate to hepatocytes incubated in the presence of excess inorganic sulfate and substrate. Rates of sulfation of 1-naphthol were doubled by addition of 1 mM p-nitrophenyl sulfate leading to the conclusion that p-nitrophenyl sulfate entered hepatocytes and stimulated sulfation by increasing cofactor supply (15,33). In contrast, Norling et al. (34) added p-nitrophenyl sulfate (1 mM) to isolated hepatocytes in the presence of excess inorganic sulfate and observed a 60% inhibition of harmol sulfation. The discrepancy in the above results may be due to differences in the ability of different sulfotransferase isoenzymes to catalyze the formation of PAPS from p-nitrophenyl sulfate.

Since most substrates for sulfotransferases are also substrates for glucuronosyltransferases, sulfate and glucuronide conjugating systems compete for substrate. With isolated hepatocytes and perfused livers, half-maximal substrate concentrations for sulfation of p-nitrophenol (19), 1-naphthol (15,33), 7-hydroxycoumarin (35,36), and harmol (20) ranged from 5 to 30 μM. These half-maximal substrate concentrations for sulfation are 3- to 7-fold lower than the respective half-maximal concentrations for glucuronidation (15,19,20,33,35,36). Since maximal rates of glucuronidation are 2- to 4-fold higher than maximal rates of sulfation, substrate delivery across the plasma membrane cannot be rate limiting for sulfation during maximal rates of glucuronidation.

B. Regulation of Glucuronidation in Intact Hepatocytes

UDPglucuronosyltransferases are a family of enzymes (37–40), associated closely with the endoplasmic reticulum, which catalyze the transfer of glucuronic

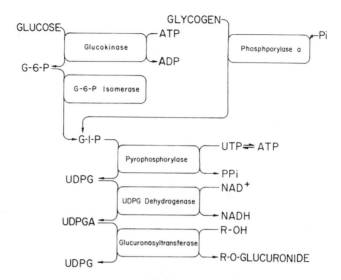

Fig. 2. Scheme depicting pathways involved in glucuronidation. G-6-P, glucose 6-phosphate; G-1-P, glucose 1-phosphate; UDPG, uridine diphosphate glucose; UDPGA, uridine diphosphate glucuronic acid; UDP, uridine diphosphate; UTP, urine triphosphate; ATP, adenosine triphosphate; NAD⁺, nicotinamide adenine dinucleotide; R-OH, substrate with phenolic or alcoholic group; Pi, inorganic phosphate; PPi, inorganic pyrophosphatate.

acid from UDPglucuronic acid (UDPGA) to a wide array of substrates (Fig. 2). Transfer of glucuronic acid to the oxygen moiety of alcohols and phenols is most common; however, nitrogen and sulfur moieties may also serve as acceptor groups. The cofactor UDPGA is synthesized intracellularly from UDPglucose and NAD⁺ by the enzyme UDPglucose dehydrogenase. UDPglucose is formed from glucose 1-phosphate and uridine triphosphate by the enzyme UDPGpyrophosphorylase (Fig. 2).

Rates of glucuronidation in intact cells may be influenced by the same factors that influence sulfation: cofactor supply, enzyme activity, and substrate availability. In a variety of hepatic preparations the concentration of UDPGA is 0.3–0.5 mM (13,19,41–46). With 1-naphthol as a substrate, the K_m of glucuronosyltransferase for UDPGA in native and detergent-activated microsomes from rat liver was 0.34 mM (47). With p-nitrophenol as a substrate, the K_m of glucuronosyltransferase for UDPGA in detergent-activated microsomes ranged from 0.16 to 0.39 mM (48–50). Since the concentration of UDPGA required for half-maximal glucuronosyltransferase activity is near the concentration of cofactor in intact cells, the concentration of cofactor may limit rates of glucuronidation *in vivo.*

The synthesis of UDPGA must be quite rapid in order to maintain maximal rates of glucuronidation. Steady-state rates of glucuronidation requiring 10 μmol/g liver/hr UDPGA have been observed in perfused livers (19,36,42,51). Under these conditions the pool of UDPGA must be resynthesized every 2 min in order to maintain intracellular concentrations above the K_m of glucuronosyltransferase. UDPGA concentrations were not altered during maximal rates of 1-naphthol glucuronidation in perfused livers (42) and hepatocytes (52), indicating that cofactor must be resynthesized rapidly. However, large doses of acetaminophen *in vivo* lowered hepatic UDPGA by 80% (13,46), suggesting that cofactor can become rate-limiting.

Rates of UDPGA synthesis, and thus glucuronidation, can be inhibited by several mechanisms. Fasting decreases maximal rates of glucuronidation in isolated hepatocytes (17,33,53), perfused livers (19,51), and *in vivo* (46) without affecting the activity of glucuronosyltransferase(s) assayed *in vitro* (19,46). Since fasting decreases the concentration of UDPGA in perfused livers (19) and *in vivo* (45,46), decreases in glucuronidation in the fasted state are most likely due to insufficient synthesis of cofactor. In line with this possibility are the observations that addition of glucose stimulates glucuronidation of *p*-nitrophenol (19), 7-hydroxycoumarin (51), 1-naphthol (33), and acetaminophen (53) in hepatocytes and perfused livers from fasted rats.

UDPGA supply may also limit glucuronidation in perfused livers from hypophysectomized rats (54). Hypophysectomy had no effect on glucuronosyltransferase activity assayed *in vitro*; however, glucuronidation of *p*-nitrophenol and 7-hydroxycoumarin in perfused livers was decreased by 50% (54). Hypophysectomy decreases hepatic glycogen *in vivo* (55) and the efflux of glucose, lactate, and pyruvate from perfused livers (54). The hypophysectomy-induced decrease in carbohydrate supply for UDPGA synthesis is consistent with the observation that glucose infusion stimulated glucuronidation in perfused livers from hypophysectomized, but not control rats (54).

Changes in the NADH redox and energy status of the cell can also affect UDPGA synthesis and glucuronidation. Ethanol inhibits glucuronidation by increasing $NADH/NAD^+$ ratios, thereby decreasing the formation of UDPGA from UDPglucose by UDPglucose dehydrogenase (56,57). A number of observations implicate energy metabolism as a key determinant of glucuronidation in intact cells. Anoxia inhibits glucuronidation by decreasing intracellular concentrations of glucose 1-phosphate and uridine triphosphate, thereby decreasing the formation of UDPglucose by UDPglucose pyrophosphorylase (53). Potassium cyanide, an inhibitor of cytochrome oxidase, and dinitrophenol, an uncoupler of oxidative phosphorylation, diminish intracellular ATP and inhibit glucuronidation in perfused rat livers (19). Further, fructose lowers hepatic ATP, UTP, and UDPglucose (58), and inhibits glucuronidation (19).

Trapping of uridine phosphates (59) and inhibition of UDPglucose pyrophosphorylase (60) by galactosamine is another mechanism whereby UDPGA concentrations may be decreased and glucuronidation inhibited (33,41,43,61). Singh and Schwartz (41) used galactosamine to lower UDPGA concentrations in isolated hepatocytes in a stepwise manner. Under these conditions a linear correlation between the cellular concentration of UDPGA and rates of glucuronidation of 3-hydroxybenzo(a)pyrene was observed (41).

Several reports suggest that glucuronidation can be stimulated by increasing UDPGA levels in the absence of induction of glucuronosyltransferase. Reinke *et al.* (19) increased hepatic UDPGA levels and rates of glucuronidation of *p*-nitrophenol by 20% by fasting and refeeding rats. Similarly, Otani *et al.* (61) increased hepatic UDPGA levels and rates of glucuronidation of 1-naphthol by 20% by orotic acid treatment.

Streptozotocin-induced diatebes increases glucuronidation in isolated hepatocytes (62) and *in vivo* (63); however, glucuronosyltransferase activity assayed with excess UDPGA *in vitro* was not altered by diabetes (62,64). Control and diabetic rats have similar UDPGA concentrations in liver (46); however, UDPGA concentrations decreased sharply in control rats given a large dose of acetaminophen, whereas UDPGA concentrations in liver from diabetic rats was not altered by acetaminophen (64). Based on these data, Price and Jollow (46) suggested that livers from diabetic rats have increased ability to synthesize UDPGA.

Although the concentration of UDPGA is an important determinant of rates of glucuronidation in intact cells, activities of glucuronosyltransferase(s) also must be taken into account. Studies with inducers of glucuronosyltransferases show a strong correlation between glucuronosyltransferase activity in microsomes and maximal rates of glucuronidation in intact cells. Reinke *et al.* (19) showed that maximal rates of glucuronidation of *p*-nitrophenol in perfused livers from phenobarbital-treated rats were 2-fold greater than rates in livers from untreated rats. Concentrations of UDPGA were not affected by phenobarbital treatment; however, glucuronosyltransferase activity was elevated 3-fold in microsomes from phenobarbital-treated rats (19). Similar data were obtained by Ullrich and Bock (52) who determined the effect of phenobarbital and 3-methylcholanthrene treatment on glucuronosyltransferase activity in microsomes, glucuronidation in intact hepatocytes, and intracellular UDPGA concentrations. 3-Methylcholanthrene and phenobarbital treatment did not alter UDPGA concentrations in isolated hepatocytes (52); however, changes in glucuronosyltransferase activity in microsome preparations correlated with maximal rates of glucuronidation of four different substrates in isolated hepatocytes (52). These correlations between the phenobarbital- and 3-methylcholanthrene-induced increases in glucuronidation and increased microsomal glucuronosyltransferase activity, in the absence of changes in UDPGA levels, indicate that glucuronosyltransferase activity is an important rate determinant of glucuronidation in intact cells (19,52).

In conclusion, several conditions can alter the relative importance of substrate availability, glucuronosyltransferase activity, and UDPGA supply as rate determinants for glucuronidation. At low concentrations of substrate, rates of glucuronidation were equivalent in perfused livers from untreated and phenobarbital-treated rats in the fasted, fed, or fasted–refed state (19,51). This implies that supply of substrate is rate-limiting under these conditions. In livers from fasted rats perfused with excess substrate, UDPGA is likely the primary rate determinant for glucuronidation since glucose stimulates glucuronidation (19,33,51,53). Both UDPGA supply and enzyme activity are probably major rate determinants for glucuronidation with excess substrate in livers from fed rats. Increases in enzyme concentration may contribute to the increases in glucuronidation observed after phenobarbital and 3-methylcholanthrene treatments (19,42). Because both hepatic UDPGA content and rates of glucuronidation are both increased by fasting–refeeding (19) and orotic acid (61), and decreased by galactosamine treatment (33,41,43,61), it appears that UDPGA generating systems are operating at near-maximal capacity during high rates of glucuronidation.

II. SULFATION AND GLUCURONIDATION IN PERIPORTAL AND PERICENTRAL REGIONS OF THE LIVER LOBULE

Microscopically, the structural repeating unit of the liver has been defined by Rappaport *et al.* (65) as the liver acinus. The acinus is the tissue supplied by terminal branches of the portal vein and hepatic artery and drained by a bile duct. The portal triad, consisting of a hepatic arteriole, portal venule, and bile duct surrounded by connective tissue, is at the center of the acinus. Arterial and venous blood flows into the sinusoids which radiate outward from the portal triad and drain into central veins at the periphery of the acinus. The microcirculatory structure of the liver acinus defines physiological differences between cells located near the portal triad (periportal area) and cells located near the central vein (pericentral area). The periportal area is exposed to blood rich in oxygen, substrates, and hormones, whereas pericentral areas are low in oxygen, substrates, and hormones. Based on differences in enzyme activites, Jungermann has proposed that gradients in oxygen tension and concentrations of hormones and substrates across the liver lobule define functional metabolic differences between cells in periportal and pericentral areas (66). Indeed, measurements with miniature oxygen electrodes in perfused liver by Thurman and colleagues indicate that gluconeogenesis is localized primarily in periportal areas (67), whereas glycolysis predominates in pericentral regions (68).

It has long been appreciated that many hepatotoxins cause zone-specific damage (69). Investigations of mechanisms of toxicity are hampered by the lack of information concerning relative rates of metabolism of xenobiotics in periportal and pericentral areas. Because of their key role in xenobiotic metabolism, sulfation and glucuronidation are likely important in the zone-specific damage caused by some hepatotoxins. For example, N-hydroxy-2-acetylaminofluorene forms an unstable, reactive sulfate conjugate which causes periportal damage (1,70). Inhibition of acetaminophen clearance via glucuronidation by pretreatment with galactosamine (71) potentiates acetaminophen-induced pericentral necrosis whereas increased glucuronidation in diabetic rats decreases acetaminophen toxicity (63,64).

A. Methods Used to Study Metabolic Heterogeneity in the Liver

A number of methods have been used to study the distribution of enzymes within the liver lobule. Sublobular localization of cell types and cell organelles has been quantitated by morphometric analysis (72). Quantitative distributions of enzyme activities and metabolites across the liver lobule have also been measured employing histochemistry (73) and microdissection of freeze-dried tissue sections followed by microchemical determination of enzyme activities (36,66,74–77). Immunohistochemical techniques that distinguish between isoenzymes have also been used (78). Data from detergent-induced release of intracellular enzymes from specific sublobular regions (79) are consistent with differences in enzyme activities measured by microdissection and histochemistry (66).

Glucuronosyltransferase activity (80) and rates of glucuronidation in perfused livers (81) have been measured after administration of zone-specific hepatotoxins to investigate glucuronidation in sublobular zones. This approach is limited because of the nonspecific effects of hepatotoxins. In addition, a number nonspecific metabolic changes occur following treatment with hepatotoxins (74).

The differential localization of maximal enzyme activities across the liver lobule does not mean a priori that the flux through metabolic pathways will be different in periportal and pericentral areas. Activities of enzymes in intact cells are controlled by the supply of substrates and cofactors and the concentrations of various activators and inhibitors. Methods that assess metabolic pathways in intact cells in or from periportal and pericentral areas are therefore advantageous. Attempts have been made to isolate viable preparations of periportal and pericentral hepatocytes. Digitonin delivered via the portal vein or vena cava has been used to destroy tissue in periportal and pericentral areas (79). Isolation of hepatocytes after destruction of periportal and pericentral regions gives cells enriched in enzyme activities characteristic of pericentral and periportal regions, respectively (85,86). Quistorff (86) has isolated hepatocytes by collagenase perfusion after selective destruction of periportal and pericentral areas by treatment with digitonin. Cells isolated from periportal areas had 2-fold higher rates of gluconeogenesis than cells from pericentral regions (86), consistent with the localization of key enzymes involved in gluconeogenesis (66). Fractions of hepatocytes having the morphology and enzyme characteristics of periportal and pericentral areas have also been separated by density (82) and size (83). Tonda and Hirata observed slight differences in the sulfation and glucuronidation of *p*-nitrophenol in fractions of hepatocytes enriched in cells from periportal and pericentral areas (84). Use of isolated hepatocytes separated into fractions enriched in periportal- and pericentral-like cells by the above methods is handicapped by the loss of cells during separation procedures. Perfusion of livers with detergents (85,86), use of calcium-free medium during cell separation (84), and periods of hypoxia during centrifugation steps (84) may compromise the metabolic integrity of the recovered hepatocytes. Thus, considerable work is required to assess the metabolic integrity of the recovered hepatocytes before comparisons between cells isolated from periportal and pericentral regions can be made with confidence.

Pang and co-workers (21,28,87) have employed a pharmacokinetic approach to study the competition between sulfation and glucuronidation pathways in perfused livers. By assuming that substrates infused into livers perfused in the anterograde (via the portal vein) and retrograde direction (via the vena cava) are preferentially conjugated by enzymes located in periportal and pericentral areas, pharmacokinetic models of conjugation in sublobular zones have been developed. This approach has been useful; however, lack of direct measurements in sublobular zones leaves conclusions derived from pharmacokinetic analysis open to question. Methodology to directly measure substrate concentrations and rates of conjugation in sublobular areas is needed to investigate regulation of conjugation in periportal and pericentral areas.

B. Development of Techniques to Quantify Metabolic Zonation Noninvasively in Perfused Liver

Microfluorometry was first used by Chance and Thorell (88) to determine the localization and kinetics of reduction of pyridine nucleotides in living cells. Subsequently, light guides (tip diameter of approximately 5 mm) containing optical fibers were used to measure NADH fluorescence and ultraviolet reflectance from the heart (89), rat cerebral cortex *in situ* (90), and perfused liver (91). Further work resulted in the development of microlight guides with tip diameters less than 200 μm allowing the measurement of NADH fluorescence and ultraviolet reflectance from small tissue areas (92).

Ji *et al.* developed a method employing a modified microlight guide to measure the NADH (93) and 7-hydroxycoumarin fluorescence (94) in periportal and pericentral regions of the liver lobule in the perfused liver noninvasively. Anterograde and retrograde perfusions of liver with India ink identified lightly pigmented regions as periportal areas and darkly pigmented spots as pericentral regions (93). A microlight guide of two strands of optical fiber with a tip of 170 μm in diameter was placed on specific regions of the liver lobule. One strand was connected to a light source and the other strand to a photomultiplier. The tissue was illuminated with light at 366 nm, and fluorescence of 7-hydroxycoumarin at 450 nm was detected from the tissue with a photomultiplier, amplified, and recorded (95).

The development of techniques to monitor fluorescence in sublobular areas with microlight guides allowed us to measure cytochrome P-450-mediated 7-ethoxycoumarin O-deethylation in periportal and pericentral areas of the liver lobule (94). A macro-light guide which collects fluorescence from many lobules was first placed on the surface of the perfused rat liver. When nonfluorescent 7-ethoxycoumarin was infused into the liver, fluorescence due to 7-hydroxycoumarin in the tissue increased in direct proportion to rates of 7-hydroxycoumarin production. The correlation between increases in fluorescence detected in the tissue, and rates of 7-ethoxycoumarin O-deethylation were used to convert fluorescence readings obtained with microlights into rates of mixed function oxidation in periportal and pericentral areas. When 7-ethoxycoumarin was infused, fluorescence increased twice as much in pericentral as in periportal areas. These fluorescence changes corresponded to local rates of mixed function oxidation of 3.6 and 7.0 μmol/g/hr in periportal and pericentral regions, respectively. This technique provided the first direct evidence that rates of mixed function oxidation are twice as great in pericentral as in periportal regions in intact livers from phenobarbital-treated rats. This method of quantifying rates of mixed function oxidation has since been utilized to study the effect of enzyme induction (94,97), nutritional state (96), and NADPH supply (96,97) on rates of 7-ethoxycoumarin O-deethylation in sublobular zones of the rat liver.

C. Sulfation Predominates in Periportal Areas of the Liver Lobule

Measurement of the disappearance of fluorescence from 7-hydroxycoumarin has been used to specifically titrate sulfotransferase and glucuronosyltransferase enzymes in periportal and pericentral areas of livers from rats pretreated with phenobarbital (35). When up to 30 μM 7-hydroxycoumarin was infused into a liver perfused in the anterograde direction, fluorescence due to free 7-hydroxycoumarin was only detected in periportal areas (Fig. 3). Since the microlight guide can detect 5 μM 7-hydroxycoumarin in the tissue, free 7-hydroxycoumarin was only present in periportal areas under these conditions. In addition, all the 7-

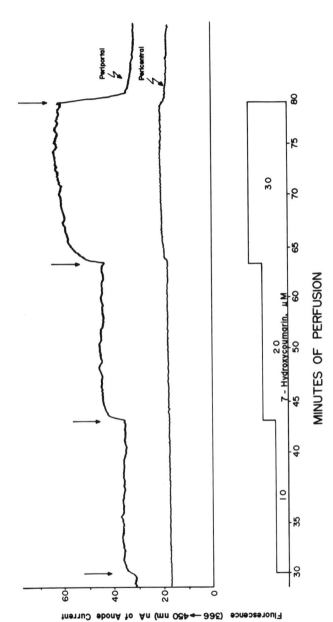

Fig. 3. Fluorescence increase upon infusion of 7-hydroxycoumarin in periportal and pericentral regions of the liver perfused in the anterograde direction. Two microlight guides were placed on periportal and pericentral regions (1 mm apart) on the left lateral lobe of the liver. Periportal and pericentral regions were identified by differences in native pigmentation as well as by differential responses to anoxia (93). 7-Hydroxycoumarin was dissolved in Krebs–Henseleit bicarbonate buffer and infused as indicated by horizontal bars (bottom) and arrows (top) (details in Ref. 35). Anterograde perfusion; fed, phenobarbital-treated rat. Reproduced by permission of Williams and Wilkins Co.

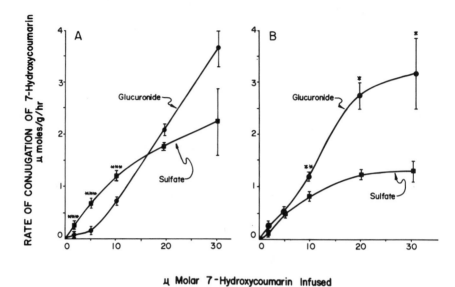

μ Molar 7-Hydroxycoumarin Infused

Fig. 4. Rates of sulfation and glucuronidation in periportal (A, anterograde perfusion) and pericentral (B, retrograde perfusion) regions of the perfused rat liver. Eight livers were perfused in the anterograde direction (see Fig. 3) and eight in the retrograde direction (35). Under these conditions, no free 7-hydroxycoumarin was detected in the effluent perfusate. Perfusate was collected and assayed for glucuronide and sulfate conjugates of 7-hydroxycoumarin. Mean ± SEM for 5–11 data points for each concentration of 7-hydroxycoumarin. $p < 0.05(*)$, $p < 0.01(**)$, and $p < 0.001(***)$ for comparisons between sulfation and glucuronidation in the same lobular region. Reproduced by permission of Williams and Wilkins Co.

hydroxycoumarin infused was recovered in the effluent perfusate as nonfluorescent sulfate and glucuronide conjugates. Thus, with low concentrations of 7-hydroxycoumarin (<30 μM) infused in the anterograde direction, substrate is conjugated exclusively in periportal areas. Similar fluorescence measurements established that low concentrations of 7-hydroxycoumarin infused in the retrograde direction are converted quanitatively into sulfate and glucuronide conjugates in pericentral areas (35).

Figure 4 depicts the rates of sulfate and glucuronide conjugation of 7-hydroxycoumarin in perfusions in the anterograde (Fig. 4A) and retrograde (Fig. 4B) directions. In periportal tissue (anterograde perfusions), sulfation predominated over glucuronidation at 7-hydroxycoumarin concentrations of 2–10 μM (Table I). At higher substrate concentrations, glucuronidation predominated. The pattern of conjugation was quite different in pericentral regions (retrograde perfusions). At low 7-hydroxycoumarin concentrations (2–5 μM), rates of sulfation and glucuronidation were equivalent; however, at higher substrate concentrations, rates of glucuronidation were significantly greater than rates of sulfation in

TABLE I

Rates of Conjugation of 7-Hydroxycoumarin during Infusion of 7-Hydroxycoumarin[a]

7-Hydroxycoumarin infused (μM)	7-Hydroxycoumarin conjugate (μmol/g/hr)			Sulfate/glucuronide
	Sulfate	Glucuronide		
Anterograde perfusion				
2	0.36 ± 0.028[b]	0.006 ± 0.005	(5)[c]	> 20[d]
5	0.71 ± 0.075[b]	0.15 ± 0.036[b]	(10)	6.11
10	1.25 ± 0.13[b]	0.64 ± 0.09[b]	(12)	1.90
Retrograde perfusion				
2	0.17 ± 0.032	0.20 ± 0.004	(5)	1.65
5	0.37 ± 0.045	0.42 ± 0.067	(9)	0.92
10	0.67 ± 0.065	1.26 ± 0.161	(7)	0.60

[a]Conditions as in Fig. 3.
[b]$p<0.01$ compares anterograde with retrograde perfusions.
[c]Mean ± SEM numbers in parentheses are the number of livers perfused.
[d]Ratio was arbitrarily defined as >20 due to the near absence of glucuronidation.

pericentral regions (Table I, Fig. 4B). Thus, at low substrate concentrations (2–10 μM). the ratio of sulfation:glucuronidation was much greater in periportal than in periportal than in pericentral regions of the liver lobule (Table I).

The apparent K_m for sulfation of 7-hydroxycoumarin in perfused liver is 7 μM (35) while the apparent K_m for glucuronidation is 26 and 47 μM in periportal and pericentral areas, respectively, under these conditions (36). Therefore, at concentrations of 7-hydroxycoumarin below 20 μM, glucuronosyltransferases would not be expected to compete effectively with sulfotransferases for this substrate. The higher rates of sulfation in periportal than pericentral regions (Table I) observed with 7-hydroxycoumarin concentrations less than 20 μM must, therefore, be due to greater concentrations of sulfotransferase(s) or PAPS in periportal areas.

D. Conjugation of 7-Hydroxycoumarin Generated from Mixed-Function Oxidation of 7-Ethoxycoumarin

O-Deethylation of 7-ethoxycoumarin is half maximal in perfused liver from phenobarbital-treated rats with about 100 μM substrate (98). Therefore, at low concentrations of 7-ethoxycoumarin (<20 μM), one would expect rates of O-deethylation to be highest in the sublobular area exposed to the highest concentration of substrate. In support of this hypothesis, we observed that perfusion

TABLE II

Rates of Conjugation of 7-Hydroxycoumarin during Infusion of 7-Ethoxycoumarin[a]

7-Hydroxycoumarin infused (μM)	7-Hydroxycoumarin conjugated ($\mu mol/g/hr$)			7-Hydroxycoumarin formed (μM)
	Sulfate	Glucuronide	Sulfate/glucuronide	
Anterograde perfusion				
5	0.32 ± 0.039[b]	0.033 ± 0.009	11.2 ± 1.68	1.74
10	0.66 ± 0.046	0.17 ± 0.054	6.6 ± 1.61	3.71
20	1.11 ± 0.094	0.58 ± 0.211	3.2 ± 0.75	7.23
Retrograde perfusion				
5	0.18 ± 0.037	0.06 ± 0.093	2.8 ± 0.39	1.61
10	0.50 ± 0.057	0.15 ± 0.021	3.4 ± 0.39	4.22
20	0.88 ± 0.141	0.41 ± 0.075	2.4 ± 0.52	7.28

[a]Mean $\pm$ SEM; $n = 4$ to 8 livers.

[b]$p < 0.05$ compares rates of sulfation in anterograde and retrograde perfusions. See Conway et al. (35) for details.

of 7-ethoxycoumarin in the anterograde direction produced fluorescence from periportal but not pericentral regions of the liver lobule (35), whereas with retrograde perfusions fluorescence was only detected in pericentral areas. At the low concentrations of 7-hydroxycoumarin formed under these conditions, conjugation occurs predominantly, if not exclusively, at the site where 7-hydroxycoumarin fluorescence was detected. The ratio of sulfate:glucuronide conjugates was much greater in periportal than pericentral areas irrespective of whether 7-hydroxycoumarin (Table I) was infused directly or generated via metabolism of 7-ethoxycoumarin (Table II).

The highly reactive sulfate ester of N-hydroxy-2-acetylaminofluorene is likely responsible for the periportal necrosis caused by N-hydroxy-2-acetylaminofluorene (1). Suppression of sulfation by exclusion of inorganic sulfate or addition of an inhibitor of sulfation, pentachlorophenol, eliminates 70–90% of the N-hydroxy-2-acetylaminofluorene binding to protein, RNA, and DNA in perfused liver (2). Pentachlorophenol pretreatment in vivo also decreased binding of N-hydroxy-2-acetylaminofluorene to protein and RNA by 50% (2). Furthermore, binding of N-hydroxy-2-acetylaminofluorene to liver in vivo was localized in periportal areas in autoradiographic studies (70). However, sulfate-dependent binding of N-hydroxy-2-acetylaminofluorene to liver slices in vitro was distributed equally in periportal and pericentral areas (70). Several factors may contribute to the discrepancy between the in vivo and in vitro binding of N-hydroxy-2-acetylaminofluorene to sublobular areas. The work discussed above exemplifies the complexities in attributing zone-specific hepatotoxicity to differences in sublobular rates of bioactivation.

Pang and Terrell (87) evaluated pharmacokinetic models of acetaminophen glucuronidation and sulfation in perfused livers from normal rats. These studies were based on the assumption that phenacetin is converted into acetaminophen via mixed function oxidation to a greater extent in pericentral than in periportal regions of the liver lobule. More acetaminophen sulfate was detected in the effluent perfusate with retrograde than with anterograde perfusions, leading the authors to suggest that sulfation occurs to a greater extent in cells in periportal than in pericentral regions (87). Pang et al. (21) also investigated the conjugation of harmol in livers perfused in anterograde and retrograde directions. The ratio of sulfate to glucuronide conjugates was greater in livers perfused in anterograde than retrograde directions, consistent with the localization of sulfation in periportal areas.

Tonda and Hirata (84) separated hepatocytes from control, phenobarbital, and 3-methylcholanthrene-treated rats into four fractions based on cell density. Sulfation of p-nitrophenol was equivalent in the fractions enriched in cells from periportal and pericentral areas, in contrast to direct measurements of sulfation in periportal and pericentral areas with microlight guide techniques (35).

Microlight guide techniques (35) have allowed direct kinetic analysis of sulfation in sublobular areas for the first time. Sulfation is 2-fold higher in periportal than pericentral areas. Since it is unlikely that the uptake of inorganic sulfate is rate limiting for sulfation, these differences are probably due to differences in sulfotransferase content and/or rates of PAPS synthesis from the precursors inorganic sulfate, ATP and UTP. Future studies should be directed at defining factors regulating PAPS synthesis and the role of PAPS in controlling rates of sulfation. The role of sulfatases in regulating net sulfate production in sublobular zones should also be evaluated.

E. Use of Microlight Guides to Study Glucuronidation in Periportal and Pericentral Regions of the Liver Lobule

Work discussed above demonstrated that rates of glucuronidation of low concentrations (<30 μM) of 7-hydroxycoumarin could be measured in specific zones of the liver lobule (35). The half-maximal concentration of 7-hydroxycoumarin for glucuronidation in perfused liver from phenobarbital-treated rats is about 60 μM (36); thus, studies discussed above were performed under conditions where substrate was likely rate limiting. Therefore, a new method was developed to measure rates of glucuronidation of 7-hydroxycoumarin in specific zones of the liver lobule as a function of substrate concentration (36). This method is based on measuring increases in surface fluorescence after glucuronidation is inhibited completely by ethanol and nitrogen. Livers from fed, phenobarbital-treated rats were perfused under normoxic conditions with sulfate-free buffer and the fluorescence of free hydroxycoumarin was monitored in periportal

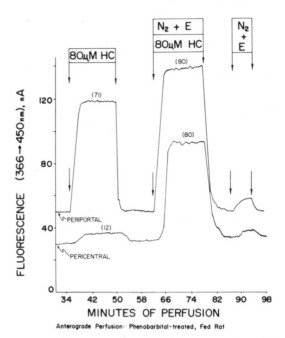

Fig. 5. Fluorescence increase upon infusion of 7-hydroxycoumarin and N_2-saturated perfusate containing 20 mM ethanol in periportal and pericentral regions of the liver perfused in the anterograde direction. Two microlight guides were placed on periportal and pericentral regions on the left lateral lobe of the liver. 7-Hydroxycoumarin (HC) and N_2-saturated perfusate containing 20 mM ethanol (N_2 + E) were infused as indicated by horizontal bars and arrows. Numbers in parentheses represent micromolar concentrations of free 7-hydroxycoumarin in the tissue calculated as described in text (details in Ref. 36). Reproduced by permission of Williams and Wilkins Co.

and pericentral regions with microlight guides (36) (Fig. 5). Subsequently, 20 mM ethanol in nitrogen-saturated perfusate was infused and fluorescence of 7-hydroxycoumarin increased to higher steady-state values (N_2 + E) (Fig. 5). All 7-hydroxycoumarin was recovered unmetabolized in the presence of nitrogen and ethanol; therefore, the 7-hydroxycoumarin fluorescence arising from the liver surface represents the infused concentration of hydroxycoumarin in the tissue (36). Maximal fluorescence changes due to 7-hydroxycoumarin during anoxia plus ethanol were used to calculate concentrations of free hydroxycoumarin in discrete areas of the liver lobule during normoxia. The decrease in free 7-hydroxycoumarin concentration in any region was assumed to be due to glucuronidation, since more than 90% of the nonfluorescent products formed in all perfusions were glucuronides (36). For example, in the experiment depicted in Fig. 5 the concentration of glucuronide conjugates formed by the periportal area was 80 − 71 = 9 μM.

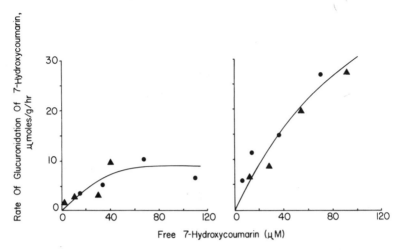

Fig. 6. Rates of glucuronidation of 7-hydroxycoumarin in periportal (left) and pericentral (right) regions of perfused livers from phenobarbital-treated rats. Concentration of glucuronide conjugates formed in each region during anterograde (●) and retrograde (▲) perfusions was derived from flourescence measurements of free 7-hydroxycoumarin in the tissue (see Fig. 5). Rates were calculated using the flow rate and the wet weight of each sublobular region (wet weight/2). Concentrations of substrate are the average of the free 7-hydroxycoumarin (μM) entering and leaving each sublobular region (details in Ref. 36). Reproduced by permission of Williams and Wilkins Co.

From the flow rate, the mass of each region, and the concentration of glucuronide formed by that region, local rates of glucuronidation were calculated. The substrate concentration in any specific region of the liver lobule was calculated as the average of the concentration of 7-hydroxycoumarin entering and leaving each region. Sublobular rates of glucuronidation in the intact liver are plotted against local 7-hydroxycoumarin concentrations in Fig. 6. Double-reciprocal analysis of data in Fig. 6 showed that maximal rates of glucuronidation in periportal regions were 9.6 μmol/g/hr; half-maximal rates were achieved with 26 μM 7-hydroxycoumarin. In contrast, maximal rates of glucuronidation by pericentral hepatocytes were 35 μmol/g/hr and were half maximal with 47 μM substrate. The differences in substrate required for half-maximal rates in the two regions were not statistically significantly different (36).

Periportal and pericentral areas were microdissected and assayed for 7-hydroxycoumarin glucuronosyltransferase activity under optimal conditions *in vitro*. Activity of glucuronosyltransferase in pericentral regions was about 3.2-fold greater than that in periportal areas. In both regions glucuronosyltransferase activity was half maximal with about 230 μM UDPGA and 54 μM hydroxycoumarin. Thus, in livers from fed, phenobarbital-treated rats, glucuronosyltransferase activity and maximal rates of glucuronidation were 3-fold higher in pericentral than periportal regions of the liver lobule (36).

Microlight guides have also been used to measure rates of glucuronidation in livers from untreated and 3-methylcholanthrene-treated rats (99). Maximal rates of glucuronidation were about 11 μmol/g/hr in both periportal and pericentral regions of livers from untreated and 3-methylcholanthrene-treated rats (99). Glucuronosyltransferase activity measured in microdissected periportal and pericentral regions did not correlate with rates of glucuronidation measured with microlight guides. For example, glucuronosyltransferase activity was 2-fold higher in pericentral than in periportal areas in perfused livers from untreated rats (77,99), while rates of glucuronidation were equivalent in both regions (99). Furthermore, 3-methylcholanthrene treatment caused a 5- to 8-fold increase in glucuronosyltransferase activity in both sublobular areas (77,99); however, it had no effect on rates of glucuronidation measured with microlight guides (99). These data illustrate the pitfalls of predicting rates of glucuronidation from assays of glucuronosyltransferase *in vitro* and suggest that further work is needed to understand the role of UDPGA in controlling rates of glucuronidation in periportal and pericentral areas.

Glucuronidation has also been studied in fractions enriched in hepatocytes isolated from periportal and pericentral areas (84). These studies agree with the relative rates of glucuronidation measured in periportal and pericentral regions of perfused livers with microlight guides (36,99). In preparations from phenobarbital-treated rats, maximal rates of glucuronidation of *p*-nitrophenol were 2-fold greater in hepatocytes isolated from pericentral than from periportal areas. However, in preparations from normal and 3-methylcholanthrene-treated rats, glucuronidation of *p*-nitrophenol was equivalent in hepatocytes from periportal and pericentral areas.

Chowdhury *et al.* (100) observed that immunohistochemical staining for glucuronosyltransferase was distributed equally across the liver lobule in liver from untreated rats. Ullrich *et al.* (101) combined immunohistochemical staining and microchemical techniques to study the effects of inducers on glucuronosyltransferase distribution and activity within the liver lobule. In livers from normal rats, pericentral areas had slightly more immunohistochemical staining and activity towards 1-naphthol (~20%) than periportal areas. 3-Methylcholanthrene treatment increased glucuronosyltransferase activity 6-fold but did not alter the relative distribution of activities in sublobular regions. In livers from phenobarbital-treated rats, periportal areas had slightly more immunohistochemical staining and activity toward 1-naphthol (~10%) than pericentral areas (101). The observation that 3-methylcholanthrene treatment increases glucuronosyltransferase activity toward 1-napthol 6-fold in microchemical determinations (101), but only increases rates of glucuronidation of 1-naphthol in perfused livers by 70% (42), illustrates further the difficulty in using assays of glucuronosyltransferase *in vitro* to predict rates of glucuronidation in periportal and pericentral areas.

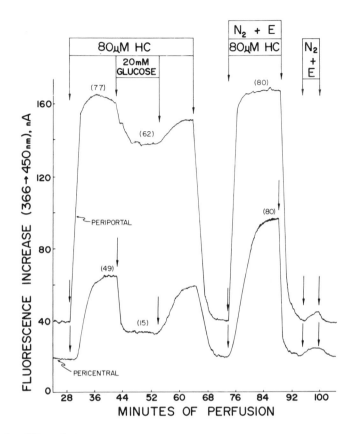

Fig. 7. Effect of glucose on 7-hydroxycoumarin fluorescence in periportal and pericentral regions of a liver perfused in the anterograde direction. Two microlight guides were placed on periportal and pericentral regions 1 mm apart on the left lateral lobe of the liver. 7-Hydroxycoumarin (HC) (80 μM), glucose (20 mM), and N_2-saturated perfusate containing 20 mM ethanol (N_2 + E) were infused as indicated by horizontal bars and arrows. Numbers in parentheses represent micromolar concentrations of free 7-hydroxycoumarin in the tissue calculated as described in text (details in Ref. 51). Anterograde perfusion; phenobarbital-treated, fasted rat. Reproduced by permission of the Biochemical Society, London.

F. Regulation of Glucuronidation by UDPGA Supply in the Fasted State

To investigate the effect of UDPGA on glucuronidation in periportal and pericentral regions of the liver lobule, the effect of glucose on sublobular rates of glucuronidation was studied (51). In perfused livers from fed, phenobarbital-

TABLE III

Effect of Glucose on Rates of Production of 7-Hydroxycoumarin Glucuronide in Periportal and Pericentral Regions of the Liver Lobule[a]

	Average free 7-hydroxycoumarin concentration during normoxic perfusion (μM)		Rate of glucuronide production (μmol g^{-1} hr^{-1})	
Additions	Periportal	Pericentral	Periportal	Pericentral
Anterograde (13)				
None	76.0 ± 0.8	59.2 ± 3.2	2.7 ± 0.7	9.5 ± 1.5
20 mM Glucose	72.6 ± 1.2	42.9 ± 2.7	5.5 ± 0.9[b]	16.6 ± 1.2[c]
Difference			2.8 ± 0.6	7.2 ± 1.3
Retrograde (8)				
None	55.7 ± 4.0	69.5 ± 2.0	2.2 ± 0.8	6.9 ± 1.3
20 mM Glucose	19.6 ± 2.7	51.1 ± 2.1	2.5 ± 1.1	18.9 ± 2.2[c]
Difference			0.3 ± 0.2	12.0 ± 1.2

[a]Concentrations of free 7-hydroxycoumarin in periportal regions during anterograde perfusions and in pericentral regions during retrograde perfusions were calculated from infused 7-hydroxycoumarin and free 7-hydroxycoumarin present in the region during normoxic perfusion from fluorescence changes (Fig. 7, this chapter; details in Ref. 51). Concentrations of free 7-hydroxycoumarin in periportal regions during retrograde perfusions and in pericentral areas during anterograde perfusions were calculated from the 7-hydroxycoumarin concentration in the "upstream" regions (i.e., periportal area during anterograde perfusion), and free 7-hydroxycoumarin detected in "downstream" regions. The rate of glucuronidation was calculated from the mass of the lobular region (one-half liver weight), the flow rate of the liver, and the concentration of glucuronide formed locally. Mean ± SEM (n).
[b]$p < 0.01$ comparing sublobular rates of glucuronidation before and after glucose infusion.
[c]$p < 0.001$ comparing sublobular rates of glucuronidation before and after glucose infusion.

treated rats, infusion of 20 mM glucose had no effect on rates of glucuronidation of 7-hydroxycoumarin (51), in agreement with earlier reports (19). Thus, in the fed state glucuronidation is not limited by the supply of glucose for the synthesis of UDPGA. However, in livers from fasted, phenobarbital-treated rats, infusion of glucose doubled rates of glucuronidation indicating that carbohydrate reserves are rate limiting for the synthesis of UDPGA and maximal rates of glucuronidation in the fasted state (19,51).

To study the regional effects of carbohydrate reserves on glucuronidation, microlight guides were used to measure rates of glucuronidation in sublobular zones (Fig. 7, Table III). Fasting decreased rates of glucuronidation by 50–70% in both regions, and glucose infusion reversed the fasting-induced decrease in rates completely indicating (Table III) that the supply of UDPGA is an important rate determinant in both regions of the liver lobule in livers from fasted rats.

Threefold higher rates were observed in pericentral areas in the absence of glucose in the fasted state (Table III). Higher rates of glucuronidation in pericentral areas are probably not due to a greater affinity of glucuronosyltransferase(s) for UDPGA, since the concentration of UDPGA required for half saturation of glucuronosyltransferase was 230 μM in both periportal and pericentral areas (36). Threefold higher rates were also observed in pericentral areas in the fed and starved state in the presence of glucose (Table III). Therefore, in the presence of excess carbohydrate, greater glucuronosyltransferase activity in pericentral areas is most likely an important rate determinant of glucuronidation.

Formation of UDPGA from glucose requires glucokinase to generate glucose 6-phosphate (102). Results presented here show that glucose stimulated glucuronidation rapidly ($t_{0.5}$ 1–2 min) in both regions of the liver lobule, indicating that glucose is phosphorylated rapidly in both regions (Fig. 7). Hexokinase is probably less important in liver than glucokinase since it is localized primarily in nonparenchymal cells (103) where there is little or no glucuronosytransferase (100). Microchemical analysis of lyophilized tissue sections from normal rats indicates that glucokinase activity is highest in pericentral regions (104). While higher activities of glucokinase in this region are consistent with our observation that glucose stimulated glucuronidation twice as much in pericentral as in periportal areas (Table III), glucokinase activity is probably not rate limiting for stimulation of glucuronidation by glucose. Matsumura and Thurman (68) observed that glucose infusion initially increased rates of O_2 uptake in periportal and pericentral regions by 23 and 15 μmol/g/hr, respectively. If we assume that this increase in respiration is due to increased ADP due to the phosphorylation of glucose, we can calculate that both sublobular areas phosphorylate glucose at rates of at least 80 μmol/g/h (51). Thus, rates of glucose phosphorylation greatly exceed the stimulation of rates of glucuronidation by glucose (Table III). The larger stimulation of glucuronidation by glucose in pericentral areas is most likely due to a higher glucuronosyltransferase activity (36) and possibly increased synthesis of UDPGA from glucose 6-phosphate in that area. This later possibility requires further study.

III. CONCLUSION

In view of the fact that many hepatotoxins damage specific zones of the liver lobule, attention is being focused on the metabolic heterogenity within the liver lobule. Because of the multiple factors regulating conjugation reactions, measurements of enzyme activity or content in sublobular zones under artificial conditions *in vitro* have limited usefulness. Further, hepatocyte subpopulations isolated from periportal and pericentral areas must be characterized metabolically before they can be used routinely. In contrast, microlight guides placed on periportal and pericentral areas of perfused livers have allowed direct kinetic analysis of sulfation and glucuronidation in sublobular zones under nearly physiological conditions for the first time. Sulfation predominates in periportal areas while glucuronidation is distributed evenly in periportal and pericentral areas in livers from untreated and 3-methylcholanthrene-treated rats. In contrast, glucuronidation predominates in pericentral areas following phenobarbital treatment. Interactions with nutrition can also be studied with microlight guides. Fasting decreases rates of glucuronidation in both sublobular zones by 50%. These new techniques should be very useful in the future in studies designed to elucidate the mechanism of zone-specific toxins.

REFERENCES

1. DeBaun, J. R., Smith, J. Y., Miller, E. C., and Miller, J. A. Reactivity *in vivo* of the carcinogen *N*-hydroxy-2-acetylaminofluorene: Increase by sulfate ion. *Science* **167**, 184–186 (1970).
2. Meerman, J. H., VanDoorn, A. B., and Mulder, G. J. Inhibition of sulfate conjugation of *N*-hydroxy-2-acetylaminofluorene in isolated perfused rat liver and in the rat *in vivo* by pentachlorophenol and sulfate ion. *Cancer Res.* **40**, 3772–3779 (1980).
3. Boberg, E. W., Miller, E. C., Miller, J. A., Poland, A., and Liem, A. Strong evidence from studies with brachymorphic mice and pentachlorophenol that l-sulfooxysafrole is the major ultimate electrophic and carcinogenic metabolite of 1-hydroxysafrole in mouse liver. *Cancer Res.* **43**, 5163–5173 (1983).
4. Kedderis, G. L., Dryoff, M. C., and Rickert, D. E. Hepatic macromolecular binding of the hepatocarcinogen 2,6-dinitrotoluene and its 2,4 isomer *in vivo*: Modulation by the sulfotransferase inhibitors pentachlorophenol and 2,6-dichloro-4-nitrophenol. *Carcinogenesis* **5**, 1199–2204 (1984).
5. Morlon, K. C., Beland, F. A., Evans, F. E., Fullerton, N. F., and Kadlubar, F. F. Metabolic activation of *N*-hydroxy-*N*-*N*'-diacetylbenzidine by hepatic sulfotransferase. *Cancer Res.* **40**, 751–757 (1980).
6. Watabe, T., Fujieda, T., Hiratsuka, A., Ishizuka, T., Hakamata, Y., and Ogura, K. The carcinogen, 7-hydroxymethyl-12-methylbenz (*a*)anthracene, is activated and covalently binds to DNA via a sulphate ester. *Biochem. Pharmacol.* **34**, 3002–3005 (1985).
7. Jakoby, W. B. Aryl and hydrosteroid sulfotransferases. *In* "Sulfate Metabolism and Sulfate Conjugation" (G. J. Mulder, J. Caldwell, G. M. J. Van Kempen, and R. J. Vonk, eds.), pp. 13–21. Taylor & Francis, London, 1982.
8. Sekura, R. D., and Jakoby, W. B. Phenol sulfotransferases. *J. Biol. Chem.* **254**, 5658–5663 (1979).
9. Chen, L. J., Thaler, M. M., Kane, R., and Bujanover, Y. Development and regulation of bile salt sulfotransferase. *n* "Sulfate Metabolism and Sulfate Conjugation" (G. J. Mulder, J. Caldwell, G. M. J. Van Kempen, and R. J. Vonk, eds.), pp. 239–245. Taylor & Francis, London, 1982.
10. Green, J. M., and Singer, S. S. Enzymatic sulfation of steroids. Study of the specific estradiol-17β sulfotransferase of rat liver cytosol that converts the estrogen to its 3-sulfate, and some elements of the endocrine control of its production. *Can. J. Biochem. Cell Biol.* **61**, 15–22 (1983).

11. Lin, J. H., and Levy, G. Sulfate depletion after acetaminophen administration and replenishment by infusion of sodium sulfate or N-acetylcysteine in rats. *Biochem. Pharmacol.* **30,** 2723–2725 (1981).

12. Krigsheld, K. R., and Mulder, G. J. The dependence of the rate of sulphate conjugation on the plasma concentration of inorganic sulphate in the rat *in vivo*. *Biochem. Pharmacol.* **31,** 3997–4000 (1982).

13. Hjelle, J. J., Hazelton, G. A., and Klaasen, C. D. Acetaminophen decreases adenosine 3'-phosphate 5'-phosphosulfate and uridine diphosphoglucuronic acid in rat liver. *Drug Metab. Dispos.* **13,** 35–41 (1985).

14. Greenberg, D. M. Biosynthesis of cysteine and cystine. *In* "Metabolic Pathways: Metabolism of Sulfur Compounds" (D. M. Greenberg, ed.), Vol. 7, Chap. 12. Academic Press, New York, 1975.

15. Schwartz, L. R. Sulfation of 1-naphthol in isolated rat hepatocytes. Dependence on inorganic sulfate. *Hoppe-Seyler's Z. Physiol. Chem.* **365,** 43–48 (1984).

16. Mulder, G. J., and Keulemans, K. Metabolism of inorganic sulfate in the isolated perfused rat liver. *Biochem. J.* **176,** 959–965 (1978).

17. Sundheimer, D. W., and Brendel, K. Factors influencing sulfation in isolated rat hepatocytes. *Life Sci.* **34,** 23–29 (1984).

18. Bracht, A., Bracht, A. K., Schwab, A., and Scholz, R. Transport of inorganic anions in the perfused rat liver. *Eur. J. Biochem.* **114,** 471–479 (1981).

19. Reinke, L. A., Belinsky, S. A., Evans, R. K., Kauffman, F. C., and Thurman, R. G. Conjugation of p-nitrophenol in the perfused rat liver: The effect of substrate concentration and carbohydrate reserves. *J. Pharmacol. Exp. Ther.* **271,** 863–870 (1981).

20. Sundheimer, D. W., and Brendel, K. Metabolism of harmol and transport of harmol conjugates in isolated rat hepatocytes. *Drug. Metab. Dispos.* **11,** 433–440 (1983).

21. Pang, K. S., Koster, H., Halsema, I. C., Scholtens, E., and Mulder, G. J. Aberrant pharmacokinetics of harmol in the perfused rat liver preparation: Sulfate and glucuronide conjugations. *J. Pharmacol. Exp. Ther.* **219,** 134–140 (1981).

22. Buch, H., Rummel, W., Pfleger, K., Eschrich, C., and Texter, N. Urinary excretion of free and conjugated sulfate following administration of N-acetyl-p-aminophenol in rats and men. *Naunyn-Schmiedeberg's Arch. Exp. Pathol. Pharmakol.* **259,** 276 (1968).

23. Galinsky, R. E., and Levy, G. Dose- and time-dependent elimination of acetaminophen in rats: Pharmacokinetic implications of cosubstrate depletion. *J. Pharmacol. Exp. Ther.* **219,** 14–20 (1981).

24. Wong, K. P., and Yeo, T. Assay of adenosine 3'-phosphate 5'-sulphatophosphate in hepatic tissues. *Biochem. J.* **181,** 107–112 (1979).

25. Roy, A. B. Sulfotransferases. *In* "Sulfation of Drugs and Related Compounds" (G. J. Mulder, ed.), pp. 83–131. CRC Press, Boca Raton, Florida, 1981.

26. Barfold, D. J., and Jones, J. G. Thiol-dependent changes in the properties of rat liver sulfotransferases. *Biochem. J.* **123,** 427–434 (1971).

27. Mulder, G. J., and Scholtens, E. The availability of inorganic sulfate in blood for sulfate conjugation of drugs in rat liver *in vivo*. *Biochem. J.* **172,** 247–255 (1978).

28. Pang, K. S., and Terrell, J. A. Conjugation kinetics of acetaminophen by the perfused rat liver preparation. *Biochem. Pharmacol.* **30,** 1959–1965 (1981).

29. Aw, T. K., and Jones, D. P. Secondary bioenergetic hypoxia: Inhibition of sulfation and glucuronidation reactions in isolated hepatocytes at low O_2 concentration. *J. Biol. Chem.* **257,** 8997–9004 (1982).

30. Lyman, S. D., and Poland, A. Effect of the brachymorphic trait in mice on xenobiotic sulfate ester formation. *Biochem. Pharmacol.* **32,** 3345–3350 (1983).

31. Roy, A. B. *In* "Handbuch der Experimentellen Pharmakologie" (B. B. Brodie and J. Gillette, eds.), Vol. 28/2, pp. 536–563. Springer-Verlag, Berlin and New York, 1971.

32. Gregory, J. D., and Lipmann, F. The transfer of sulfate among phenolic compounds with 3', 5' diphosphosadenosine as coenzyme. *J. Biol. Chem.* **229**, 1081–1090 (1957).

33. Schwartz, L. R. Modulation of sulfation and glucuronidation of 1-naphtol in isolated rat liver cells. *Arch. Toxicol.* **44**, 137–145 (1980).

34. Norling, A., Moldeus, P., Anderson, B., and Hanninen, O. Passage of glucuronides and sulphates into isolated hepatocytes and action on conjugation reactions. *In* "Conjugation Reactions in Drug Biotransformation" (A. Aitio, ed.), pp. 303–311. Elsevier/North-Holland, Amsterdam, 1978.

35. Conway, J. G., Kauffman, F. C., Ji, S., and Thurman, R. G. Rates of sulfation and glucuronidation of 7-hydroxycoumarin in periportal and pericentral regions of the liver lobule. *Mol. Pharmacol.* **22**, 507–516 (1982).

36. Conway, J. G., Kauffman, F. C., Tsukada, T., and Thurman, R. G. Glucuronidation of 7-hydroxycoumarin in periportal and pericentral regions of the liver lobule. *Mol. Pharmacol.* **25**, 487–494 (1984).

37. Jayle, M. F., and Pasqualini, J. R. Implication of conjugation of endogenous compounds-steroids and thyroxine. *In* "Glucuronic Acid: Free and Combined" (G. J. Dutton, ed.), pp. 507–543. Academic Press, New York, 1966.

38. Falany, C. N., and Tephly, T. R. Separation, purification and characterization of three iso-enzymes of UDP-glucuronyltransferase from rat liver microsomes. *Arch. Biochem. Biophys.* **227**, 248–258 (1983).

39. Bock, K. W., Jasting, D., Lilienblum, W., and Pfeil, H. Purification of rat liver microsomal UDP-glucuronyltransferase. Separation of two enzyme forms inducible by 3-methylcholanthrene and phenobarbital. *Eur. J. Biochem.* **98**, 19–26 (1979).

40. Weatherill, P. J., and Burchell, B. The separation and purification of rat liver UDP-glucuronyltransferase activities toward testosterone and estrone. *Biochem. J.* **189**, 377–380 (1980).

41. Singh, J., and Schwartz, L. R. Dependence of glucuronidation rate on UDP-glucuronic acid levels in isolated hepatocytes. *Biochem. Pharmacol.* **30**, 3252–3254 (1981).

42. Bock, K. W., and White, I. N. H. UDP-glucuronyltransferase in perfused rat liver and microsomes: Influence of phenobarbital and 3-methylcholanthrene. *Eur. J. Biochem.* **46**, 451–459 (1974).

43. Abou-El-Makarem, M. M., Otani, G., and Bock, K. W. Glucuronidation of 1-naphthol and bilirubin by intact liver and microsomal fractions: Influence of the uridine diphosphate glucuronic acid content. *Biochem. Soc. Trans.* **3**, 881–883 (1975).

44. Watkins, J. B., and Klassen, C. D. Chemically-induced alteration of UDP-glucuronic acid concentration in rat liver. *Drug. Metab. Dispos.* **11**, 37–40 (1983).

45. Felsher, B. F., Carpio, N. M., and VanCouvering, K. Effect of fasting and phenobarbital on hepatic UDP-glucuronic acid formation in the rat. *J. Lab. Clin. Med.* **93**, 414–427 (1979).

46. Price, V. F., and Jollow, D. J. Role of UDPGA flux in acetaminophen clearance and hepatotoxicity. *Xenobiotica* **14**, 553–559 (1984).

47. Lucier, G. W., McDaniel, O. S., and Matthews, H. B. Microsomal rat liver UDP-glucuronyltransferase: Effects of piperonyl butoxide and other factors on enzyme activity. *Arch. Biochem. Biophys.* **145**, 520–530 (1971).

48. Winsnes, A. Kinetic properties of different forms of hepatic UDP-glucuronyltransferase. *Biochim. Biophys. Acta* **284**, 394–405 (1972).

49. Bock, K. W., Frohling, W., Remmer, H., and Rexer, B. Effects of phenobarbital and 3-methylcholanthrene on substrate specificity of rat liver microsomal UDP-glucuronyltransferase. *Biochim. Biophys. Acta* **327**, 46–56 (1973).

50. Antoine, B., Magdalon, J., and Siest, G. Kinetic properties of UDP-glucuronosyltransferase(s) in different membranes of rat liver cells. *Xenobiotica* **14**, 575–579 (1984).

51. Conway, J. G., Kauffman, F. C., and Thurman, R. G. Effect of glucose on 7-hydroxy-

coumarin in glucuronide production in periportal and pericentral regions of the liver lobule. *Biochem. J.* **226,** 749–756 (1985).

52. Ullrich, D., and Bock, K. W. Glucuronide formation of various durgs in liver microsomes and in isolated hepatocytes from phenobarbital- and 3-methylcholanthrene-treated rats. *Biochem. Pharmacol.* **33,** 97–101 (1984).

53. Aw, T. Y., and Jones, D. P. Control of glucuronidation during hypoxia: Limitation by UDP-glucose pyrophosphorylase. *Biochem. J.* **219,** 707–712 (1984).

54. Al-Turk, W. A., and Reinke, L. A. Diminished conjugation of products of mixed-function oxidation in perfused livers from hypophysectomized rats. *Pharmacology* **27,** 74–84 (1983).

55. Anderson, C. H., and Chatterson, R. T. Effect of hypophysectomy on estrogen conjugation and on plasmas and tissue concentrations of tritium after administration of ³H-estradiol-17. *Steroids* **28,** 785–803 (1976).

56. Moldeus, P., Anderson, B., and Norling, A. Interaction of ethanol oxidation with glucuronidation in isolated hepatocytes. *Biochem. Pharmacol.* **27,** 2583–2588 (1978).

57. Aw, T. Y., and Jones, J. P. Intracellular inhibition of UDP-glucose dehydrogenase during ethanol oxidation. *Chem.-Biol. Interact.* **43,** 283–288 (1983).

58. Burch, H. B., Lowry, O. H., Meinhardt, L., Max, P., and Chyu, K. Effect of fructose, dihydroxyacetone, glycerol, and glucose on metabolites and related compounds in liver and kidney. *J. Biol. Chem.* **245,** 2902–2102 (1970).

59. Keppler, D. O. R., Rudigier, J. F. M., Bischoff, E., and Decker, K. F. The trapping of uridine phosphates by D-galactosamine, D-glucosamine and 2-deoxy-D-galactose. *Eur. J. Biochem.* **17,** 246–253 (1970).

60. Keppler, D., and Decker, K. Studies on the mechanism of galactosamine hepatitis: Accumulation of galactosamine-1-phosphate and its inhibition of UDP-glucose pyrophosphorylase. *Eur. J. Biochem.* **10,** 219–225 (1969).

61. Otani, G., Abou-El-Makarem, M. M., and Bock, K. W. UDP-glucuronyltransferase in perfused rat liver and in microsomes: Effects of galactosamine and carbon tetrachloride on the glucuronidation of 1-naphthol and bilirubin. *Biochem. Pharmacol.* **25,** 1293–1297 (1976).

62. Eacho, P. I., Sweeny, D., and Weiner, M. Effects of glucose and fructose on conjugation of *p*-nitrophenol in hepatocytes of normal and streptozotocin diabetic rats. *Biochem. Pharmacol.* **30,** 2616–2619 (1981).

63. Price, V. F., and Jollow, D. J. Increased resistance of diabetic rats to acetaminophen-induced hepatotoxicity. *J. Pharmacol. Exp. Ther.* **220,** 504–513.

64. Price, V. F., and Jollow, D. J. Resistance of diabetic rats to acetaminophen hepatotoxicity. Role of co-substrates. *Fed. Proc., Fed. Am. Soc. Exp. Biol.* **40,** Abstr. 724 (1981).

65. Rappaport, A. M., Borrowy, Z. J., Lougheed, W. M., and Lotto, W. N. Subdivision of hexagonal liver lobules into a structural and functional unit: Role in hepatic physiology and pathology. *Anat. Rec.* **119,** 11–38 (1954).

66. Jungermann, K., and Katz, N. Functional hepatocellular heterogeneity. *Hepatology* **2,** 385–394 (1982).

67. Matsumura, T., Kashiwagi, T., Meren, H., and Thurman, R. G. Gluconeogenesis predominates in periportal regions of the liver lobule. *Eur. J. Biochem.* **144,** 409–415 (1984).

68. Matsumura, T., and Thurman, R. G. Predominance of glycolysis in pericentral regions of the liver lobule. *Eur. J. Biochem.* **140,** 229–234 (1984).

69. Rappaport, A. M. Physio-anatomical basis of toxic liver injury. *In* "Toxic Injury of the Liver" (E. Farber and M. Fischer, eds.), pp. 1–58. Dekker, New York, 1978.

70. Shirai, T., and King, C. M. Sulfotransferases and deacetylase in normal and tumor-bearing liver of CD rats: autoradiographical studies with *N*-hydroxy-2-acetylaminofluorene and *N*-hydroxy-4-acetylaminobiphenyl *in vitro* and *in vivo*. *Carcinogenesis* **3,** 1385–1391 (1982).

71. Smith, C., and Jollow, D. J. Potentiation of acetaminophen-induced liver necrosis in hamsters by galactosamine. *Pharmacologist* **18**, 156 (1976).

72. Loud, A. V. Quantitative stereological description of the ultrastructure of normal rat liver parenchymal cells. *J. Cell Biol.* **37**, 27–46 (1968).

73. Shank, R. E., Morrison, G., and Cheng, C. H. Cell heterogeneity within the hepatic lobule (quantitative histochemistry). *J. Histochem. Cytochem.* **7**, 237–239 (1959).

74. Morrison, G. R., Brock, F. E., and Karl, I. E. Quantitative analysis of regenerating and degenerating areas within the lobule of the carbon tetrachloride-injured liver. *Arch. Biochem. Biophys.* **111**, 448–464 (1965).

75. Welsch, F. A. Changes in distribution of enzymes within the liver lobule during adaptive increases. *J. Histochem. Cytochem.* **20**, 107–111 (1972).

76. Ghosh, A. H., Finegold, D., White, W., Zawalich, K., and Matchinsky, F. M. Quantitative histochemical resolution of oxidation-reduction and phosphate potentials within the simple hepatic acinus. *J. Biol. Chem.* **10**, 5476–5481 (1982).

77. Tsukada, T., Kauffman, F. C., and Thurman, R. G. The effect of inducing agents on the distribution and kinetic properties of UDP-glucuronosyltransferase in periportal and pericentral zones of rat liver. *Fed. Proc., Fed. Am. Soc. Exp. Biol.* **42**, Abstr. 3642 (1983).

78. Baron, J., and Kawabata, T. T. Intratissue distribution of activating and detoxicating enzymes. *In* "Biological Basis of Detoxification" (J. Caldwell and W. B. Jakoby, eds.), pp. 105–135. Academic Press, New York, 1983.

79. Quistroff, B., Grunnet, N., and Cornell, N. W. Digitonin perfusion of rat liver. A new approach in the study of intra-acinar and intracellular compartments in the liver. *Biochem. J.* **226**, 289–291 (1985).

80. Desmond, J. R., Kupfer, A., Schenker, S., and Branch, R. A. The differential localization of various drug metabolizing systems within the rat liver lobule as determined by the hepatotoxins allyl alcohol, carbon tetrachloride and bromobenzene. *J. Pharmacol. Exp. Ther.* **217**, 127–132 (1981).

81. Branch, R. A., Cotham, R., Johnson, R., Porter, J., Desmond, P. V., and Schenker, S. Periportal localization of lorazepam glucuronidation in the isolated perfused rat liver. *J. Lab. Clin. Med.* **102**, 805–812 (1983).

82. Tonda, K., Hasegawa, T., and Hirata, M. Effects of phenobarbital and 3-methylcholanthrene pretreatments on monooxygenase activities and proportions of isolated rat hepatocyte sub-populations. *Mol. Pharmacol.* **23**, 235–243 (1983).

83. Sumner, I. G., Freedman, R. B., and Lodola, A. Characterization of hepatocyte sub-populations generated by centrifugal elutriation. *Eur. J. Biochem.* **134**, 539–545 (1983).

84. Tonda, K., and Hirata, M. Glucuronidation and sulfation of *p*-nitrophenol in isolated rat hepatocyte subpopulations. Effects of phenobarbital and 3-methylcholanthrene pretreatment. *Chem.-Biol. Interact.* **47**, 277–287 (1983).

85. Lindros, K. O., and Penttila, K. E. Digitonin-collagenase perfusion for efficient separation of periportal or perivenous hepatocytes. *Biochem. J.* **228**, 757–760 (1985).

86. Quistorff, B. Gluconeogenesis in periportal and perivenous hepatocytes of rat liver, isolated by a new high-yield digitonin/collagenase technique. *Biochem. J.* **229**, 221–226 (1985).

87. Pang, K. S., and Terrell, J. A. Retrograde perfusion to probe the heterogeneous distribution of hepatic drug metabolizing enzymes in rats. *J. Pharmacol. Exp. Ther.* **216**, 339–346 (1981).

88. Chance, B., and Thorell, B. Localization and kinetics of reduced pyridine nucleotides in living cells by microfluorometry. *J. Biol. Chem.* **234**, 3044–3050 (1959).

89. Chance, B., Oshino, N., Sugano, T., and Mela, L. Factors in oxygen delivery to tissue. *Microvasc. Res.* **8**, 276–282 (1974).

90. Mayevsky, A., and Chance, B. A new long-term method for the measurement of NADH fluorescence in intact rat brain with chronically implanted cannula. *In* "Oxygen Transport to Tissue" (H. I. Bicher and D. F. Bruly, eds.), pp. 239–244. Plenum, New York, 1974.

91. Scholz, R., Thurman, R. G., Williamson, J. R., Chance, B., and Bucher, T. Flavin and pyridine nucleotide oxidation-reduction changes in perfused rat liver. *J. Biol. Chem.* **244**, 2317–2324 (1969).
92. Ji, S., Chance, B., Nishiki, K., Smith, T., and Rich, T. Micro-light guides: A new method for measuring tissue fluorescence and reflectance. *Am. J. Physiol.* **236**, C144–C156 (1979).
93. Ji, S., Lemasters, J. J., and Thurman, R. G. A noninvasive method to study metabolic events within sublobular regions of hemoglobin-free perfused liver. *FEBS Lett.* **114**, 349 (1980).
94. Ji, S., Lemasters, J. J., and Thurman, R. G. A fluorometric method to measure sublobular rates of mixed-function oxidation in the hemoglobin-free perfused rat liver. *Mol. Pharmacol.* **19**, 513–516 (1981).
95. Lemasters, J. J., Ji, S., and Thurman, R. G. New micro-methods for studying sublobular structure and function in the isolated, perfused rat liver. *In* "Regulation of Hepatic Metabolism: Intra- and Intercellular Compartmentation" (R. G. Thurman, F. C. Kauffman, and K. Jungermann, eds.), pp. 159–183. Plenum, New York, 1986.
96. Belinsky, S. A., Kauffman, F. C., Ji, S., Lemasters, J. J., and Thurman, R. G. Stimulation of mixed-function oxidation of 7-ethoxycoumarin in periportal and pericentral regions of the perfused rat liver by xylitol. *Eur. J. Biochem.* **137**, 1–6 (1983).
97. Belinsky, S. A., Kauffman, F. C., and Thurman, R. G. Reducing equivalents for mixed function oxidation in periportal and pericentral regions of the liver lobule in perfused livers from normal and phenobarbital-treated rats. *Mol. Pharmacol.* **26**, 574–581 (1984).
98. Belinsky, S. A. and Thurman, R. G., unpublished observations.
99. Conway, J. G., Kauffman, F. C., Tsukada, T., and Thurman, R. G. Rates of production of glucuronides in periportal and pericentral regions of the lobule in livers from untreated and 3-methylcholanthrene-treated rats. *Toxicologist* **5**, Abstr. 676 (1985).
100. Chowdhury, J. R., Novikoff, P. M., Chowdhury, N. R., and Novikoff, A. B. Distribution of UDP glucuronosyltransferase in rat tissue. *Proc. Natl. Acad. Sci. U.S.A.* **82**, 2990–2994 (1985).
101. Ullrich, D., Fischer, G., Katz, N., and Bock, K. W. Intralobular distribution of UDP-glucuronosyltransferase in livers from untreated, 3-methylcholanthrene- and phenobarbital-treated rats. *Chem.-Biol. Interact.* **48**, 181–190 (1984).
102. Dutton, G. J. The biosynthesis of glucuronides. *In* "Glucuronidation of Drugs and Other Compounds" (G. J. Dutton, ed.), pp. 69–78. CRC Press, Boca Raton, Florida, 1980.
103. Dileepan, K. N., Wagle, S. R., Hoffman, F., and Decker, K. Distribution profile of glucokinase and hexokinase in parenchymal and sinusoidal cells of rat liver during development. *Life Sci.* **24**, 89–96 (1979).
104. Fischer, W., Ick, M., and Katz, N. Reciprocal distribution of hexokinase and glucokinase in the periportal and pericentral zone of the liver acinus. *Hoppe-Seyler's Z. Physiol. Chem.* **363**, 375–380 (1982).

3

Toxicology Studies in Cultured Hepatocytes from Various Species

CHARLENE A. MCQUEEN AND GARY M. WILLIAMS

Naylor Dana Institute
American Health Foundation
Valhalla, New York 10595

I. INTRODUCTION

During the past 10 years, the use of hepatocytes as a cell system for *in vitro* studies has steadily increased (Borek and Williams, 1980; Harris and Cornell, 1983). This was made possible by the development and availability of methods to isolate and maintain intact and functioning liver cells (see reviews in Sirica and Pitot, 1980; Thurman and Kaufman, 1980; Suolinna, 1982; McQueen and Williams, 1985). Early procedures used hyaluronidase and collagenase digestion followed by mincing (Howard *et al.*, 1967; Howard and Pesch, 1968; Berry and Friend, 1969). Subsequently, it was found that perfusion of the liver, first with a chelating agent and then with collagenase, was a highly effective method for dissociation (Seglen, 1972). This two-step perfusion of the liver is currently the most commonly used method and is preferred for the preparation of hepatocytes for monolayer culture (Williams *et al.*, 1977). This perfusion procedure has been used to isolate hepatocytes from several species including rat (Laishes and Williams, 1976; Dougherty *et al.*, 1980; McQueen *et al.*, 1981; Reese and Byard, 1981; Maslansky and Williams, 1982, 1985a), mouse (Dougherty *et al.*, 1980; Klaunig *et al.*, 1981, 1984; McQueen *et al.*, 1981, 1983; Reese and Byard, 1981; Maslansky and Williams; 1982, 1985a), hamster (McQueen *et al.*, 1981, 1983; Maslansky and Williams, 1982, 1985a; Kornbrust and Barfknecht, 1984), guinea pig (Reese and Bayard, 1981; Maslansky and Williams, 1985a), rabbit (Reese and Byard, 1981; McQueen and Williams, 1983; Maslansky and

51

THE ISOLATED
HEPATOCYTE

Williams, 1982, 1985a), and human (Reese and Bayard, 1981; Strom *et al.*, 1982; Guguen-Guillouzo *et al.*, 1982; Blaaubuer *et al.*, 1985).

Isolated hepatocytes can be used in suspension or monolayer culture. Monolayer cultures have several advantages: (1) the cells have time to recover from alterations sustained during the isolation procedure; (2) monolayer cultures initially consist of virtually all viable cells whereas suspensions consist of both viable and nonviable cells; (3) monolayer cultures can be maintained for longer periods of time than suspensions; and (4) monolayer cultures permit study of effects on functions requiring organization of cells.

II. HEPATOCYTE PRIMARY CULTURES: METHODS

A. Isolation of Hepatocytes

The techniques to be described were developed to provide a rapid, efficient method for isolation in high yield and viability of functioning hepatocytes suitable for monolayer culture. These techniques have been reviewed in detail (Maslansky and Williams, 1985a; McQueen and Williams, 1985). Young adult animals are the best source for hepatocytes. The method originally described for rat (Williams *et al.*, 1977) has been modified for use with mouse, hamster, guinea pig, and rabbit (McQueen *et al.*, 1981, 1982, 1983; Maslansky and Williams, 1981, 1982, 1985a). In this procedure, a peristaltic pump is used to perfuse the liver *in situ* with two solutions (I and II): I consists of 0.5 mM ethylene glycol bis(β-aminoethylether)-N, N'-tetraacetic acid (EGTA) in Hanks' balanced salt solution without Ca^{2+} and Mg^{2+}, and II consists of 100 units prescreened collagenase/ml Williams medium E (WME) buffered with 10 mM N-2-hydroxyethylpiperazine-N'-1-ethanesulfonic acid (HEPES) and adjusted to pH 7.35 with 1 N sodium hydroxide. Both solutions are sterilized by filtration before use and maintained at 37°C. The animal is anesthetized with an intraperitoneal injection of 50 mg pentobarbital/kg body weight. A ventral midline incision is made from the pubic bone to the xiphisternum and the portal vein is cannulated. A 21-gauge butterfly needle is used for rat, hamster, and guinea pig, a 25-gauge needle for mouse, and a 2.5-mm internal diameter Teflon cannula for rabbit. Solution I is initially perfused at a low flow rate, and the subhepatic inferior vena cava is severed to prevent excessive swelling of the liver. At this point, there should be uniform blanching of the liver. The incision is extended through the diaphragm, the thoracic vena cava is cut, and the perfusate allowed to run to waste. The subhepatic vena cava is then clamped to close the system, and the pump speed is increased to 40 ml/min for rat, 8 ml/min for mouse, 25 ml/min for hamster, 50 ml/min for guinea pig, and 70 ml/min for rabbit. Solution I is perfused for approximately 4 min.

Solution II is perfused for 10 min at 20, 15, 5, 30, or 50 ml/min for rat, hamster, mouse, guinea pig, or rabbit, respectively. The liver is covered to keep it moist and a 40-w bulb is positioned above the liver to keep it warm.

The liver is removed to a sterile dish containing WME. Following removal of any extraneous tissue, the liver is transferred to a dish containing either WME or solution II and the capsule is stripped away from the ventral side. Hepatocytes are detached from the fibrovascular supporting tissue by gentle brushing with a 1-in. hog-bristle paint brush or a wide-tooth stainless-steel dog hair comb. The cell suspension is transferred with a wide-bore pipet into 50-ml centrifuge tubes and the volume brought to 50 ml with WME. The cells are sedimented at 50 g for 2.5 min, then resuspended in WME containing 50 μg gentamycin/ml and calf serum. A serum concentration of 10% is used for rat, guinea pig, and rabbit, and 1% is used for mouse and hamster. An additional 4 mM $CaCl_2$ and $MgSO_4$ is added to all media for hamster hepatocytes. Cell viability is determined by trypan blue exclusion.

Greater than 90% of the cells obtained from rat, hamster, guinea pig, and rabbit are viable, while a viability of 85% or greater is seen with mouse.

B. Initiation of Monolayer Cultures

Monolayer cultures can be initiated in tissue culture flasks (25 or 75 cm^2), Petri dishes (100 mm), or directly onto plastic or glass coverslips (glass coverslips must be prepared by boiling in distilled water). Monolayer cultures can also be initiated on collagen membranes or collagen-coated dishes (Michalopoulos et al., 1978; Seglen and Fossa, 1978; Oldham et al., 1980; Sirica et al., 1980; Gebhart and Jung, 1982).

For DNA repair studies, an aliquot containing 5×10^5 viable cells/ml of WME with the appropriate serum concentration is seeded into a 6-well dish containing a coverslip and 2 ml of WME plus serum. A seeding density of 1 to 2 $\times 10^6$ cells is used for 25 cm^2 flasks for cytotoxicity or metabolism studies.

After an attachment interval of 2 hr (3 hr if glass coverslips are used) at 37°C in a humidified 5% CO_2 incubator, the cultures are washed and refed with WME.

C. Autoradiographic DNA Repair Assay

Following cell attachment, washed cultures are simultaneously exposed to 10 μCi methyl [^{3}H]thymidine (60–80 Ci/mmol) and the test chemical (Williams, 1976, 1977). Triplicate coverslips are done for each concentration of each chemical as well as the controls. An established genotoxic chemical, a structurally related nongenotoxic analogue, solvent, and untreated cultures serve as controls.

After 18 hr, the cultures are washed by dipping the coverslips in three successive beakers of phosphate-buffered saline, pH 7.4. To allow for better quantification of grains, the cells are swollen by immersing the coverslips in 1% sodium citrate for 8–10 min. The cells are fixed by three 30-min changes in ethanol:glacial acetic acid (5:1). The coverslips are air dried and mounted, cell side up on glass slides. The slides are then dipped in NTB emulsion which is prewarmed for 1 hr at 45°C, dried overnight in the dark, then placed in a light-tight box for 9–10 days at 4°C.

The slides are developed in D19 (Kodak) for 4 min and placed in a stop bath of acidified tap water for 30 sec. The slides are then immersed for 10 min in fixer (Kodak) followed by a 10-min tap water wash. The developed slides are stained with hemotoxylin and eosin.

Grain counts are determined in both the nucleus and the adjacent cytoplasm. Net nuclear grain counts are determined by subtracting the highest cytoplasmic count from the nuclear count. The mean and standard deviation of the average net nuclear counts of triplicate samples are determined. Currently, a value of greater than five is considered as positive; however, as statistical criteria for a positive response are being developed (Casciano and Gaylor, 1983), this value may be too conservative.

It has been suggested that rather than net nuclear counts, dose–response curves should be generated for both the nucleus and the cytoplasmic background (Lonati-Galligani *et al.*, 1983). This approach can result in substantial variation in replicate cultures and may lead to a different conclusion from that when the data is analyzed by the method of Williams (Rossberger and Andrae, 1985).

D. Cytotoxicity Assays

Following cell attachment, the washed cultures are exposed to the test chemical (McQueen and Williams, 1982). At intervals up to 24 hr, the culture medium is removed and centrifuged for 5 min at 1000 g to remove floating cells. The supernatant can be stored at $-10°C$. Lactate dehydrogenase activity in the culture medium is determined by measuring the lactate-dependent reduction of NAD to NADH (Amador *et al.*, 1963).

Viability of hepatocytes is determined by the *in situ* trypan blue uptake method (Williams *et al.*, 1977; Maslansky and Williams, 1982). Trypan blue is added to the cultures for 5 min and the cells are fixed with formalin. The number of cells that exclude dye as well as those that take up trypan blue are determined by light microscopy. These data are also used to determine cell survival, i.e., the number of cells remaining attached to the culture vessel.

In order to determine protein synthesis by the hepatocytes, 0.1 µCi of a ^{3}H-labeled amino acids mixture is added to the culture during the exposure to the test chemical. After 18 hr, the cells are pooled, lysed with 1% sodium lauryl sulfate and precipitated with 20% trichloroacetic acid (TCA). The TCA precipitates are collected on glass fiber filters, washed, and counted.

III. USES OF HEPATOCYTE PRIMARY CULTURES FOR XENOBIOTIC STUDIES

Monolayer cultures of hepatocytes can be utilized to study chemical biotransformation, cytotoxicity, and genotoxicity.

A. Chemical Biotransformation Studies

Cultured hepatocytes have been shown to retain enzymes that are expressed in the intact organ (Decad et al., 1977; Schmeltz et al., 1978; Dybing et al., 1979; Shirkey et al., 1979; McQueen et al., 1982; Maslansky and Williams, 1982; Miller et al., 1983). In this laboratory, enzyme preservation has been demonstrated in a comparison of the activities of phase I and II enzymes and P-450 levels in rat liver, freshly isolated cells, and monolayer cultures of hepatocytes immediately after cell attachment and after 20 hr in culture (Croci and Williams, 1985). The P-450 content of liver from male F-344 rats was 0.786 nmol/mg protein. After 20 hr in culture, the level decreased approximately 50%. The loss of P-450 differs according to culture conditions (Paine and Hockin, 1980; Nelson and Acosta, 1982; Lake and Paine, 1982; Holme et al., 1983), sex (Croci and Williams, 1985) and species (Maslansky and Williams, 1982). When hepatocytes from several species were cultured in Williams medium E, rabbit hepatocytes retained approximately 75% of the initial P-450 content while mouse hepatocytes showed the greatest decrease (Maslansky and Williams, 1982). Rat hepatocytes maintained for 4 days in supplemented Hams F_{12} medium lost 90% of the initial P-450 while human hepatocytes retained 50% after 6 to 8 days (Guillouzo et al., 1985).

Supplementation of culture medium with pyridines, particularly metyrapone, has resulted in maintenance of P-450 content (Villa et al., 1980; Paine et al., 1980; Lake and Paine, 1982). Similar results were observed when cystine and cysteine were omitted from the culture medium (Paine and Hockin, 1980; Nelson and Acosta, 1982). When enzyme activities were determined, however, low activities were noted in cultures in cystine- and cysteine-free medium (Lake and Paine, 1982). This was in contrast to the effect of metyrapone; cells cultured with this ligand had enzyme activities equal to or higher than the initial cell preparations. Thus, although P-450 content can be maintained, enzyme activities are not necessarily representative of the initial preparation or intact liver (Lake and Paine, 1982).

The mechanisms that result in a decrease in P-450 content, as well as the mechanism of action of metyrapone in cultured hepatocytes, have been investigated (Paine et al., 1980; Hockin and Paine, 1983). The decrease in P-450 content can be partially offset by supplementation of the culture medium with 5-aminolevulinic acid, a heme precursor (Paine and Hockin, 1980), or the addition of exogenous heme (Engelmann and Fierer, 1983). Although this observation suggested that cultured hepatocytes might have a defect in heme synthesis, the activity of 5-aminoevulinic acid synthetase was the same in freshly isolated cells and in those cultured for 24 hr (Hockin and Paine, 1983), while an increase in activity was observed for heme oxygenase, an enzyme involved in P-450 degradation. Metyrapone had no effect on 5-aminolevulinic acid synthetase but did lessen the increase of heme oxygenase, although not completely. It has been proposed that metyrapone acts primarily by binding to P-450 and protecting it from degradation (Hockin and Paine, 1983).

Although the procedures for isolation and culture of hepatocytes can result in a decrease in P-450 content, related enzyme activities are not always affected (Croci and Williams, 1985). The activity of aryl hydrocarbon hydroxylase, a mixed function oxidase, was found to remain fairly constant (Table I). The activities of two phase II enzymes, glutathione transferase and UDPglucuronyltransferase were stable during this period (Table I). It was interesting to note that in this study the level of UDPglucuronic acid decreased by approximately 95% during the isolation procedure. Cells used at this point, for example as suspensions, would probably need supplementation with exogenous substrate in order for efficient glucuronidation to occur. In contrast, with monolayer culture,

TABLE I

Enzyme Activities in Monolayer Cultures of Male F-344 Rat Hepatocytes[a]

	Product per minute per milligram protein (nmol)			
Enzyme	Liver	Isolated hepatocytes	Monolayer cultures (0 hr.)	Monolayer cultures (20 hr)
AHH[b]	1.01 ± 0.09	0.84 ± 0.11	0.84 ± 0.17	0.76 ± 0.05
GT[c]	1089	1187	1173	1023
UDPGT[d]	5.3 ± 0.2	4.9 ± 0.7	5.5 ± 0.6	6.9 ± 0.3

[a]Data from Croci and Williams (1985).
[b]Aryl hydrocarbon hydroxylase.
[c]Glutathionine-transferase; substrate, 1,2-dichloro-4-nitrobenzene.
[d]UDP-glucuronyltransferase; substrate, phenolphthalein.

UDPglucuronic acid returned to approximately 60% of the original value (Croci and Williams, 1985).

The preservation of enzyme activities in cultured hepatocytes has also been reflected in metabolism studies. Rat hepatocytes biotransformed benzo(a)pyrene continuously during 24 hr in culture (Schmeltz *et al.*, 1978). Phenols, diols, and quinones were identified. Both activation and detoxification pathways were preserved. This was in contrast to a subcellular fraction which generally lacks the capacity for detoxification. Monolayer cultures have also been shown to be capable of metabolizing 2-acetylaminofluorene (2-AAF) (Leffert *et al.*, 1977). In studies in this laboratory, a log linear rate of disappearance of 2-acetylaminofluorene was observed in hepatocytes from several species. Rabbit hepatocytes had the most rapid half-life while hepatocytes from rat and mouse had comparable rates (Table II). This rapid metabolism by rabbit hepatocytes may account for the lack of liver tumor induction by 2-AAF in rabbit.

Sex-, age-, and strain-dependent variations in biotransformation also occur in monolayer cultures of hepatocytes. Hepatocytes isolated from male and female F-344 rats differed in their ability to conjugate 2-AAF metabolites (McQueen *et al.*, 1986). When hepatocyte cultures from male rats were exposed to 10^{-5} M 2-AAF, the ratio of sulfate to glucuronide conjugates was about 6. This decreased slightly to 4 in cultures exposed to 10^{-4} M 2-AAF. The sulfate to glucuronide ratio for cells from females was 0.3 at 10^{-5} M 2-AAF and 0.7 at 10^{-4} M.

A difference in glucuronidation was noted between hepatocytes from 4- and 10-week-old Long–Evans rats. Generally, more glucuronides were detected in cultured cells from the older animals (Table III). Age-related changes in glu-

TABLE II

Half-Life ($t1/2$) of 2-Acetylaminofluorene in Monolayer Cultures of Hepatocytes

Species	Strain	Concentration (M)	$t1/2$ (hours)
Rat[a]	F344	10^{-4}	24.7 ± 6.6[d]
		10^{-5}	19.5 ± 7.8
Mouse[b]	C57/B16	10^{-4}	13.6[e]
		10^{-5}	27.7
Rabbit[c]	New Zealand White	10^{-4}	9.5 ± 5.3[d]
		10^{-5}	2.5 ± 2.2

[a]$n = 4$.
[b]$n = 2$.
[c]$n = 3$.
[d]Mean $\pm$ SD.
[e]Mean.

TABLE III

Effect of Age on Metabolism of 2-Acetylaminofluorene (2-AAF) in Hepatocytes from Long–Evans Rats

Concentration (M)	Age (weeks)	Parent (%)[a]				
		AAF	AF[b]	AAF-OH[c]	Sulfates	Glucuronides
5×10^{-6}	4	20.5	2.8	61.7	54.2	11.5
	4	22.2	1.3	61.8	57.3	11.3
5×10^{-6}	10	17.0	2.2	65.1	40.7	31.2
	10	3.2	1.0	73.6	58.8	55.0
10^{-5}	4	33.2	8.2	42.6	35.0	8.7
	4	19.3	2.6	61.0	43.1	14.4
10^{-5}	10	21.3	3.8	49.0	24.1	19.2
	10	14.1	3.0	60.6	31.5	31.6
10^{-4}	4	47.6	19.8	23.9	6.0	1.2
	4	44.7	17.4	23.2	6.8	1.0
10^{-4}	10	46.9	5.9	19.9	4.8	2.6
	10	46.0	7.3	21.7	13.1	3.3

[a]Cultures were incubated for 20 hr.
[b]2-Aminofluorene.
[c]Ring and N-hydroxylated 2-acetylaminofluroene.

curonidation have been reported, and the developmental pattern observed is substrate dependent (Basu *et al.*, 1971; Henderson, 1971; Scragg *et al.*, 1983).

B. Chemical Cytotoxicity Studies

Xenobiotic cytotoxicity has been monitored in monolayer cultures by numerous end points which measure membrane integrity or cellular function (Grisham *et al.*, 1978; Anuforo *et al.*, 1978; Jauregui *et al.*, 1981; McQueen *et al.*, 1982). These indicators of cytotoxicity can be utilized with hepatocytes of several species (Anuforo *et al.*, 1978; McQueen *et al.*, 1982; Harman and Fisher, 1983; Green *et al.*, 1984). In this laboratory, four indicators of toxicity have been utilized: leakage of lactate dehydrogenase (an intracellular enzyme), trypan blue exclusion, cell survival, and inhibition of protein synthesis. The sensitivity and usefulness of each end point was found to be dependent on the chemical tested. For example, no increase in enzyme release was observed in cultures exposed to 10^{-4} M 2-aminofluorene (2-AF) (Table IV). This concentration of 2-AF did not significantly decrease the number of viable cells; however, there was a slight decrease in the total number of cells that survived. A 70% decrease in protein synthesis was observed.

TABLE IV

Cytotoxicity of 2-Aminofluorene in Monolayer Cultures
of Rat Hepatocytes[a]

Concentration (M)	Control (%)			
	LDH[b] release	Trypan blue exclusion	Cell survival	Protein synthesis
10^{-5}	84	97	87	102
10^{-4}	108	93	84	67
10^{-3}	208	87	77	30

[a]Data from McQueen and Williams (1982, and unpublished observations).
[b]Lactate dehydrogenase.

It is important to note that culture conditions can alter the susceptibility of hepatocytes to cytotoxicity. Hepatocytes cultured in Williams medium E, which contains vitamin E, were sensitive to chemical toxicity only in the presence of extracellular Ca^{2+} (Schanne et al., 1979), while hepatocytes maintained in medium that lacked vitamin E were more susceptible to toxicity in the absence of Ca^{2+} (Smith et al., 1981). Recently, it was demonstrated that the addition of vitamin E had a protective effect only when added to Ca^{2+}-free medium, and it was suggested that the presence or absence of vitamin E may explain the earlier conflicting reports (Fariss et al., 1985).

C. Chemical Genotoxicity Studies

One aspect of the cellular effects of a chemical can be detected by measurement of interaction of reactive derivatives with DNA. Damage to DNA in monolayer cultures of hepatocytes can be measured directly as DNA binding (Leffert et al., 1977), by adduct formation (Poirier et al., 1980), and by determining strand breaks using alkaline elution (Bradley et al., 1982) and alkaline sucrose gradients (Schwarz et al., 1980). DNA damage can also be measured indirectly as DNA repair by autoradiography (Williams, 1976, 1977; Williams et al., 1982; Probst et al., 1981), liquid scintillation methods (Oldham et al., 1980; Althaus et al., 1982; Olson et al., 1983), or density gradient centrifugation (Andrae and Schwarz, 1981; McQueen and Williams, 1981; Yeager and Miller, 1978). Some difficulties with false-positive results due to the background have been reported with the liquid scintillation technique (Gupta et al., 1985).

The determination of chemically induced DNA repair in hepatocytes was developed to take advantage of the metabolic capability of liver and the specificity of DNA repair as an indicator of DNA damage. The autoradiographic measurement of DNA repair allows the cells undergoing replicative DNA synthesis to be readily distinguished from those in repair, thus eliminating the need for inhibitors such as hydroxyurea which must be used when liquid scintillation methods are employed. Additionally, DNA repair is a response that is not mimicked by cytotoxicity. At cytotoxic concentrations of a chemical, DNA repair is actually inhibited. This is in contrast to what is observed with strand breaks which can occur under cytotoxic conditions.

The use of the autoradiographic measurement of DNA repair was developed into a screening test, originally using rat hepatocytes (Williams, 1976, 1977). The test has been shown to reliably identify genotoxic chemicals of a wide variety of structural classes (Williams, 1980a,b; Probst et al., 1981); it has been recommended by working groups (IARC, 1980; Mitchell et al., 1983; National Research Council, 1983) and accepted by a number of testing and regulatory agencies (USFDA, 1982; Andrae, 1984; National Toxicology Program, 1984). In our laboratory, a wide variety of structural types of chemical have been tested (Williams, 1977, 1980a,b, 1983; McQueen and Williams, 1983). All of the noncarcinogens in these groups were negative while approximately 90% of the carcinogens were positive (McQueen and Williams, 1983).

Although DNA repair is induced by chemical carcinogens, the number of cells responding is different from that observed in cultures exposed to ultraviolet light. Damage repair can be induced by UV irradiation in 100% of the cells (Maslansky and Williams, 1985b). Cultures exposed to chemicals, however, generally show less than 100% of the cells in repair. This difference in response to radiation vs chemically induced DNA damage may reflect the metabolic heterogeneity of hepatocytes.

A carcinogen can fail to elicit DNA repair in hepatocytes for several reasons. Chemical carcinogens of the epigenetic type that do not interact with DNA (Weisburger and Williams, 1981) will, of course, not produce DNA damage. Thus, when carcinogens do not elicit DNA repair, it indicates a possible nongenotoxic mode of action. A carcinogen that inhibits DNA repair will also be negative in this assay, which requires DNA repair for a positive response.

For some chemicals, the lack of a positive response is due to the fact that the active metabolite is not produced by cultured hepatocytes. One example of this is cycasin, which requires cleavage to the agylcone, methylazoxymethanol, to be genotoxic. This can be performed by intestinal bacteria; therefore, when the hepatocyte primary culture (HPC)/DNA repair test is supplemented with bacterial β-glucosidase, cycasin produced DNA damage (Williams et al., 1981a).

Differences between mammalian species in metabolism and susceptibility to chemical carcinogens have been described (Miller et al., 1964; Irving, 1979; Weisburger and Weisburger, 1973). Hepatocytes from a variety of species can now be used to measure DNA repair. Known carcinogens and known (or presumed) noncarcinogens of six structural classes have been tested in mouse, rat, and hamster hepatocytes (McQueen and Williams, 1983). The carcinogens were positive in all species and the noncarcinogens were negative in mouse and rat hepatocytes. However, two presumed noncarcinogens, aflatoxin G_2 and pyrene, induced DNA repair in hamster hepatocytes. These results suggest that these two chemicals may be carcinogenic in hamsters.

Both qualitative and quantitative species differences have been noted in DNA repair in hepatocytes (Table V). In cultures of B6C3F$_1$ mouse hepatocytes, aflatoxin B$_1$ induced maximum DNA repair at a concentration 10–100 times higher than that required for the other species tested. This was also observed in hepatocytes isolated from C3H mice (Mori et al., 1984). This reflects the relative resistance of the mouse to the tumorigenic effect of aflatoxin B$_1$ (Wogan, 1973).

The value of multispecies testing is apparent with a chemical such as safrole, which is negative in rat hepatocytes but positive if mouse or hamster hepatocytes are used (Table V). Species-specific results have also been seen with procarbazine and other hydrazine derivatives (McQueen and Williams, 1985).

D. Other Studies

Cultured hepatocytes have been utilized to activate chemicals for *in vitro* assays to monitor genotoxic and epigenetic effects. Hepatocytes from several species including humans have been used in bacterial and mammalian cell mutagenesis and transformation assays (San and Williams, 1977; Langenbach et al., 1978; Poiley et al., 1979; Bartsch et al., 1980; Tong et al., 1981; Strom and Michalopoulos, 1982; Ved Brat and Williams, 1982). In these systems, DNA-

TABLE V

Species Differences in Genotoxicity of Aromatic Amines[a]

	Concentration inducing maximum repair			
Chemical	Rat (F-344)	Mouse (C57/B16)	Hamster (Syrian Golden)	Rabbit (New Zealand White)
2-Acetylaminofluorene (*M*)	10^{-4}	10^{-3}	10^{-4}	10^{-3}
Aflatoxin B$_1$ (M)	10^{-6} to 10^{-7}	10^{-4} to 10^{-5}	10^{-6}	10^{-6}
Methylene bis-2-choloraniline (*M*)	10^{-5}	5×10^{-5}	10^{-5}	10^{-3} to 10^{-4}
Safrole (percentage v/v)	Negative	10^{-4}	10^{-3}	NT[b]

[a]Data from McQueen et al. (1981, 1983, and unpublished observations).
[b]NT, Not tested.

damaging metabolites are formed by the hepatocytes and transferred to the cocultured cells, and genotoxic effects are determined in the target cells.

Monolayer cultures of hepatocytes can also be used to investigate processes requiring cell-to-cell contact. For example, promoters, one class of epigenetic carcinogens (Williams, 1980b), are agents that have the ability to inhibit intracellular communication (Murray and Fitzgerald, 1979; Yotti et al., 1979). An assay to measure this effect was originally developed with fibroblasts (Yotti et al., 1979). It has now been extended to a liver-derived system (Williams, 1980b; Williams et al., 1981b) to take advantage of the metabolic capacity of hepatocytes. Phenobarbital (Williams, 1980b) and pesticides such as DDT, chlordane, heptachlor and polybrominated biphenyls (Williams et al., 1981b; Telang et al., 1982; Williams et al., 1984) have been shown to markedly inhibit intracellular communication in liver cultures.

IV. CONCLUSIONS

Primary cultures of hepatocytes offer a useful system in which to study the genotoxicity, cytotoxicity, and metabolism of xenobiotics. Hepatocytes from a variety of species can be utilized for these studies, thus permitting identification of species-specific responses.

REFERENCES

Althaus, F. R., Lawrence, S. D., Sattler, G. L., Longfellow, D. G., and Pitot, H. C. (1982). Chemical quantification of unscheduled DNA synthesis in cultured hepatocytes as an assay for the rapid screening of potential carcinogens. Cancer Res. 42, 3010–3015.

Amador, E., Dorfman, L. E., and Walker, W. E. C. (1963). Serum lactic dehydrogenase activity: An analytical assessment of current assays. Clin. Chem. 9, 391–399.

Andrae, U. (1984). The DNA repair test with isolated hepatocytes. In "Critical Evaluation of Mutagenicity Testing" (R. Bas, V. Gloclain, P. Grosdanoff, D. Henschler, B. Kolbey, D. Muller, and N. Neubert, eds.), pp. 371–378. MMV Med. Verlag, Munich.

Andrae, U., and Schwarz, L. R. (1981). Induction of DNA repair synthesis in isolated rat hepatocytes by 5-diazouracil and other DNA damaging compounds. Cancer Lett. 13, 187–193.

Anuforo, D. C., Acosta, D., and Smith, R. V. (1978). Hepatotoxicity studies with primary cultures of rat liver cells. In Vitro 14, 981–987.

Bartsch, H., Malaveille, C., Camus, A. M., Martel-Planche, G., Hautefeuille, A., Sabadie, N., Barbin, A., Kuroki, T., Drevon, C., Piccoli, C., and Montesano, R. (1980). Validation and comparative studies on 180 chemicals with S. typhimurium strains and V79 Chinese hamster cells in the presence of various metabolizing systems. Mutat. Res. 76, 1–50.

Basu, T. K., Dickerson, J. W. T., and Parke, D. V. W. (1971). Effect of development on the activity of microsomal drug-metabolizing enzymes in rat liver. Biochem. J. 124, 19–24.

Berry, M. N., and Friend, D. S. (1969). High yield preparations of isolated rat parenchymal cells; a biochemical and fine structural study. *J. Cell Biol.* **43**, 506–520.

Blaaubuer, B. J., von Holsteijn, I., von Graft, M., and Paine, A. J. (1985). The concentration of cytochrome *P*-450 in human hepatocyte culture. *Biochem. Pharmacol.* **34**, 2405–2408.

Borek, C., and Williams, G. M. (1980). Differentiation and carcinogenesis in liver cell culture. *Ann. N.Y. Acad. Sci.* **349**, 1–429.

Bradley, M. O., Dysart, G., Fitzsimmons, K., Harbach, P., Lewin, J., and Wolf, G. (1982). Measurement by filter elution of DNA single- and double-strand breaks in rat hepatocytes: Effects of nitrosamines and γ-irradiation. *Cancer Res.* **42**, 2569–2597.

Casciano, D. A., and Gaylor, D. W. (1983). Statistical criteria for evaluating chemicals as positive or negative in the hepatocyte/DNA repair assay. *Mutat. Res.* **122**, 81–86.

Croci, T., and Williams, G. M. (1985). Activities of several phase I and phase II xenobiotic biotransformation enzymes in cultured hepatocytes from male and female rats. *Biochem. Pharmacol.* **34**, 3029–3035.

Decad, G. M., Hsieh, D. P. H., and Byard, J. L. (1977). Maintenance of cytochrome P-450 and metabolism of aflatoxin B_1 in primary hepatocyte cultures. *Biochem. Biophys. Res. Commun.* **78**, 279–287.

Dougherty, K. K., Spilman, S. D., Green, C. E., Steward, A. R., and Byard, J. L. (1980). Primary cultures of adult mouse and rat hepatocytes for studying the metabolism of foreign compounds. *Biochem. Pharmacol.* **29**, 2117–2124.

Dybing, E., Soderlund, E., Timm-Haug, L., and Thorgeirsson, S. S. (1979). Metabolism and activation of 2-acetylaminofluorene in isolated rat hepatocytes. *Cancer Res.* **39**, 3268–3275.

Engelmann, G. L., and Fierer, J. A. (1983). Use of exogeneous heme to maintain cytochrome P-450 content in primary cultures of adult rat hepatocytes. *In* "Isolation, Characterization, and Use of Hepatocytes" (R. A. Harris and N. W. Cornell, eds.), pp. 117–122. Elsevier, Amsterdam.

Fariss, M. W., Pascoe, G. A., and Reed, D. J. (1985). Vitamin E reversal of the effect of extracellular calcium on chemically induced toxicity in hepatocytes. *Science* **227**, 751–753.

Gebhart, R., and Jung, W. (1982). Biliary secretion of sodium fluorescein in primary cultures of adult rat hepatocytes. *J. Cell Sci.* **56**, 233–244.

Green, C. E., Dabbs, J. E., and Tyson, C. A. (1984). Metabolism and cytotoxicity of acetaminophen in hepatocytes isolated from resistant and susceptible species. *Toxicol. Appl. Pharmacol.* **76**, 139–149.

Grisham, J. W., Charlton, R. K., and Kaufman, D. G. (1978). *In vitro* assay of cytotoxicity with cultured liver: accomplishments and possibilities. *Environ. Health Perspect.* **25**, 161–171.

Guguen-Guillouzo, C., Campion, J. P., Brissot, P., Glaise, D., Launois, B., Bourel, M., and Guillouzo, A. (1982). High yield preparation of isolated human adult hepatocytes by enzymatic perfusion of the liver. *Cell Biol. Int. Rep.* **6**, 625–628.

Guillouzo, A., Beaune, P., Gascoin, M. N., Begue, J. M., Campion, J. P., Guengerich, F. P., and Guguen-Guillouzo, C. (1985). Maintenance of cytochrome *P*-450 in cultured adult human hepatocytes. *Biochem. Pharmacol.* **34**, 2991–2995.

Gupta, R. C., Goel, S. K., Earley, K., Singh, B., and Reddy, J. (1985). [32]P—Postlabeling analysis of peroxisome proliferator—DNA adduct formation in rat liver *in vivo* and hepatocytes *in vitro*. *Carcinogenesis* **6**, 933–936.

Harman, A. W., and Fisher, L. J. (1983). Hamster hepatocytes in culture as a model for acetaminophen toxicity: Studies with inhibitors of drug metabolism. *Toxicol. Appl. Pharmacol.* **71**, 330–341.

Harris, R. A., and Cornell, N. W. (1983). "Isolation, Characterization, and Use of Hepatocytes." Elsevier, New York.

Henderson, P. T. (1971). Metabolism of drugs in rat liver during the perinatal period. *Biochem. Pharmacol.* **20**, 1225–1232.

Hockin, L. J., and Paine, A. J. (1983). The role of 5-aminolevulinate synthetase, haem oxygenase, and ligand formation in the mechanism of maintenance of cytochrome P-450 concentration in hepatocyte culture. *Biochem. Pharmacol.* **210**, 855–857.

Holme, J. A., Soderlund, E., and Dybing, E. (1983). Drug metabolizing activities of isolated rat hepatocytes in monolayer culture. *Acta Pharmacol. Toxicol.* **52**, 348–356.

Howard, R. B., and Pesch, L. A. (1968). Respiratory activity of intact, isolated parenchynal cells from rat liver. *J. Biol. Chem.* **243**, 3105–3109.

Howard, R. B., Christensen, A. K., Gibbs, F. A., and Pesch, L. A. (1967). The enzymatic preparation of isolated, intact parenchymal cells from rat liver. *J. Cell Biol.* **35**, 675–684.

International Agency for Research on Cancer (1980). Long-term and short-term screening assays for carcinogens: A critical appraisal. *IARC Monogr.* Suppl. 2.

Irving, C. C. (1979). Species and tissue variation in the metabolic activation of aromatic amines. *In* "Carcinogens: Identification and Mechanisms" (A. C. Griffin and C. B. Shaw, eds.), pp. 221–227. Raven, New York.

Jauregui, J. O., Hayner, N. T., Driscoll, J. L., Williams-Holland, R., Lipsky, M. H., and Galletti, P. M. (1981). Trypan blue dye uptake and lactate dehydrogenase in adult rat hepatocytes— freshly isolated cells, cell suspensions and primary monolayer cultures. *In Vitro* **17**, 1100–1110.

Klaunig, J. E., Goldblatt, P. J., Hinton, D. E., Lipsky, M. M., Chasko, J., and Trump, B. F. (1981). Mouse liver cell culture I. Hepatocyte isolation. *In Vitro* **17**, 913–925.

Klaunig, J. E., Goldblatt, P. J., Hinton, D. E., Lipsky, M. M., and Trump, B. F. (1984). Carcinogen induced unscheduled DNA synthesis in mouse hepatocytes. *Toxicol. Pathol.* **12**, 119–125.

Kornbrust, D. J., and Barfknecht, T. R. (1984). Comparison of rat and hamster primary culture/DNA repair assays. *Environ. Mutagen.* **6**, 1–11.

Laishes, B. A., and Williams, G. M. (1976). Conditions affecting primary cultures of functional adult rat hepatocytes. 1. The effect of insulin. *In Vitro* **12**, 521–532.

Lake, B. G., and Paine, A. J. (1982). The effect of hepatocyte culture conditions on cytochrome P-450 linked drug metabolizing enzymes. *Biochem. Pharmacol.* **31**, 2141–2144.

Langenbach, R., Freed, H. J., and Huberman, E. (1978). Liver cell mediated mutagenesis of mammalian cells by liver carcinogens. *Proc. Natl. Acad. Sci. U.S.A.* **75**, 2864–2867.

Leffert, H. L., Moran, T., Boorstein, R., and Koch, K. S. (1977). Procarcinogen activation and hormal control of cell proliferation in differentiated primary adult rat liver cell cultures. *Nature (London)* **267**, 58–61.

Lonati-Galligani, M., Lohman, P. H. M., and Berends, F. (1983). The validity of the auto-radiographic method for detecting DNA repair synthesis in rat hepatocytes in primary cultures. *Mutat. Res.* **113**, 145–160.

McQueen, C. A., and Williams, G. M. (1981). Characterization of DNA repair elicited by carcinogens and drugs in the hepatocyte primary culture/DNA repair test. *J. Toxicol. Environ. Health* **8**, 463–477.

McQueen, C. A., and Williams, G. M. (1982). Cytotoxicity of xenobiotics in adult rat hepatocytes in primary culture. *Fund. Appl. Toxicol.* **2**, 139–144.

McQueen, C. A., and Williams, G. M. (1983). The use of cells from rat, mouse, hamster, and rabbit in the hepatocyte primary culture/DNA repair test. *Ann. N.Y. Acad. Sci.* **407**, 119–130.

McQueen, C. A., and Williams, G. M. (1985). Methods and modifications of the hepatocyte primary culture/DNA repair test. *In* "Handbook of Carcinogenic Testing" (H. A. Milman and E. K. Weisburger, eds.), pp. 116–129. Noyes Publications, Park Ridge, New Jersey.

McQueen, C. A., Maslansky C. J., Crescenzi, S. B., and Williams, G. M. (1981). The genotoxicity of 4,4′-methylene-bis-2-chloroaniline in rat, mouse, and hamster hepatocytes. *Toxicol. Appl. Pharmacol.* **58**, 231–235.

McQueen, C. A., Maslansky, C. J., Glowinski, I. B., Crescenzi, S. B., Weber, W. W., and Williams, G. M. (1982). Relationship between the genetically determined acetylator phenotype and DNA damage induced by hydralazine and 2-aminofluorene in cultured rabbit hepatocytes. *Proc. Natl. Acad. Sci. U.S.A.* **79,** 1269–1272.

McQueen, C. A., Kreiser, D. M., and Williams, G. M. (1983). The hepatocyte primary culture (HPC)/DNA repair assay using mouse and hamster hepatocytes. *Environ. Mutagen.* **5,** 1–8.

McQueen, C. A., Miller, M. J., and Williams, G. M. (1986). Sex differences in the biotransformation of 2-acetylaminofluorene in cultured rat hepatocytes. *Cell Biol. Toxicol.* **2,** 271–281.

Maslansky, C. J., and Williams, G. M. (1981). Evidence for an epigenetic mode of action in organochlorine pesticide hepatocarcinogenicity: a lack of genotoxicity in rat, mouse, and hamster hepatocytes. *J. Toxicol. Environ. Health* **8,** 121–130.

Maslansky, C. J., and Williams, G. M. (1982). Primary cultures and the levels of cytochrome P-450 in hepatocytes from mouse, rat, hamster, and rabbit liver. *In Vitro* **18,** 663–693.

Maslansky, C. J., and Williams, G. M. (1985a). Methods for the initiation and use of hepatocyte primary cultures from various rodent species to detect metabolic activation of carcinogens. *In* "*In Vitro* Models for Cancer Research" (M. Webber and L. Sekely, eds.), pp. 43–60. CRC Press, Boca Raton, Florida.

Maslansky, C. J., and Williams, G. M. (1985b). Ultraviolet light-induced DNA repair synthesis in hepatocytes from species of differing longevities. *Mech. Aging Dev.* **29,** 191–203.

Michalopoulos, G., Sattler, G. L., O'Connor, L., and Pitot, H. C. (1978). Unscheduled DNA synthesis induced by procarcinogens in suspensions and primary cultures of hepatocytes on collagen membranes. *Cancer Res.* **38,** 1866–1871.

Miller, E. C., Miller, J. A., and Enomoto, M. (1964). The comparative carcinogenicity of 2-acetylaminofluorene and its *N*-hydroxy metabolite in mice, hamsters, and guinea pigs. *Cancer Res.* **24,** 2018–2013.

Miller, M. S., Haung, M. T., Williams, G. M., Jeffrey, A. M., and Conney, A. H. (1983). Effects of betamethasone on the *in vitro* and *in vivo* 2-hydroxylation of biphenyl in rats. *Drug Metab. Dispos.* **11,** 556–561.

Mitchell, A. D., Casciano, D. A., Meltz, M. L., Robinson, D. E., San, R. H. C., Williams, G. M., and Von Halle, E. S. (1983). Unscheduled DNA synthesis tests. A report of the U.S. Environmental Protection Agency Gene-Tox Program. *Mutat. Res.* **123,** 363–410.

Mori, H., Kawai, K., Ohbayashi, F., Kuniyasu, T., Yamazaki, M., Hamasak, T., and Williams, G. M. (1984). Genotoxicity of a variety of mycotoxins, in the hepatocyte primary culture/DNA repair test using rat and mouse hepatocytes. *Cancer Res.* **44,** 2918–2923.

Murray, A. W., and Fitzgerald, D. J. (1979). Tumour promoters inhibit metabolic cooperation in coculture of epidermal and 3T3 cells. *Biochem. Biophys. Res. Commun.* **91,** 395–401.

National Research Council (1983). "Identifying and Estimating the Genetic Impact of Chemical Mutagens." Natl. Acad. Press, Washington, D.C.

National Toxicology Program (1984). "Report on the NTP Ad Hoc Panel on Chemical Carcinogenesis Testing and Evaluation." U.S. Government Printing Office, Washington, D.C.

Nelson, K. F., and Acosta, D. (1982). Long term maintenance and induction of cytochrome P-450 in primary cultures of rat hepatocytes. *Biochem. Pharmacol.* **31,** 2211–2214.

Oldham, J. W., Casciano, D. A., and Cave, M. D. (1980). Comparative induction of unscheduled DNA synthesis by physical and chemical agents in non-proliferating primary cultures of rat hepatocytes. *Chem.-Biol. Interact.* **29,** 303–314.

Olson, M. J., Casciano, D. A., and Pounds, J. G. (1983). A method for rapid, sensitive quantitation of short patch DNA repair in cultured rat hepatocytes. *Mutat. Res.* **119,** 381–386.

Paine, A. J., and Hockin, L. J. (1980). Nutrient imbalance causes the loss of cytochrome *P*-450 in liver cell culture: Formulation of culture media which maintain *P*-450 at *in vivo* concentrations. *Biochem. Pharmacol.* **29,** 3215–3217.

Paine, A. J., Villa, P., and Hockin, L. J. (1980). Evidence that ligand formation is a mechanism underlying the maintenance of cytochrome P-450 in rat liver cell culture. *Biochem. J.* **188,** 937–939.

Poiley, J. A., Raineri, R., and Pienta, R. J. (1979). Use of hamster hepatoytes to metabolize carcinogens in an *in vitro* bioassay. *JNCI, J. Natl. Cancer Inst.* **63,** 519–523.

Poirier, M. C., Williams, G. M., and Yuspa, S. H. (1980). Effect of culture conditions, cell type, and species of origin on the distribution of acetylated and deacetylated deoxyguanosine C-8 adducts of *N*-acetoxy-2-acetylaminofluorene. *Mol. Pharmacol.* **18,** 581–587.

Probst, G. S., McMahon, R. E., Hill, L. E., Thompson, C. Z., Epp, J. K., and Neal, S. B. (1981). Chemically-induced unscheduled DNA synthesis in primary rat hepatocyte cultures. *Environ. Mutagen.* **3,** 11–32.

Reese, J. A., and Byard, J. L. (1981). Isolation and culture of adult hepatocytes from liver biopsies. *In Vitro* **17,** 935–940.

Rossberger, S., and Andrae, V. (1985). DNA repair synthesis induced by *N*-hydroxyurea, acetohydroxamic acid, and *N*-hydroxyurethane in primary rat hepatocyte cultures: Comparative evaluation using the autoradiographic and the bromodeoxyuridine density-shift method. *Mutat. Res.* **145,** 201–207.

San, R. H. C., and Williams, G. M. (1977). Rat hepatocyte primary cell culture-mediated mutagenesis of adult rat liver epithelial cells by procarcinogens. *Proc. Soc. Exp. Biol. Med.* **156,** 534–538.

Schanne, F. A. X., Kane, A. B., Young, E. E., and Farber, J. L. (1979). Calcium dependence of toxic cell death: A final common pathway. *Science* **206,** 700–702.

Schmeltz, I., Tosk, J., and Williams, G. M. (1978). Comparison of the metabolic profiles of benzo(*a*)pyrene obtained from primary cell cultures and subcellular fractions derived from normal and methylcholanthrene induced rat liver. *Cancer Lett. (Shannon, Irel.)* **5,** 81–89.

Schwarz, M., Appel, K. E., and Rickart, R. (1980). Carcinogen metabolism and carcinogen-induced strand breaks in DNA of isolated liver cells. *Arch. Toxicol.* **44,** 157–166.

Scragg, I., Pollard, M., Burchell, B., and Dutton, G. J. (1983). The temporary postnatal decline in glucuronidation of certain phenols by rat liver. *Biochem. J.* **214,** 533–537.

Seglen, P. O. (1972). Preparation of rat liver cells. II. Effects of ions and chelators on tissue dispersion. *Exp. Cell Res.* **76,** 25–30.

Seglen, P. O., and Fossa, J. (1978). Attachment of rat hepatocytes *in vitro* to substrate of serum protein, collagen, or concanavalin A. *Exp. Cell Res.* **116,** 199–206.

Shirkey, R. J., Kao, J., Fry, J. R., and Bridges, J. W. (1979). A comparison of xenobiotic metabolism in cells isolated from rat liver and small intestinal mucosa. *Biochem. Pharmacol.* **28,** 1461–1466.

Siricia, A. E., and Pitot, H. C. (1980). Drug metabolism and effects of carcinogens in cultured hepatic cells. *Pharmacol. Rev.* **31,** 205–228.

Siricia, A. E., Hwang, C. G., Sattler, G. L., and Pitot, H. C. (1980). Use of primary cultures of adult rat hepatocytes on collagen gel-nylon mesh to evaluate carcinogen-induced unscheduled DNA synthesis. *Cancer Res.* **40,** 3259–3267.

Smith, M. T., Thor, H., and Orrenius, S. (1981). Toxic injury to isolated hepatocytes is not dependent on extracellular calcium. *Science* **213,** 1257–1259.

Strom, S., and Michalopoulos, G. (1982). Mutagenesis and DNA binding of benzo(*a*)pyrene in cocultures of rat hepatocytes and human fibroblasts. *Cancer Res.* **42,** 4519–4524.

Strom, S. C., Jirtle, R. L., Jones, R. S., Novicki, D. L., Rosenberg, M. R., Novotny, A., Irons, G. P., McLain, J. R., and Michalopoulous, G. (1982). Isolation, culture, and transplantation of human hepatocytes. *JNCI, J. Natl. Cancer Inst.* **68,** 771–778.

Suolinna, E. M. (1982). Isolation and culture of liver cells and their use in the biochemical research of xenobiotics. *Med. Biol.* **60,** 237–254.

Telang, S., Tong, C., and Williams, G. M. (1982). Epigenetic membrane effects of a possible tumor promoting type on cultured liver cells by the nongenotoxic organochlorine pesticides chlordane and heptachlor. *Carcinogenesis* **3**, 1175–1178.

Thurman, R. G., and Kaufman, F. C. (1980). Factors regulating drug metabolism in intact hepatocytes. *Pharmacol. Rev.* **31**, 229–251.

Tong, C., Fazio, M., and Williams, G. M. (1981). Rat hepatocyte-mediated mutagenesis of human cells by carcinogenic polycyclic aromatic hydrocarbons but not organochlorine pesticides. *Proc. Soc. Exp. Biol. Med.* **167**, 572–575.

U.S. Food and Drug Administration (1982). "Toxicological Principles for the Safety Assessment of Direct Food Additives and Color Additives Used in Foods." U.S. Gov. Print. Off., Washington, D.C.

Ved Brat, S., and Williams, G. M. (1982). Hepatocyte-mediated production of sister chromatid exchange in cocultured cells by acrylonitrile; evidence for extracellular transport of a stable reactive intermediate. *Cancer Lett.* **17**, 213–216.

Villa, P., Hockin, L. J., and Paine, A. J. (1980). The relationship between the ability of pyridine and substituted pyridines to maintain cytochrome P-450 and inhibit protein synthesis in rat hepatocyte cultures. *Biochem. Pharmacol.* **29**, 1773–1777.

Weisburger, J. H., and Weisburger, E. K. (1973). Biochemical formation and pharmacological, toxicological and pathological properties of hydroxylamines and hydroxamic acids. *Pharmacol. Rev.* **25**, 1–66.

Weisburger, J. H., and Williams, G. M. (1981). Carcinogen testing. Current problems and new approaches. *Science* **214**, 401–407.

Williams, G. M. (1976). Carcinogen-induced DNA repair in primary rat liver cell cultures; a possible screen for chemical carcnogens. *Cancer Lett. (Shannon, Irel.)* **1**, 231–236.

Williams, G. M. (1977). The detection of chemical carcinogens by unscheduled DNA synthesis in rat liver primary cell cultures. *Cancer Res.* **37**, 1845–1851.

Williams, G. M. (1980a). The detection of chemical mutagens/carcinogens by DNA repair and mutagenesis in liver cultures. *Chem. Mutagens* **6**, 61–79.

Williams, G. M. (1980b). Classification of genotoxic and epigenetic hepatocarcinogens using liver culture assays. *Ann. N.Y. Acad. Sci.* **349**, 273–282.

Williams, G. M. (1983). Genotoxic and epigenetic carcinogens: their identification and significance. *Ann. N.Y. Acad. Sci.* **407**, 328–333.

Williams, G. M., Bermudez, E., and Scaramuzzino, D. (1977). Rat hepatocyte primary cultures. III. Improved dissociation and attachment techniques and the enhancement of survival by culture medium. *In Vitro* **13**, 809–817.

Williams, G. M., Laspia, M. F., Mori, H., and Hirona, I. (1981a). Genotoxicity of cycasin in the hepatocyte primary culture/DNA repair test supplemented with β-glucosidase. *Cancer Lett.* **12**, 329–333.

Williams, G. M., Telang, S., and Tong, C. (1981b). Inhibition of intracellular communication between liver cells by the liver tumor promoter 1,1,1-trichloro-2,2-bis(*p*-chlorophenyl)ethane (DDT). *Cancer Lett.* **11**, 339–344.

Williams, G. M., Laspia, M. F., and Dunkel, V. C. (1982). Reliability of the hepatocyte primary culture/DNA repair test. *Cancer Lett.* **6**, 199–306.

Williams, G. M., Tong, C., and Telang, S. (1984). Polybrominated biphenyls are nongenotoxic and produce an epigenetic membrane effect in cultured cells. *Environ. Res.* **34**, 310–320.

Wogan, G. N. (1973). Aflatoxin carcinogenesis. *Methods Cancer Res.* **7**, 309–344.

Yeager, J. D., and Miller, J. A. (1978). DNA repair in primary cultures of rat hepatocytes. *Cancer Res.* **38**, 4385–4395.

Yotti, L. P., Chang, C. C., and Trosko, J. E. (1979). Elimination of metabolic cooperation in Chinese hamster cells by a tumour promoter. *Science* **206**, 1089–1091.

4

Cytochrome *P*-450-Dependent Monooxygenase Systems in Mouse Hepatocytes

KENNETH W. RENTON

Department of Pharmacology
Dalhousie University
Halifax, Nova Scotia, Canada B3H 4H7

I. INTRODUCTION

The capacity of the liver to metabolize drugs and foreign chemicals by the mixed function oxidase system involves a complex relationship between the levels of enzymes which are present in the liver, the influence of agents which can act as inducers or inhibitors, and a large number of extrahepatic influences including blood flow, circulating hormone levels, and oxygen concentration (Vessel, 1982). Most of the information on drug biotransformation and its regulation has been obtained from studies using liver homogenates and subcellular fractions or by using *in vivo* models of drug elimination in the intact animal. The development of procedures to study drug biotransformation in the isolated hepatocyte bridges the important gap between these *in vitro* and *in vivo* study systems. Dose–response relationships and the effect of inhibitors on the overall pathways for the biotransformation of drugs and chemicals are easily studied in isolated cell systems. Major advantages in using isolated cells are that the interrelationship between various types of hepatic drug biotransformation can be studied without the influence of complicating extrahepatic factors and investigations can be carried out using specific treatment conditions which would cause major problems to the health of the intact animal. In addition, the interrelationships between the cytochrome *P*-450-based system and other drug metabolizing path-

THE ISOLATED
HEPATOCYTE

ways which exist *in vivo* are maintained in the isolated hepatocyte and can be studied under the influence of a variety of easily controlled conditions. Overall, the isolated hepatocyte provides a model of true *in vivo* drug biotransformation in combination with the inherent simplicity of an *in vitro* system.

Although most of the studies concerning drug metabolism in isolated heptocytes have been carried out using the rat as an animal model, several other species, including mouse, hamster, dog, ferret, and human, have been utilized to provide hepatic cells. The use of rat hepatocytes to study drug biotransformation has recently been reviewed by Smith and Orrenius (1984). The mouse has been particularly valuable for the study of drug biotransformation; however, isolated hepatocytes from this species have not been widely used, presumably because of the technical difficulties involved in preparing hepatic cells from such a small species. This chapter will present a method for preparing isolated mouse hepatic cells in high yields using a method which has few technical difficulties (Renton *et al.*, 1978) and will review some of the applications in which mouse cells have been utilized to study the metabolism of drugs by cytochrome *P*-450-dependent monooxygenase enzyme system.

II. METHODS USED TO STUDY DRUG BIOTRANSFORMATION IN ISOLATED HEPATIC CELLS

A. Preparation of Mouse Hepatocytes

Attempts to isolate hepatocytes using the proteolytic enzymes, collagenase and hyaluronidase, to dissociate liver slices produces a poor yield of cells which have a low percentage of viability (Fry and Bridges, 1979). It appears that a high yield of viable cells can only be produced using techniques which involve liver perfusion with solutions of collagenase (Berry and Friend, 1969). In the rat, liver perfusion is normally accomplished by cannulating the hepatic portal vein and perfusing the liver with collagenase in a recirculating system (Moldeus *et al.*, 1978). In the mouse this blood vessel is small and fragile, and perfusion by this route is technically very difficult, although Klaunig *et al.* (1981) have produced excellent yields of viable cells by this method. We have developed a method (Renton *et al.*, 1978) in which the mouse liver is perfused in a retrograde direction using the vena cava as the site for cannula insertion. This method also differs from that used in the rat in that the perfusion fluid is not recycled. This procedure, which is relatively easy to carry out, produces a high yield of viable hepatocytes from the mouse and has also been used in our laboratory for the preparation of cells from other small animal species such as the hamster and the newly hatched chick.

Mice (20–30 g) are anesthetized with sodium pentobarbital (100 mg/kg), and the perfusion apparatus is assembled as illustrated in Fig. 1. The reservoirs of the solutions are maintained at 37°C in a water bath or the perfusion fluid can be warmed to 37°C by placing a narrow-guage condenser coil between the pump and the cannula. The mouse is placed in a tray to collect the waste perfusate. The calcium-free buffer (adjusted to pH 7.4) contains 2.4 g HEPES, 8.3 g NaCl, and 0.5 g KCl in 1 liter double-distilled water, and the dissociating media was the same buffer containing 0.5 mg/ml collagenase. The collagenase utilized is crude collagenase type I (Sigma, St. Louis, MO #0130). Although we have found that this type of enzyme has been the most successful for the purpose of preparing

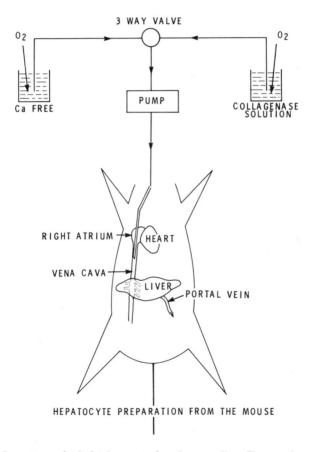

Fig. 1. Apparatus used to isolate heptocytes from the mouse liver. The procedure was described by Renton *et al.* (1978) and provides a retrograde perfusion of the liver via a cannula inserted through the right atrium into the anterior portion of the inferior vena cava.

hepatocytes, it has been our experience that some batches of collagenase produced cells of poor viability and that highly purified types of collagenase produced poor yields.

Following anesthesia the abdomen is opened, the intestines displaced to the left, and the portal vein exposed. The chest is then opened and the inferior vena cava is cannulated in a retrograde manner via an incision made in the right atrium using a 20-gauge plastic cannula connected to the variable speed pump. During the cannulation procedure, a flow rate of 0.5 ml/min of calcium-free buffer is established which keeps the incision site in the atria free of blood during the cannulation procedure. The cannula is pushed down into the vena cava until it rests just above the liver, and it is tied in place. By using a cannula size which forms a tight fit in the vena cava, it is possible to select a cannula which does not have to be tied in place. This saves time and allows the entire procedure to be completed in less than 5 min, which is important to protect the viability of the cells. The hepatic portal vein is then cut to allow the perfusion fluid to escape. The perfusion rate is then immediately increased to 10 ml/min, and the liver clears of blood within a few seconds. The perfusion fluid is allowed to run to waste. Following perfusion with calcium-free media for 2 min, the flow is switched to the media containing collagenase and perfusion is continued for a further 10 min. During the perfusion with collagenase solution, the liver becomes swollen and fluid leaks from the surface. For some batches of collagenase, shorter perfusion times are required to prevent excessive cell damage and a resulting poor viability. The flow is then switched back to the calcium-free media for a further 2 min to flush the liver free of collagenase.

The liver is excised, placed in a petri dish containing calcium-free media, broken apart, and then raked gently with a pair of blunt forceps. The buffer containing the isolated cells is decanted into a 50-ml centrifuge tube. This is repeated three times, and the resulting pooled batches of cells are allowed to cool on ice and then centrifuged at 50 g for 2 min. The sedimented cells are resuspended in 30 ml of buffer and then recentrifuged. This washing procedure is repeated three times before the cells are finally resuspended in 10 ml of Leibovitz L-15 cell culture media. Other media can be used to resuspend the cells depending on the individual requirements of each experiment. Cells are counted in a hemocytometer and viability is determined using trypan blue exclusion. A typical yield from a 1.5 g liver is about 100×10^6 cells with a viability of 95%.

Klaunig et al. (1981) compared the yield and viability of hepatocytes prepared by the retrograde perfusion technique described above with that obtained by two variations of the more difficult portal perfusion procedure (Table I). Although there was no difference in yield or viability between the retrograde method and an intermittent portal perfusion, it is interesting to note that continuous portal perfusion produced very poor yields. The technical ease by which the retrograde perfusion can be carried out makes this a very attractive procedure with which to

TABLE I

The Comparison of the Yields and Viabilities of Mouse Liver Cells Prepared by Different Isolation Methods[a]

Liver perfusion method	Total cell yield ($\times 10^6$)	Viability (%)
Retrograde perfusion via inferior vena cava.	69.3 ± 14.2 (30)	92.2 ± 2.0
Intermittent perfusion via portal vein.	77.4 ± 18.2 (50)	94.0 ± 2.0
Continuous perfusion via portal vein.	22.8 ± 5.6 (15)	86.6 ± 6.0

[a]The values are reported as the mean ± SD for the number of preparations indicated in parenthesis. Data adapted from Klaunig *et al.* (1981).

prepare hepatocytes from small animal species or from young animals of larger species, such as the rat. Klaunig *et al.* (1981) also describes a number of other factors, such as collagenase concentration, centrifugal force, and washing procedures, which can affect the viability, yield, and morphology of isolated hepatic cells in the mouse. It has been our experience that the influence of many of these factors is highly variable and that individual investigators should establish their own conditions to provide a maximum yield while maintaining a high proportion of viable cells.

B. Separation of Liver Cell Types

The preparation described above yields a mixture of parenchymal and nonparenchymal cells which can be further separated into various cell types for use in certain types of experimental procedures. Fractionation of this cell mixture can be achieved by a modification of the method first described by Skilleter and Price (1978) to yield a preparation which contains predominantly hepatocytes and a preparation which contains predominantly Kupffer cells (Peterson and Renton, 1984).

The mixture of hepatic cells is suspended in 50 ml of ice-cold calcium-free buffer (see Section II,A) in a 50-ml graduated cylinder packed in ice and allowed to settle by gravity for 20 min. The heavier parenchymal cells settle to form a layer on the bottom of the cylinder, and the upper layer of buffer contains predominantly nonparenchymal cells. The cells are then processed further through various low-speed centrifugation steps as outlined in Fig. 2. The resulting cell fractions are finally suspended in Leibovitz L-15 media which can be supplemented with fetal calf serum and antibiotics if required. The entire pro-

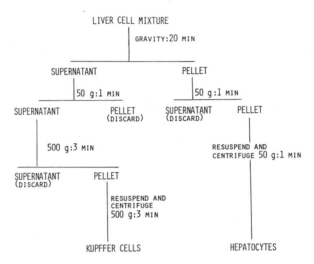

Fig. 2. Scheme for the separation of parenchymal and nonparenchymal cells from a cell mixture isolated from the mouse liver.

cedure should be carried out under aseptic conditions if the cells are to be used for long-term incubations or introduced into a cell culture system. The hepatocyte fraction contains predominantly parenchymal cells as identified by size and appearance. The nonparenchymal cell fraction contains predominantly Kupffer cells which are identified by their size and their ability to take up carbon particles (Munthe-Kaas *et al.*, 1975), but other cell types of similar size are also present in this fraction. Most preparations from a single liver contained about 100×10^6 cells/g liver for the hepatocyte fraction and about 30×10^6 cells for the Kupffer cell fraction. This method provides hepatocyte cell fractions which are relatively free of other cell types and produces a Kupffer cell fraction which, although it contains other cells is relatively free of parenchymal cells, as shown in Table II. Cell fractions can be used to study the cellular distribution of cytochrome *P*-450-dependent monooxygenase in the liver and to investigate the interrelationships between cell types and the influence of factors released from Kupffer cells on the mixed function oxidase which is contained predominantly in the parenchymal cells.

The distribution of aryl hydrocarbon hydroxylase obtained in liver cell fractions is illustrated in Table III. The activity in hepatocytes is about 4-fold greater than that found in Kupffer cells or mouse peritoneal macrophages. This distribution in the mouse is different from that reported in the rat by Cantrell and Bresnick (1972) who reported a 13-fold difference in the activity of aryl hydrocarbon hydroxylase in the two cell types. The method used by these investigators

TABLE II

The Cross-Contamination of Cell Types in Preparations of
Hepatocytes and Kupffer Cells[a]

	Number of cells/g liver	Viability (%)
Hepatocyte fraction		
Hepatocytes	109.6×10^6	82
Kupffer cells	1.2×10^6	100
Kupffer cell fraction		
Hepatocytes	1.2×10^6	0
Kupffer cells	30.0×10^6	92

[a]Mice were treated intravenously with colloidal carbon particles, and cell fractions prepared as described in the text. Kupffer cells were identified by the presence of carbon within the cell. The viability of the cells was determined by the exclusion of trypan blue. The data illustrated are from a preparation with a liver weight of 2.5 g and is representative of most cell fractions prepared in this way. Data reproduced with permission from Peterson and Renton (1984), © by *Am. Soc. for Pharmacology and Experimental Therapeutics*.

to prepare Kupffer cells involved the digestion of the hepatocytes with 0.1% Pronase solutions to leave the intact Kupffer cells. This process may have destroyed mixed function oxidase activity in the Kupffer cells and may account for the extremely low levels of mixed function oxidase which were reported to be present in Kupffer cells of the rat compared to the values which we have found in the mouse. The other explanation for the discrepancy in the two studies is that a true species difference exists for the distribution of this enzyme in cells of various types in the liver. In any study involving the distribution of mixed function oxidase between cell types, it is important to ensure that enzymatic

TABLE III

Comparison of Aryl Hydrocarbon Hydroxylase Activity in
Different Cell Types[a]

Cell type	Aryl hydrocarbon hydroxylase activity (nmol 3-OHBP/mg protein/hr)[b]
Hepatocytes	1.12 ± 0.19
Kupffer cells	0.30 ± 0.09
Peritoneal macrophages	0.37 ± 0.05
Hepatic microsomes	5.14 ± 0.56

[a]Each value represents the mean ± SE for four preparations.
[b]3-OHBP, 3-hydroxybenzo(a)pyrene.

activity found in any cell type is not due to contamination with another cell type. In the method of cell fractionation described above, it is important to ensure that the activity attributed to Kupffer cells is not due to the contamination of this cell fraction by hepatocytes which contain a greater amount of the enzyme system. In our studies we have calculated that in most preparations of Kupffer cells about 4% of the cells present in the fraction could be identified as hepatocytes which would contribute only about 3% to the total amount of fluorescent metabolites produced from benzo(a)pyrene. The activity of mixed function oxidase found in Kupffer cells is, therefore, a true reflection of the level of this enzyme present in the reticuloendothelial cells of the liver.

C. Drug Biotransformation in Isolated Cell Fractions

The measurement of mixed function oxidase and related drug biotransformation pathways can usually be carried out in the isolated cell system using modifications of the procedures used in the whole animal or in isolated microsomes. Freshly isolated cells can be used directly or introduced into a primary cell culture system. It must be recognized, however, that when drug biotransformation pathways are studied using isolated cells, the compounds are metabolized via pathways such as glucuronidation, sulfation, and acetylation in addition to oxidation by mixed function oxidase and that the pattern of drug metabolites formed in isolated cells will be more like those achieved in the intact animal rather than those reported for isolated microsomes.

Hemoprotein Levels

The main hemoprotein associated with the mixed function oxidase system is cytochrome P-450, which is usually measured by the absorbance of its reduced carbon monoxide difference spectrum. Cytochrome P-450 measurement in hepatocytes can be carried out using a modification of the method first described for microsomes by Omura and Sato (1964). The cells are centrifuged and washed with isotonic saline and then solubilized by adding 50 μl of 4% Lubrol to 1.95 ml of phosphate buffer (0.1 M, pH 7.4) containing about 1×10^6 hepatocytes/ml (Peterson and Renton, 1984). A few crystals of sodium dithionite are then added to reduce the cytochrome, and the sample is divided equally into two cuvettes. A baseline spectra is determined in a double-beam spectrophotometer before carbon monoxide is bubbled into the sample cuvette for 30 sec and the spectrum is redetermined. Cytochrome P-450 concentration is obtained from the absorbance difference between 450 and 490 nm using an extinction coefficient of 91 mM^{-1} cm^{-1}. The assay can also be carried out in intact cell preparations which have not been solubilized with detergent; however, the cells settle in the cuvette at a rapid rate and great care must be taken to carry out the spectral scans before this occurs. This problem can also be overcome by using a cuvette stirring device (Instech Laboratories Inc., Horsham, PA) or by suspending the cells in an isopycnic solution of Ficoll 400 (Jones *et al.*, 1979).

An example of the use of these methods to determine the effect of interferon on cytochrome *P*-450 in isolated cells is illustrated in Fig. 3. A cloned hybrid interferon was incubated for 19 hr with hepatocytes attached to collagen boats, and cytochrome *P*-450 was determined after solubilization of the cells with detergents. In this experiment interferon caused a loss of cytochrome *P*-450 but had no effect on the total protein content of the cells. This type of experiment has been used by us to demonstrate that the effect of interferon on hepatic drug biotransformation results from a direct effect of this agent within the hepatocyte and rules out the possibility that the effect was caused by an indirect action of interferon in other cell types (Renton, 1983).

A major difficulty in determining cytochrome *P*-450 in solubilized or whole cells is interference from the cytochromes which are present in the mitochondria.

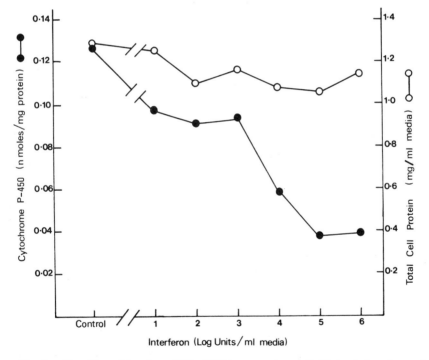

Fig. 3.   The effect of interferon (IFN-αCON₁) on cytochrome *P*-450 levels in cultures of isolated liver cells. Isolated hepatocytes were cultured on collagen boats using L-15 as media as described by Renton *et al.* (1978). Interferon was incubated with the cells for 19 hr before solubilization with detergent was carried out and cytochrome *P*-450 was determined.

The absorbance maximum of cytochrome P-450 is decreased due to the negative contribution of the trough at 445 nm of cytochrome a_3 (Jones et al., 1979). The error produced in cytochrome P-450 estimation is about 10% for control hepatocytes but only 2–3% in heptocytes prepared from animals treated with phenobarbital or 3-methylcholanthrene. If required, a correction to the value for the cytochrome P-450 level can be made by following the procedures originally described by Jones et al. (1979). The concentration of cytochromes $a + a_3$ is estimated from the change in absorbance (630–605 nm) following reduction of the oxidized forms using antimycin A to ensure complete oxidation. Under a nitrogen atmosphere the endogenous reduction of cytochrome $a + a_3$ is determined using an extinction coefficient of 13.1 mM^{-1} cm^{-1}. The cytochrome a_3 concentration is estimated to be 50% of the total for cytochrome $a + a_3$ concentration. The correction to cytochrome P-450 is derived from the contribution of cytochrome a_3 at the absorbance difference (450–490 nm) which is used to estimate cytochrome P-450 concentration. This is determined from the extinction coefficient for cytochrome a_3 at this absorbance difference which is 36 mM^{-1} cm^{-1}. A second method which can be used to overcome the difficulty of contaminating mitochondrial cytochromes is to prepare microsomes from the cell preparations. This, however, requires a large number of cells and for most experiments is not a practical option. A minor difficulty which may be encountered in determining cytochrome P-450 in isolated cells is the high turbidity often present in these preparations. This can usually be overcome if cell concentrations are kept below 1.5×10^6 cells/ml media or if the cells are solubilized with detergents.

The stability of cytochrome P-450 in isolated cells is also a serious concern in most of the species which have been studied (Jones et al., 1979). In the mouse we have shown that, in heptocytes incubated for 6 hr, no loss of aryl hydrocarbon hydroxylase was apparent, but on further incubation mixed function oxidase was rapidly lost until only 35 and 24% remained at 16 and 21 hr respectively (Peterson and Renton, 1984). When mouse cells were introduced into cultures, similar losses have been reported (Renton et al., 1978), and the half life of mouse cytochrome P-450 in such a system has been estimated at about 20 hr (Maslansky and Williams, 1982). A similar half-life was reported for the rat, but in the hamster and the rabbit the half-life of cytochrome P-450 is considerably longer. The losses of cytochrome P-450 in the rat can apparently be prevented by the addition of high concentrations of nicotinamide to the media (Paine et al., 1979). Although we have used this procedure successfully in rat hepatocytes, we have been unable to prevent losses in the mouse by the same means.

The determination of cytochrome b_5 presents some special problems due to the presence of a number of mitochondrial cytochromes. In cells obtained from induced animals, cytochrome b_5 can be determined using tert-butylhydroperoxide to produce a difference spectrum in the presence of a mitochondrial uncoupler (Jones *et al.*, 1979). In hepatocytes obtained from uninduced animals, a much more accurate determination can be obtained using a methyl viologen-induced difference spectrum obtained in the presence of oxygen (Smith and Orrenius, 1984).

III. APPLICATIONS

A. The Influence of Cells of the Reticuloendothelial System on Drug Biotransformation in Hepatocytes

Although most of the drug biotransformation capacity of the liver is contained in the hepatocytes, the status of the nonparenchymal cells of the reticuloendothelial system (RES) appears to play an important role in the maintenance of cytochrome *P*-450 levels within the hepatocyte. Drug metabolism in the liver is decreased when the RES is either activated or depressed (Wooles and Munson, 1971). Several different agents, including methyl palmitate, thorium dioxide, and pyran copolymers, all of which are known to stimulate the RES, are capable of prolonging barbiturate anesthesia (Wooles and Munson, 1971). Al-Tuwaijri and DiLuzio (1982) also demonstrated that pentobarbital sleeping time was increased and blood levels of pentobarbital were elevated when the RES was stimulated. When the Kupffer cells of the liver are loaded with particulate matter, such as colloidal carbon, the metabolism and toxicity of carbon tetrachloride are reduced (Stenger *et al.*, 1969). Peterson and Renton (1984, 1986) have shown that drug biotransformation in the liver is lost when animals are treated with dextran or latex particles, that these agents are taken up soley in the Kupffer cells of the liver, and that none of these agents are found in the hepatocytes. It appears that whenever animals are treated with agents which can be phagocytosed, the levels of drug biotransformation in the liver decrease (Renton, 1983). If the depression of cytochrome *P*-450 by immunoactive agents is mediated by activation of the liver reticuloendothelial cells, this hypothesis would require some type of communication between the Kupffer cells, which contain a small proportion of the mixed function oxidase system, and the hepatocytes, which contain the bulk of the cytochrome *P*-450 in the liver. In a series of recent experiments Peterson and Renton (1984, 1986) used isolated liver cells to demonstrate that Kupffer cells could release a factor which depressed drug biotransformation in adjacent hepatocytes and thus provided an explanation for the action of immunoactive agents on the mixed function oxidase enzyme system of the liver.

Hepatocytes and Kupffer cells were isolated in highly pure fractions by the methods described earlier in this chapter. When these two cell types were recombined and incubated in the presence of dextran sulfate or latex beads, cytochrome *P*-450 and drug biotransformation were depressed in the cell mixture as shown in Fig. 4 supporting the idea that the primary action of dextran sulphate and latex particles was in the Kupffer cell rather than directly on the hepatocyte. In this type of experiment the two cell types remained in physical contact with each other, and Kupffer cells could have influenced hepatocyte function in this way. Neither of these agents had any effect on cytochrome *P*-450 or drug biotransformation when incubated with hepatocytes alone. Unequivocal proof that this effect was mediated by a factor released from Kupffer cells was demonstrated by incubating the liver cell fractions in a Marbrook vessel in which the two cell types were separated by a semipermeable membrane (Fig. 5). When dextran or latex particles were added to the chamber containing Kupffer cells, drug biotransformation and cytochrome *P*-450 were lost from the hepatocytes contained in the adjacent chamber (Figs. 6 and 7). The factor which was released from Kupffer cells is most likely released by all cells of the reticuloendothelial system, as peritoneal macrophages could be substituted for Kupffer cells in the upper chamber and produce the same result in hepatocytes maintained on the other side of the membrane in the Marbrook vessel. By using such a procedure in isolated

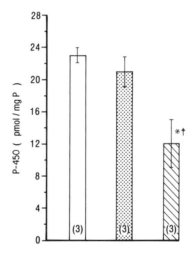

Fig. 4. Cytochrome *P*-450 levels in hepatocytes incubated with Kupffer cells and dextran sulfate. Kupffer cells (1 ml of 7.5×10^6 cells/ml) were incubated for 30 min with 50 µg/ml dextran and then added to 1 ml of hepatocytes (14×10^6 cells/ml) and incubated for a further 30 min. Similar incubations were carried out with hepatocytes alone or with dextran and hepatocytes alone. Each value is the mean of 3 separate experiments. Asterisk, Significantly different from control; $p < 0.05$. Double dagger, Significantly different from incubation mixture containing dextran sulfate, $p < 0.05$. [Reproduced with permission from Peterson and Renton (1984), © by *Am. Soc. for Pharmacology and Experimental Therapeutics.*]

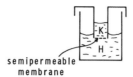

semipermeable
membrane

Fig. 5.   Marbrook vessel. This vessel has an inner chamber separated from an outer chamber by a semipermeable membrane with a molecular weight cutoff of 12,000. The inner chamber contained Kupffer cells (K) and the outer chamber contained hepatocytes (H). [Reproduced with permission from Peterson and Renton (1984), © by *Am. Soc. for Pharmacology and Experimental Therapeutics.*]

cells, we were able to demonstrate the mechanism by which certain immunoactive agents lowered mixed function oxidase in the liver. These experiments would have been impossible to carry out *in vivo,* as a direct effect of these agents on the hepatocyte could never have been ruled out. Using isolated cell systems it was shown, for the first time, that drug biotransformation in the parenchymal cells of the liver is influenced by indirect actions originating in adjacent cell types in the same organ.

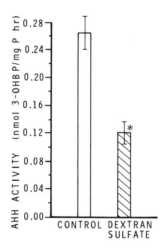

Fig. 6.   The activity of aryl hydrocarbon hydroxylase in mouse hepatocytes incubated with Kupffer cells and dextran sulphate in a double chambered Marbrook vessel. One ml of Kupffer cells (15×10^6 cells) and dextran sulfate (50 μg/ml) were contained in the upper chamber and separated from 10 ml hepatocytes (70×10^6 cells) by a semipermeable membrane. Incubations were carried out for 16 hr at 37°C. Vessels containing both cell types but without dextran acted as controls. Each value is the mean ± SE of six determinations. Asterisk, significantly different from control, $p < 0.05$. [Reproduced with permission from Peterson and Renton (1984) © by *Am. Soc. for Pharmacology and Experimental Therapeutics.*]

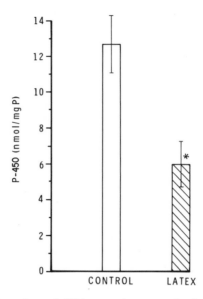

Fig. 7. The level of cytochrome *P*-450 in mouse hepatocytes incubated with Kupffer cells and latex particles in a Marbrook vessel. Conditions are identical to those described in Fig. 6. Latex concentration in the upper chamber was 1 mg/ml media. [Reproduced with permission from Peterson and Renton (1986), copyright Pergamon Press.]

The use of a similar experimental procedure in Marbrook vessels could be used to investigate the influence of other cell types or the effect of any mediator secreted by another cell type on hepatic drug biotransformation in the liver. Such an experimental procedure will allow us to determine if the loss of drug biotransformation during viral and bacterial infections is caused by the release of a factor from immunocompetent cells. This type of experiment will also be of particular value in investigations concerning the influence of hormones released by a variety of cell types on the mixed function oxidase system of the liver.

B. The Alteration of Drug Biotransformation during Hypoxia

Oxygen is required for mixed function oxidase systems directly as a substrate for xenobiotic oxidations, and therefore low oxygen concentration within the cell can potentially lead to an impairment in the liver to metabolize and excrete drugs. Jones (1981) suggested that the oxygen concentration in the liver could be rate limiting for the biotransformation of a number of different compounds in the intact animal. Such a limitation caused by the oxygen concentration in the cell will depend on the K_m O_2 relative to the oxygen concentration which is available within the cell. It is likely that different isozymes of cytochrome *P*-450 will have different affinities for oxygen and that depending on the substrate and the isozyme in question some biotransformation pathways will be affected more than others. The limiting effect of oxygen concentration on the metabolism of hexobarbital and pentobarbital have been studied *in vivo* (Cumming and Mannering, 1970), but experiments of this type are obviously complicated by a number of factors, such as cardiac output, vasodilatation, and hemoconcentration, which are altered by changes in oxygen tension in the intact animal. The severity of the hypoxia and the range of oxygen concentrations which can be produced are also limited by the survival of the animal. Another major problem with experiments *in vivo* is the inability of the investigator to define the degree of hypoxia which occurs at the cellular level during experimental procedures, as regional hypoxia may severely compromise the conclusions which can be made. Organ perfusion experiments can eliminate many of the limitations imposed in the whole animal; however, the use of portal vein perfusion produces an unnatural oxygen gradient across the organ, and the use of hemoglobin-free perfusion media produces large oxygen gradients between the perfusion fluid and the tissue. The use of isolated hepatocytes eliminates most of the problems outlined above and allows the investigator to vary the oxygen tension in live, respiring cells to a degree which is not possible in the intact animal or in perfused liver systems.

In microsomes it has been reported that the K_m value for oxygen varies from 0.7 μM (0.5 Torr) to 200 μM (140 Torr) depending on the substrate and the type of cytochrome *P*-450 which was involved (Bernhardt *et al.*, 1973; Jones, 1981). In the normal liver the oxygen concentration is about 35 μM (Kessler *et al.*, 1973), which indicates that any substrate with a K_m value of greater than about 15 μM will be oxidized at a rate which is limited by the availability of oxygen. Jones (1981) has suggested that about 25% of all drugs may fall into such a category in the normally oxygenated liver. In conditions which produce hypoxia in the liver, the rate of metabolism of a greater number of drugs will be limited by the available oxygen concentration.

In recent experiments Nakatsu (1985) has investigated the metabolism of theophylline in isolated heptocytes to test the hypothesis (Jones, 1981) that oxygen concentration could be rate limiting for the metabolism of certain drugs at oxygen concentrations which are close to those found *in vivo*. Theophylline is an important drug which has a high K_m and is used at high plasma concentrations in therapeutics (Hendeles and Weinberger, 1983). Isolated mouse hepatocytes were incubated at 37°C with theophylline (15 μg/ml) in round-bottomed flasks which were gassed with various concentrations of oxygen. The gas phase also contained 5% CO_2 and nitrogen to make up the remaining volume. Samples were removed from the flasks at various time intervals, and the metabolism of theophylline was determined by the disappearance of substrate using HPLC analysis. The metabolism of theophylline by hepatocytes in different concentrations of oxygen are illustrated in Fig. 8. These experiments indicate that there is a high correlation between the rate of metabolism of theophylline and the oxygen concentration. Theophylline biotransformation in isolated hepatocytes was severely impaired at concentrations of oxygen which were below those occurring in the normal liver. In the same study, Nakatsu (1985) demonstrated that a similar relationship existed when the experiments were carried out in perfused livers, and in preliminary experiments the clearance of theophylline was depressed by 27 and 44% in animals maintained in a 17 and 13% oxygen atmosphere, respectively. These experiments demonstrate that the oxygen limitation hypothesis of Jones (1981) can be studied in isolated hepatic cells and that the conclusions obtained from such an experimental system are directly applicable to the metabolism of drugs *in vivo*. In therapeutics the impairment of theophylline elimination which occurs in diseases such as respiratory illness or congestive heart failure could be explained on the basis of poor liver oxygenation which is likely to occur in conditions that are known to cause some degree of hypoxia (Hendeles and Weinberger, 1983).

The experiments of Nakatsu (1985) with theophylline in hepatocytes illustrate

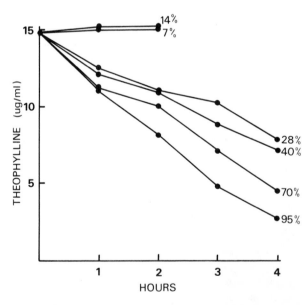

Fig. 8.   The effect of oxygen availability on the metabolism of theophylline in isolated mouse hepatocytes. Hepatocytes were incubated with theophylline as described in the text using gas mixtures containing the percentage of oxygen indicated at the right. Results are expressed as the amount of theophylline remaining in the incubation mixtures at various time intervals. [Adapted from Nakatsu (1985.]

an experimental system that can be used to predict which drugs will be dependent on oxygen concentrations which are normally present in the liver. This system will allow the identification of drugs that will have decreased elimination rates during disease processes which cause hypoxia, and screening experiments can play a role in decreasing the number of drug reactions occurring in conditions in which the delivery of oxygen is impaired. It will also be interesting to investigate drug interactions between agents which have different oxygen requirements or to determine if drug interactions are likely during the use of agents which are known to impair blood flow and oxygen delivery to the liver. At the theoretical level the use of isolated heptocytes can be used to test the hypothesis of Bernhardt *et al.* (1973) which suggested that the nature of a drug substrate will determine the affinity of the different isozymes of cytochrome *P*-450 toward oxygen.

C. The Use of Hepatocytes in Drug Toxicity Studies

Hepatocytes are ideally suited to study the effects of chemical toxicity because not only are isolated cells able to inactivate or activate drugs to reactive species but the cells themselves can act as the target for the toxicity and provide a measure of the effect in question. The most often used indicator of cellular damage in intact cells is the measurement of plasma membrane leakiness (Schanne *et al.*, 1979). This can be carried out either by measuring the exclusion of trypan blue from the cells or by determining the latency of a cytosolic enzyme such as lactic acid dehydrogenase (Moldeus, 1978). In some cases specific biochemical or morphological alterations within the cell can be measured as an assessment of a particular drug-induced cellular toxicity. A major advantage in using isolated cells for this toxicological assessment is that concentrations of the toxic agent and the conditions in which it is used can be closely controlled in an environment which is similar to that found *in vivo* but without the restraints caused by using toxic substances in an intact animal. Several investigators have used isolated liver cells from various species to investigate the toxicity of acetaminophen, which serves as an excellent example of how isolated cells can be utilized to study toxic events caused by the metabolism of drugs and chemicals.

At concentrations of acetaminophen which would be encountered during therapeutic use, the drug is predominantly conjugated as a sulfate and a glucuronide (Mitchell *et al.*, 1973). A small amount of acetaminophen is oxidized by the mixed function oxidase system to a highly reactive benzoquinone, which is normally inactivated by glutathione, and excreted as the mercapturic acid derivative. Following the ingestion of large doses of acetaminophen, glutathione is depleted and the excess reactive metabolite escapes to react irreversibly with cell macromolecules, which has been postulated to result in hepatic necrosis and organ death (Mitchell *et al.*, 1974). Most of the studies on acetaminophen have been carried out *in vivo*; however, by using isolated hepatocytes it is possible to avoid the problems of absorption, tissue binding, and excretion and allows reaction sequences and kinetics of the reactions to be carried out in a more controlled manner. This method has been particularly valuable in determining the mechanism of action of agents which protect the liver against acetaminophen toxicity.

Moldeus (1978) has compared the toxicity of acetaminophen in hepatocytes isolated from both the mouse and the rat. Although glucuronidation rates in both species were similar, the rate of sulphate conjugation was approximately 10-fold higher in the rat than in the mouse. On the other hand, the formation of the glutathione conjugate was about 10-fold higher in mouse hepatocytes. Even when cells were used from phenobarbital-induced rats, the rate of glutathione conjugate synthesis was only about 50% of that found in untreated mouse cells. When rat cells from phenobarbital-treated animals were incubated with 10 m*M* acetaminophen for 5 hr, the formation of sulfate and glucuronide was linear for 2 hr, but the formation of the glutathione conjugate decreased after 1 hr (Fig. 9). This loss of conjugation capacity coincided with the depletion of glutathione levels; however, even at the end of 5 hr there was no loss in cell permeability as measured by NADH penetration. This contrasts sharply with the loss in cell integrity which occurs within 3 hr in the mouse. After 5 hr of incubation, over half of the cells were damaged (Fig. 10). In the mouse cells, the rate of glutathione conjugation was more rapid than in the rat, but glutathione levels were

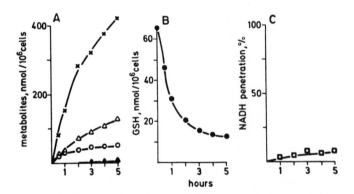

Fig. 9.   Acetaminophen metabolite formation (A), glutathione levels (B), and cell viability (C) in rat hepatocytes from phenobarbital-induced rats. Incubations were carried out using 10^6 cells/ml and 10 m*M* acetaminophen. x, Glucuronide formation; △, sulfate conjugation; ○, glutathione conjugation; ▲, cysteine conjugation; ●, glutathione level; □, cell viability as measured by NADH penetration. [Reproduced with permission from Moldeus (1978), copyright Pergamon Press.]

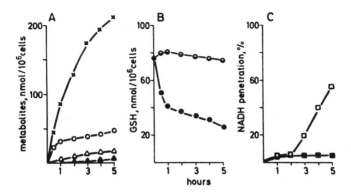

Fig. 10. Acetaminophen metabolite formation (A), glutathione levels (B), and cell viability (C) in mouse hepatocytes from noninduced mice. Conditions and key are identical to those described in Fig. 9. [Reproduced with permission from Moldeus (1978), copyright Pergamon Press.]

not depleted to levels lower than that found in the rat cells. This is perhaps surprising as the mouse cells were much more susceptible to damage than the rat cells, but it does agree with previous studies in the whole animal that demonstrated that the mouse was much more susceptible to the hepatotoxic effects of acetaminophen compared to the rat (Davis *et al.,* 1974).

Massey and Racz (1981, 1983) have used the sensitivity of mouse hepatocytes to investigate the mechanism by which *N*-acetylcysteine protects the liver against the necrosis caused by large doses of acetaminophen. These investigators demonstrated that when *N*-acetylcysteine was incubated with acetaminophen in isolated hepatocytes, no protective effect was observed as measured by LDH latency (Fig. 11). When *N*-acetylcysteine was preincubated with the cells prior to the addition of acetaminophen, this compound prevented damage to the integrity of the cells. The requirement for a preincubation period with *N*-acetylcysteine in order to produce protection against acetaminophen toxicity indicates that *N*-acetylcysteine is not acting as a direct nucleophile for conjugation. Such a suggestion is confirmed by the absence of an acetaminophen–*N*-acetylcysteine conjugate in the deproteinized incubation media obtained from cells incubated with both of these compounds (Fig. 12). Hepatocytes which were incubated with *N*-acetylcysteine and acetaminophen produced increasing amounts of both cysteine and glutathione conjugates, and the covalent binding of [^{3}H]acetaminophen was decreased. From the experiments in isolated cells, these investigators concluded that *N*-acetylcysteine did not protect hepatic cells from acetaminophen damage by covalently binding to the reactive metabolite but acted indirectly to increase the availability of glutathione. This could occur by the metabolism of *N*-acetylcysteine within the hepatocytes to form cysteine which is a direct precursor of glutathione.

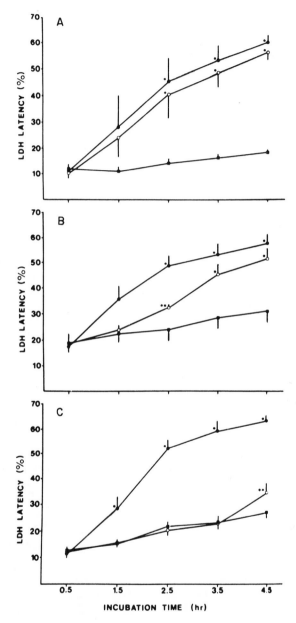

Fig. 11. The effect of acetaminophen and N-acetylcysteine on the integrity of the plasma membrane of mouse hepatocytes. ■, Control; ●, acetaminophen; ○, acetaminophen + N-acetylcysteine. (A) 1 mM acetaminophen and 1 mM N-acetylcysteine (no preincubation). (B) 1 mM acetaminophen and 1 mM N-acetylcysteine (45 min preincubation). (C) 1 mM acetaminophen and 2 mM N-acetylcysteine (no preincubation). [Reproduced from Massey and Racz (1981).]

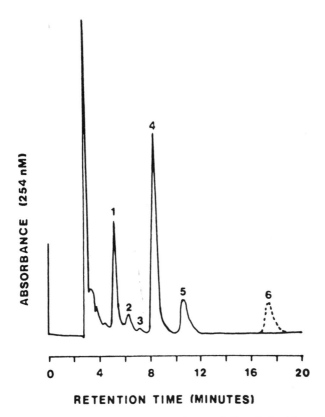

Fig. 12. An HPLC chromatogram of acetaminophen metabolites obtained from a deproteinized incubation mixture of acetaminophen and *N*-acetylcysteine. 1, Glucuronide; 2, sulfate; 3, cysteine; 4, acetaminophen; 5, GSH conjugate; 6, acetaminophen–*N*-acetylcysteine conjugate standard (none found in incubation mixtures). [Reproduced from Massey and Racz (1981).]

A major advantage in using mouse hepatocytes to study acetaminophen toxicity is the possible use of ultrastructural changes to monitor toxicity in the isolated cells. Walker *et al.* (1983) have demonstrated that isolated mouse hepatocytes exposed to acetaminophen demonstrate cytoplasmic lesions and bleb formation on the cell wall surface which correlates with other features of acetaminophen toxicity such as glutathione depletion. These authors were able to identify cells which had undergone a terminal hydropic degeneration and had a very unstructured appearance which was serious enough to classify the cells as being dead. Although similar changes in morphology can be demonstrated for bromobenzene toxicity in glutathione-depleted rats (Smith and Orrenius, 1984), the cells from this species are resistant to the effects of acetaminophen. It can be concluded that a preparation of rat hepatocytes is a poor preparation with which to study acetaminophen toxicity and its treatment and that isolated cells from the mouse would be the model system of choice.

IV. CONCLUSIONS

Isolated parenchymal cells from the liver are useful experimental models to study the metabolism of drugs and the toxic consequences of drug biotransformation. Of particular value is the ability to reproduce conditions which are essentially identical to those occurring *in vivo* but without the restraints imposed by the whole animal or the limitations which occur during the use of subcellular fractions. To date, isolated cell systems have been widely used for studies in drug biotransformation, but a major limitation of the system remains the loss of cytochrome *P*-450 and drug biotransformation which occurs on prolonged incubation of the cells or following the introduction of the cells to culture. The usefulness of isolated cells could be greatly expanded if a method were discovered which would protect the integrity of mixed function oxidase for a longer period of time. Nevertheless, the isolated hepatocyte is a useful addition to the various techniques used to study drug biotransformation and provides us with a model similar to the *in vivo* situation combined with the simplicity and control which is provided by *in vitro* systems.

REFERENCES

Al-Tuwaijri, A., and DiLuzio, N. R. (1982). Modifications of pentobarbital metabolism by macrophage stimulants and depressants. *nt. J. Immunopharmacol.* **4**, 327.

Bernhardt, F. H., Erdin, N., Staudinger, H., and Ullrich, V. (1973). Interactions of substrates with a purified 4-methoxybenzoate monooxygenase system from *Pseudomonas putida*. *Eur. J. Biochem.* **35**, 126–139.

Berry, M. N., and Friend, D. S. (1969). High yield preparation of isolated rat liver parenchymal cells: A biochemical and fine structure study. *J. Cell Biol.* **43**, 506–520.

Cantrell, E., and Bresnick, E. (1972). Benzpyrene hydroxylase activity in isolated parenchymal and non-parenchymal cells of rat liver. *J. Cell Biol.* **52**, 316–321.

Cumming, J. F., and Mannering, G. J. (1970). Effect of phenobarbital administration on the oxygen requirement for hexobarbital metabolism in the intact rat. *Biochem. Pharmacol.* **19**, 973–978.

Davis, D. C., Potter, W. Z., Jollow, D. J., and Mitchell, J. R. (1974). Species differences in hepatic glutathione depletion, covalent binding and hepatic necrosis after acetaminophen. *Life Sci.* **14**, 2099–2109.

Fry, J. R., and Bridges, J. W. (1979). Use of primary hepatocyte cultures in biochemical toxicology. *In* "Reviews of Biochemical Toxicology" (E. Hodgson, J. Bend, and R. M. Philpot, eds.), pp. 201–247. Elsevier/North-Holland, New York.

Hendeles, L., and Weinberger, M. (1983). Theophylline. A state of the art review. *Pharmacotherapy* **3**, 2–44.

Jones, D. P. (1981). Hypoxia and drug metabolism. *Biochem. Pharmacol.* **30,** 1019–1023.

Jones, D. P., Orrenius, S., and Mason, H. S. (1979). Hemoprotein quantitation in isolated hepatocytes. *Biochim. Biophys. Acta* **576,** 17–29.

Kessler, M., Lang, H., Sinagowitz, E., Rink, R., and Hoper, D. (1973). Homeostasis of oxygen supply in liver and kidney. *In* "Oxygen Transport to Tissues" (D. F. Bruley and H. I. Bicker, eds.), pp. 351–360. Plenum, New York.

Klaunig, J. E., Goldblatt, P. J., Hinton, D. E., Lipsky, M. M., Chako, J., and Trump, B. F. (1981). Mouse liver cell culture. I. Hepatic isolation. *In Vitro* **17,** 913–925.

Maslansky, C. J., and Williams, G. M. (1982). Primary culture and the levels of cytochrome P-450 in hepatocytes from mouse, rat, hamster and rabbit liver. *In Vitro* **18,** 683–693.

Massey, T., and Racz, W. J. (1981). Effects of *N*-acetylcysteine on metabolism, covalent binding, and toxicity of acetaminophen in isolated mouse hepatocytes. *Toxicol. Appl. Pharmacol.* **60,** 220–228.

Massey, T., and Racz, W. J. (1983). Acetaminophen toxicity and its prevention in isolated mouse hepatocyte suspensions. *In* "Isolation, Characterization and Use of Hepatocytes" (R. A. Harris and N. W. Cornell, eds.), pp. 379–384. Elsevier/North-Holland, New York.

Mitchell, J. R., Jollow, D. J., Potter, W. J., Davis, D. C., Gillette, J. R., and Brodie, B. B. (1973). Acetaminophen induced hepatic necrosis I. Role of drug metabolism. *J. Pharmacol. Exp. Ther.* **187,** 185–194.

Mitchell, J. R., Thorgeirson, S. S., Potter, W. Z., Jollow, D. J., and Keiser, H. (1974). Acetaminophen induced hepatic necrosis: Protective role of glutathione in man and rationale for therapy. *Clin. Pharmacol. Ther.* **16,** 676–684.

Moldeus, P. (1978). Paracetamol metabolism and toxicity in isolated hepatocytes from rat and mouse. *Biochem. Pharmacol.* **27,** 2859–2863.

Moldeus, P., Jones, D. P., and Orrenius, S. (1978). Isolation and use of liver cells. *Methods Enzymol.* **52,** 60–71.

Munthe-Kaas, A. C., Berg, T., Seglen, P. O., and Seljelid, R. (1975). Mass isolation and culture of rat Kupffer cells. *J. Exp. Med.* **141,** 1–10.

Nakatsu, K. (1985). Limitation of theophylline elimination by reduced oxygen availability in mouse hepatocytes and rat isolated livers. *Can. J. Physiol. Pharmacol.* **63,** 903–907.

Omura, T., and Sato, R. (1964). The carbon monoxide binding pigment of liver microsomes. I. Evidence for its hemoprotein nature. *J. Biol. Chem.* **239,** 2370–2378.

Paine, A. J., Williams, L. J., and Legg, R. F. (1979). Apparent maintenence of cytochrome P-450 by nicotinamide in primary cultures of rat hepatocytes. *Life Sci.* **24,** 2185–2192.

Peterson, T. C., and Renton, K. W. (1984). Depression of cytochrome P-450 dependent drug biotransformation in hepatocytes after the activation of the RES by dextran sulphate. *J. Pharmacol. Exp. Ther.* **229,** 299–304.

Peterson, T. C., and Renton, K. W. (1986). Kupffer cell factor mediated depression of hepatic parenchymal cell cytochrome P-450. *Biochem. Pharmacol.* **35,** 1491–1497.

Renton, K. W. (1983). Relationships between the enzymes of detoxication and host defence mechanisms. *In* "Biological Basis of Detoxication" (J. Caldwell and W. B. Jakoby, eds.), pp. 307–324. Academic Press, New York.

Renton, K. W., Deloria, L. B., and Mannering, G. J. (1978). Effects of poly rI.rC and mouse interferon preparations on cytochrome P-450 dependent mono-oxygenase systems in cultures of mouse hepatocytes. *Mol. Pharmacol.* **14,** 672–681.

Schanne, F. A. X., Kane, A. B., Young, E. E., and Farber, J. L. (1979). Calcium dependence of toxic cell death; A final common pathway. *Science* **206,** 700–702.

Skilleter, D., and Price, R. J. (1978). The uptake and subsequent loss of beryllium by rat liver parenchymal and non-parenchymal cells after the intra-venous administration of particulate and soluble forms. *Chem.-Biol. Interact.* **20,** 383–396.

Smith, M. T., and Orrenius, S. (1984). Studies on drug metabolism and drug toxicity in isolated mammalian cells. *In* "Drug Metabolism and Drug Toxicity" (J. Mitchell and M. Horning, eds.), pp. 71–98. Raven, New York.

Stenger, R. J., Petrelli, M., Segal, A., Williams, J. N., and Johnson, E. (1969). Modification of carbon tetrachloride hepatotoxicity by prior loading of the RES with carbon particles. *Am. J. Pathol.* **57,** 689–706.

Vessel, E. S. (1982). On the significance of host factors that affect drug disposition. *Clin. Pharmacol.* **31,** 1–7.

Walker, R. M., McElligott, T. F., Massey, T. E., and Racz, W. J. (1983). Ultrastructural effects of acetaminophen in isolated mouse hepatocytes. *Exp. Mol. Pathol.* **39,** 163–175.

Wooles, W. R., and Munson, A. E. (1971). The effects of stimulants and depressants of RES activity on drug metabolism. *Proc. Reticuloendothel. Soc.* **9,** 108–119.

5

Control of Hepatocyte
Proliferation *in Vitro*

NOREEN C. LUETTEKE AND
GEORGE K. MICHALOPOULOS

Department of Pathology
Duke University Medical Center
Durham, North Carolina 27710

I. INTRODUCTION

Primary cultures of hepatocytes have been employed extensively in studies of hepatic metabolism, differentiation and growth. The major limitation of this *in vitro* system has been a failure to maintain functional, replicating hepatocytes in culture for prolonged periods of time. The ability to clone hepatocytes would be particularly advantageous to investigations in liver toxicology and carcinogenesis. Discovery of the factors or conditions that permit or promote hepatocyte proliferation *in vitro* would also contribute to the understanding of normal and neoplastic growth processes *in vivo*. This chapter reviews the current information available on hepatocyte growth factors. The relevance of this research to liver regeneration and carcinogenesis is also discussed.

Previous *in vivo* experiments involving cross circulation of parabiotic rats (Moolten and Bucher, 1967) or transplantation of liver tissue (Grisham *et al.*, 1964) and cells (Jirtle and Michalopoulos, 1982) to extrahepatic sites have indicated that the stimuli for liver regeneration following partial hepatectomy are blood-borne. Therefore, numerous *in vitro* studies have focused on the role of serum and the interaction of its defined or unknown components in the regulation of hepatocyte growth. Serum induces fetal, neonatal, and adult rat hepatocytes in primary culture to synthesize DNA and undergo mitoses (Paul *et al.*, 1972; Draghi *et al.*, 1980; Michalopoulos *et al.*, 1982). Homologous rat serum is

93

THE ISOLATED
HEPATOCYTE

usually more potent than other sera (Strain *et al.*, 1982), suggesting a possible species specificity of some relevant factor (Draghi *et al.*, 1980). However, at high concentrations, even normal rat serum inhibits hepatocyte DNA synthesis *in vitro* (Michalopoulos *et al.*, 1982). Several studies have shown that, at all percentages tested, serum from two-thirds partially hepatectomized rats (collected at 24–48 hr postoperative) is more effective in stimulation of hepatocyte proliferation than serum from control or sham-operated rats (Paul *et al.*, 1972; Michalopoulos *et al.*, 1982). Furthermore, partially hepatectomized rat serum contains little or no inhibitory activity even when added to hepatocyte cultures at greater than 50% concentration (Michalopoulos *et al.*, 1982). Thus, it has been hypothesized that in the regenerating liver mature hepatocytes may be released from quiescence by a relative decrease in circulating levels of growth inhibitor(s) and/or induced to proliferate by a relative increase in circulating levels of growth stimulator(s). To date, most of the research of hepatocyte growth in primary culture has been directed toward the identification of positive growth factors which presumably could initiate or potentiate liver regeneration.

II. HEPATOCYTE ISOLATION AND CULTURE

Much of the work described here utilized cultured rat hepatocytes in bioassays of DNA synthesis. Hepatocytes for primary culture can be isolated from rodent livers by a two-step collagenase perfusion method (Seglen, 1976; Berry and Friend, 1969). The data presented in the figures were generated from experiments performed with rat hepatocytes isolated and cultured as follows. However, this is only one of many published modifications of the basic procedures. Young adult Fischer rats (150–250 gm, usually male) are anesthetized with an intraperitoneal injection (50 mg kg^{-1} body weight) of sodium pentobarbital. Perfusions are performed in a sterilized hood using a peristaltic pump, autoclaved tubing, and sterile solutions maintained at 37°C. A catheter is inserted in the liver via the inferior vena cava above the kidneys and flow (15 ml/min) is directed out of the severed portal vein by clamping the major vessels above the diaphragm. The first step circulates 250 ml of calcium-free salt solution (0.5 g/liter KCl, 8.3 g/liter NaCl, 2.5 g/liter HEPES) through the liver *in situ*. The second perfusate is 250 ml of the same buffer plus 0.64 g/liter $CaCl_2$ and 0.5 g/liter collagenase (Worthington type I, 100–250 U/mg). After the digestion, the softened liver is carefully excised, placed in ice-cold sterile salt solution, and gently minced to disperse the cells. Undigested tissue fragments are removed by filtration through 100-μm nylon mesh. The 50-ml suspension is centrifuged at 50 g at 4°C for 2–3 min to enrich the yield of viable parenchymal hepatocytes. After three such washes with ice-cold buffer, the preparation usually contains 95–98% hepatocytes with >85% viability as determined by trypan blue exclusion.

Hepatocytes are suspended in warm Eagle's minimum essential medium with Earles's salts and nonessential amino acids supplemented with 1 mM pyruvate, 50 μg/ml gentamicin, 10^{-7} M insulin, and 5% HyClone fetal bovine serum to enhance attachment. They are plated at a density of 100,000 cells/ml for each 35-mm plastic culture dish. Tissue culture dishes, with or without autoclaved 22 $\times$ 22 mm glass coverslips, are coated in advance with rat-tail collagen (Michalopoulos *et al.*, 1982). Cultures are incubated at 37°C in a 90% humidity, 7% CO_2 atmosphere. Two to four hours after plating, the medium is replaced with 1 ml of a formulation identical to the above with the exception of serum, the addition of radiolabel, and the substances to be tested. DNA synthesis is expressed as tritiated thymidine (specific activity 40 mCi/mmol) incorporation, which can be measured by liquid scintillation counting (added concentration 5 μCi/ml) or autoradiography (10 μCi/ml). Cultures are routinely incubated in the experimental medium for 48 hr and harvested accordingly. The plates (for liquid scintillation counting) or coverslips (for autoradiography) are washed six times in saline. For liquid scintillation counting, cells are solubilized with NaOH and macromolecules are precipitated with cold trichloroacetic acid (Michalopoulos *et al.*, 1984). Data are usually expressed as dpm/culture, but can also be quantitated as dpm/unit DNA or protein. For autoradiography, cells on coverslips are fixed with 10% buffered formalin, dehydrated with 95% ethanol, mounted on microscope slides, and coated with nuclear track emulsion (Michalopoulos *et al.*, 1982). After 1 week exposure at -20°C, slides are developed, stained, and examined under high power of a light microscope. Hepatocyte nuclei in random fields are counted, and data are expressed as a percentage labeling index. For detailed description of other procedures, such as chromatography, the reader should consult the specific references cited.

III. EPIDERMAL GROWTH FACTOR

Epidermal growth factor (EGF) is the only hepatotrophic hormone which has been fully characterized. EGF is a heat- and acid-stable, nondialyzable, single 6 kDa polypeptide consisting of 53 amino acid residues (St. Hilaire and Jones, 1982). The sequences of mouse (Carpenter and Cohen, 1979), rat (Simpson *et al.*, 1985), and human (Gregory, 1975) EGF are known. Murine EGF lacks lysine, phenylalanine, and alanine. Human EGF lacks threonine and phenylalanine. EGF has six cysteine residues which form three intramolecular disulfide linkages that are required for biological activity. Mouse EGF retains activity when cleaved by trypsin (St. Hilaire and Jones, 1982).

Isolation of complementary DNA clones has recently revealed the primary structure of the messenger RNA encoding mouse EGF (Gray *et al.*, 1983; Scott *et al.*, 1983). The 4.7-kilobase mRNA predicts the sequences of a 130 kDa precursor protein (preproEGF) of about 1200 amino acids. The mature EGF corresponds to residues 977–1029 which may be cleaved from the precursor or processing intermediate(s) by an EGF-binding arginine esteropeptidase activity (Carpenter and Cohen, 1979). The sequence of preproEGF also contains seven regions upstream which share significant homology with EGF, particularly with respect to the pattern of cysteine residues (Scott *et al.*, 1983). As yet, it is not known if these seven sequences represent other biologically active polypeptides. PreproEGF includes two stretches of hydrophobic amino acids which may serve as an amino terminal signal sequence and a carboxy terminal transmembrane sequence. In this regard the EGF precursor structurally resembles a membrane protein and indeed shows limited homology to the low-density lipoprotein receptor (Pfeffer and Ullrich, 1985).

Epidermal growth factor or related proteins have been found in various tissues and body fluids, predominantly in submaxillary salivary glands, duodenal Brunner's glands, saliva, gastric juice, milk, and urine (St. Hilaire and Jones, 1982; Carpenter and Cohen, 1979). Plasma levels in the adult mouse are approximately 1 ng/ml. EGF protein and mRNA have recently been detected in relative abundance in the rodent kidney using EGF antibodies (Rall *et al.*, 1985; Skov Olsen *et al.*, 1984) or a preproEGF cDNA probe (Rall *et al.*, 1985). Despite the extensive body distribution of EGF, its expression in normal adult rodent liver could not be confirmed (Rall *et al.*, 1985).

Epidermal growth factor interacts in a specific, saturable manner with high-affinity receptors on the surface of responsive cells. The hepatic EGF receptor has been purified and characterized as a 170 kDa integral membrane glycoprotein with intrinsic basal kinase activity (Cohen *et al.*, 1982). Binding of EGF to the extracellular domain stimulates the autophosphorylation of the intracellular kinase domain at specific tyrosine residues (Downward *et al.*, 1984). The receptor also possesses tyrosine kinase activity toward exogenous substrates. The EGF receptor has been shown to be phosphorylated on specific serine and threonine residues by protein kinase C, with variable concomitant effects on EGF binding (Hunter *et al.*, 1984). This may represent an important biochemical mechanism in the heterologous regulation of EGF action by other hormones [e.g., platelet-derived growth factor (PDGF)] or tumor promoters (e.g., phorbol esters). Receptor binding of iodinated EGF has been studied in rat liver membrane preparations (Benveniste and Carson, 1985) and in rat hepatocyte primary cultures (Moriarty and Savage, 1980). Scatchard analyses of ^{125}I-labeled EGF binding at 4°C to isolated intact hepatocytes indicate a single class of high-affinity sites on the order of 10^5 receptors per cell with dissociation constants in the nanomolar range (Moriarty and Savage, 1980; Cruise *et al.*, 1986). At 37°C, *in vivo* (St. Hilaire and Jones, 1982; Dunn and Hubbard, 1984) and *in vitro* (Moriarty and Savage, 1980; Carpentier *et al.*, 1981) hepatocytes internalize receptor-bound EGF by endocytosis. This ligand-induced internalization accounts for the down-regulation of hepatic EGF receptors observed in primary culture (Moriarty and Savage, 1980). Experiments with perfused rat livers suggest that some hepatic EGF receptors are conserved and recycled back to the cell membrane (Dunn and Hubbard, 1984). Hepatocytes degrade EGF within lysosomal structures (Dunn and Hubbard, 1984; Carpentier *et al.*, 1981). *In vivo*, a small portion of EGF internalized at the hepatocyte sinusoidal surface must be processed through a direct, nonlysosomal pathway because both intact and degraded proteins are secreted into bile (St. Hilaire *et al.*, 1983; Burwen *et al.*, 1984). Supraphysiological intraportal doses of ^{125}I-labeled EGF are completely cleared by the liver in a single pass and are found within a steep portal-to-central lobular concentration gradient (St. Hilaire *et al.*, 1983). Thus, in addition to dual intracellular transport routes, hepatocytes have a high capacity and efficiency for the uptake of circulating EGF.

Hepatic sequestration may account, in part, for the finding that continuous intraperitoneal infusion of EGF into normal adult rats stimulated DNA synthesis and mitosis in liver but not in other tissues examined (Bucher *et al.*, 1978). The hepatotrophic response was substantially augmented by combination of EGF with insulin and/or glucagon. EGF, supplemented to serum-free media, can stimulate DNA synthesis and mitosis in primary cultures of normal adult rat hepatocytes (McGowan *et al.*, 1981; Richman *et al.*, 1976; Enat *et al.*, 1984). Insulin and glucagon alone are ineffective in initiation of hepatocyte DNA synthesis but can synergistically enhance the EGF-induced growth (McGowan *et al.*, 1981). DNA synthesis stimulated by EGF is dose dependent (1–100 ng/ml), saturable, and peaks within the second to third day of continuous exposure (Richman *et al.*, 1976; Enat *et al.*, 1984). Under optimal culture conditions, labeling indices up to 75–80% can be observed following the addition of $10^{-7} M$ insulin and 10–20 ng/ml EGF. Other variables which influence hormonal responsiveness of hepatocytes *in vitro* include cell density (Seglen, 1976; Enat *et al.*, 1984), substratum (Enat *et al.*, 1984), and medium composition. For example, improved survival and proliferation has been described for primary cultures maintained at low density on rat liver biomatrix in defined serum-free medium containing EGF, insulin, and other factors (Enat *et al.*, 1984). Addition of pyruvate (20–40 mM) to serum-free medium enhances hepatocyte DNA synthesis, particularly in the presence of EGF and insulin (McGowan and Bucher, 1983). The responsiveness of hepatocytes to EGF is strongly dependent upon the presence of proline. Proline (0.2–1.0 mM) alone does not stimulate DNA synthesis in primary cultures of adult rat hepatocytes but is necessary for maximum stimulation by EGF (Houck and Michalopoulos, 1985; Nakamura *et al.*, 1984b). There is some evidence that proline may be required for collagen biosynthesis, even by hepatocytes cultured on rat tail type I collagen-coated plates. Addition of *cis*-4-hydroxyproline or other inhibitors of collagen synthesis to proline-containing media caused dose-dependent suppression of EGF-induced hepatocyte DNA synthesis (Nakamura *et al.*, 1984a). It remains unknown why endogenous collagen synthesis should be a prerequisite for DNA synthesis, but the extracellular matrix is believed to be important for hepatocyte growth and function *in vitro*. Hepatocytes in primary culture have been shown to produce and degrade collagen (Diegelmann *et al.*, 1983; Tseng *et al.*, 1983). Such activity may create a proline "sink" as proline is converted to hydroxyproline and diverted from incorporation into other proteins or utilization in metabolism. Although hepatocytes can synthesize proline from other intermediates such as glutamate, the exogenous proline may serve to alleviate the need of hepatocytes *in vitro* for this otherwise nonessential amino acid.

IV. NOREPINEPHRINE

Norepinephrine is another factor which can interact with EGF in the stimulation of hepatocyte DNA synthesis. Previous *in vivo* studies of the effects of denervation or administration of adrenergic agents on hepatocyte proliferation following partial hepatectomy have suggested a role for catecholamines in liver regeneration (Morley and Royse, 1981). However, it was not known if, or how, catecholamines acted directly on hepatocytes to promote their growth or if they acted indirectly by modulating synthesis or secretion of growth factors in other tissues. For example, treatment of rodents with α-adrenergic agonists resulted in the release of EGF from submaxillary glands, with subsequent increased concentrations of EGF measured in saliva (Olsen *et al.*, 1984) and plasma (Byyny *et al.*, 1974). Since the development and improvement of hepatocyte culture and more selective adrenergic drugs, the regulation of hepatocyte function by catecholamines has been examined at the cellular level. Recently, it was demonstrated that norepinephrine and epinephrine stimulate DNA synthesis in adult rat hepatocytes cultured in serum-free medium supplemented with insulin ($10^{-7} M$) and EGF (10 ng/ml) (Cruise *et al.*, 1985). The hepatotrophic response was dose dependent, becoming significant between 10^{-7} and $10^{-6} M$, and reaching a maximum at $10^{-4} M$ norepinephrine (Fig. 1a). Norepinephrine exerts this effect through interaction with α_1-adrenergic receptors, because the response was in-

---

Fig. 1. Dose response of norepinephrine stimulation of hepatocyte DNA synthesis in the presence of insulin ($10^{-7} M$) and EGF (10 ng/ml). Tritiated thymidine incorporation was measured by liquid scintillation counting (left scale, solid lines) and autoradiography (right scale, dashed lines). (b) Effect of increasing doses of adrenergic blockers on the stimulation of hepatocyte DNA synthesis by $10^{-5} M$ norepinephrine. Prazosin (circles), yohimbine (squares), or propanolol (single triangle) was added to primary cultures simultaneously with norepinephrine. [Reproduced from Cruise *et al.* (1985).]

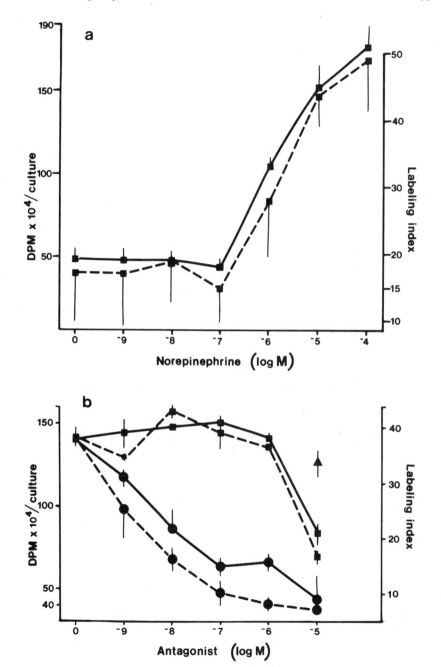

hibited, in a dose-dependent manner, by the α_1-specific antagonist prazosin (Fig. 1b) (Cruise et al., 1985). The enhancement of hepatocyte DNA synthesis by norepinephrine is strictly dependent upon the presence of EGF (Cruise and Michalopoulos, 1985). These two hormones appear to interact synergistically. Norepinephrine is most effective when incubated with hepatocytes for 12 hr or more, beginning 2–4 hr after plating. Hepatocytes were unresponsive to norepinephrine added to cultures 24 hr after plating (Cruise and Michalopoulos, 1985). This time restriction correlates with the rapid decline in α_1-adrenergic receptors (60% loss within 24 hr, measured by radiolabeled prazosin binding) on hepatocytes in primary culture (K. R. Schwarz et al., 1985). EGF reeptor numbers also drop (albeit less drastically) during primary culture of hepatocytes (Cruise and Michalopoulos, 1985; Lin et al., 1984) yet the cells retain responsiveness to EGF. DNA synthesis was still augmented when hepatocytes were preincubated with only norepinephrine for the first 24 hr and then exposed to EGF for the second 24 hr. However, maximal labeling indices were obtained when hepatocytes were incubated for the full 48 hr with norepinephrine, insulin, and EGF (Cruise and Michalopoulos, 1985). Again, insulin seems to play a permissive role in the stimulation of hepatocyte growth.

Further investigations into the effects of norepinephrine and EGF on adult rat hepatocytes have revealed a possible mechanism for their interaction. Exposure of primary cultures at 37°C to norepinephrine caused a time- and dose-dependent decrease in ^{125}I-labeled EGF binding assayed at 4°C (Cruise et al., 1986). The greatest reduction in EGF binding, 35–45%, was observed with hepatocytes preincubated for 1 hr with 10^{-5} M norepinephrine. EGF binding remained about 25% lower than controls in cultures exposed to norepinephrine for 24 hr, so the effect is not transient (Cruise and Michalopoulos, 1985). Norepinephrine inhibition of EGF binding to hepatocytes was not competitive, but was antagonized by 10^{-7} M prazosin (Cruise et al., 1986). Thus, both the stimulation of EGF-induced DNA synthesis and the inhibition of EGF-receptor binding are mediated through the interaction of norepinephrine with α_1-adrenergic receptors on hepatocytes. Scatchard analysis of saturation curves of ^{125}I-labeled EGF binding to hepatocyes demonstrated that preincubation for 1 hr with 10^{-5} M norepinephrine reduced EGF receptor number by about 40% (from 100,000 to 60,000 sites per cell) without significantly altering receptor affinity (K_D 1.5 nM) (Fig. 2a,b) (Cruise et al., 1986). The precise mechanism of such heterologous

Fig. 2. Effect of 1 hr preincubation with 10^{-5} M norepinephrine on the binding of ^{125}I-labeled EGF to hepatocytes. (a) Saturation binding curves of control (solid curve, circles) and norepinephrine-treated (dashed curve, squares) triplicate cultures. Binding was performed for 2 hr at 4°C with hepatocytes plated in 24 multiwell dishes at a density of 100,000 cells per well. (b) Scatchard plot of data. Estimated K_d values are 1.40 and 1.53 nM for control (circles) and norepinephrine-treated (squares) hepatocytes, respectively. [Reproduced from Cruise et al. (1986).]

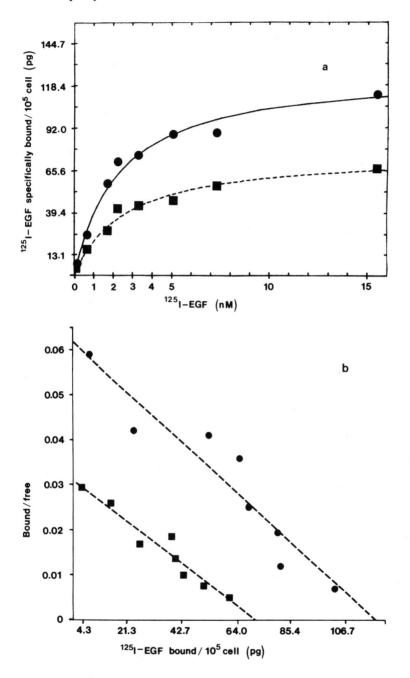

regulation remains undefined. However, several ligands (e.g., norepinephrine, PDGF, vasopressin, bombesin), which bind to their own distinct receptors but also alter binding at EGF receptors, can stimulate phosphatidyl inositol turnover and/or calcium mobilization (Tolbert *et al.*, 1980; Brown *et al.*, 1984). The formation of diacylglycerol, like the presence of 12-*O*-tetradecanoylphorbol-13-acetate (TPA), can activate protein kinase C which has been shown to phosphorylate and modulate EGF receptors (Hunter *et al.*, 1984; Fearn and King, 1985).

Several pieces of experimental evidence plus some lines of logical reasoning support a role for the interaction of norepinephrine and EGF in liver regeneration. Comparison of ^{125}I-labeled EGF binding (Earp and O'Keefe, 1981) and EGF-dependent phosphorylation (Rubin *et al.*, 1982) in hepatic membranes prepared at various times after sham operation or partial hepatectomy indicated a progressive decrease in EGF receptor number and tyrosine kinase activity during liver regeneration. The reduction in EGF binding was detectable at 8 hr and maximal at 36 hr posthepatectomy. Theoretically, such a decline in EGF receptors could be due to down-regulation by EGF (or homologous ligands) and/or heterologous regulation by norepinephrine (or other hormones). Therefore, a rise in the circulating levels or norepinephrine and/or EGF following liver resection would be expected. Preliminary data show that serum concentrations of norepinephrine and epinephrine increase within hours after surgery. At all times tested, the levels remained higher in partially hepatectomized rats than in sham-operated rats (Cruise and Michalopoulos, 1987). It is possible that the stress of any surgical manipulation can induce the release of catecholamines. Perhaps blood levels in two-thirds hepatectomized animals remain higher for longer due to deficient monoamine oxidase activity or insufficient metabolic capacity in the liver remnant. Higher circulating concentrations of catecholamines might stimulate the α-adrenergic receptor-mediated secretion of EGF from salivary and/or duodenal glands. If absorbed intact from the gastrointestinal tract, EGF could be enriched in portal blood perfusing the liver. A rise in serum levels of EGF following partial hepatectomy has not been documented. As stated earlier, even supraphysiological intraportal doses of EGF are cleared within a single pass through the normal adult rat liver (St. Hilaire *et al.*, 1983). So a gradual increase in serum concentrations of EGF may not be detectable, especially if peripheral or systemic blood rather than portal blood, is collected and assayed. EGF may be more concentrated in regenerating liver if far fewer hepatocytes must sequester greater amounts. Of course, these proposed events are currently speculative. Preliminary results indicate that prazosin, given in multiple doses to rats prior to partial hepatectomy, can delay or decrease the first wave of hepatocyte DNA synthesis during liver regeneration (Cruise and Michalopoulos, 1987). Surgical and chemical denervation experiments are in progress to determine if local release of norepinephrine at hepatic nerve terminals is crucial. Earlier *in vitro* studies suggest that blood-borne norepinephrine is relevant in liver regeneration. As previously mentioned, serum from partially hepatectomized rats was found to stimulate DNA synthesis in primary cultures of adult rat hepatocytes better than serum from normal rats (Michalopoulos *et al.*, 1982). In the same report, various hormones were tested for the ability to diminish or abolish this difference in hepatotrophic activity. Addition of EGF (10 ng/ml) and insulin (10^{-7} M) slightly increased hepatocyte-labeling indices observed in cultures supplemented with either 50% partially hepatectomized rat serum or 50% control rat serum. But the difference in effectiveness between the sera was maintained. Only the combination of norepinephrine (10^{-4} M) with EGF and insulin abolished this difference in hepatocyte DNA synthesis by elevating the labeling index obtained with normal rat serum to nearly equal that obtained with partially hepatectomized rat serum. However, the levels of norepinephrine measured in hepatectomized rat serum were only slightly higher than those measured in normal rat serum. Therefore, it was proposed that norepinephrine might interact with other serum components.

V. FACTORS DERIVED FROM SERUM OR PLATELETS

Comparable increases in hepatocyte DNA synthesis were achieved by supplementing primary cultures with either serum or platelet-poor plasma from partially hepatectomized rats (Michalopoulos *et al.*, 1982). However, depletion of platelets did diminish, but did not abolish, the hepatotrophic activity in normal rat serum (Michalopoulos *et al.*, 1982; Strain *et al.*, 1982). Rat platelet lysates were found to restore the activity of platelet-poor plasma-derived rat serum to that of normal rat serum (Russell *et al.*, 1984a). Partial purification and characterization of the hepatotrophic activity from rat platelets indicated that it was due to a 65-kDa, heat and acid-labile, trypsin- and dithiothreitol-sensitive, cationic protein distinct from platelet-derived growth factor (Russell *et al.*, 1984a,b). This hepatocyte growth factor from rat platelets could account for about half of the stimulatory activity of normal rat serum. A similar heat- and acid-labile hepatotrophic activity was found to be secreted by rat platelets during their aggregation in response to thrombin (Paul and Piasecki, 1984). Neither the rat platelet releasate nor rat plasma alone contained substantial hepatotrophic activity, but together they interacted synergistically to stimulate DNA synthesis in primary cultures of adult rat hepatocytes (Paul and Piasecki, 1984). It has not been shown if the serum concentration or activity of the platelet-derived hepatocyte growth factor(s) increased following liver injury or surgery.

When untreated serum from normal or partially hepatectomized rats was subjected to gel filtration at neutral pH and the fractions were bioassayed for stimulation of DNA synthesis in primary cultures of adult rat hepatocytes, two regions of activity were separated (Fig. 3) (Michalopoulos *et al.*, 1984). A large-molecular-weight peak of activity eluting from Sephadex G-100 in the void volume (>150,000) has been empirically termed hepatopoetin A (HPTA). The second peak of activity, denoted hepatopoetin B (HPTB), can be recovered in the very low molecular weight fractions (<1000) from Sephadex G-100 or G-15. Physicochemical characterization demonstrated that hepatopoetin A activity is acid and heat (60°C) labile, but resistant to dialysis in neutral, isotonic buffers. HPTA is very unstable and can only be stored for more than 1 week at 4°C in complete media or frozen at −20°C in 25% glycerol or in lyophilized form (Thaler and Michalopoulos, 1985). In contrast, hepatopoetin B is acid and heat stable, but lost through dialysis in 1-kDa cutoff membranes (Michalopoulos *et al.*, 1984). The biochemical nature of HPTB remains undefined, but its hydrophilic, basic molecular behavior are consistent with its possible identification as a small peptide or modified amino acid. Hepatopoietin B does not comigrate in gel filtration with vasopressin (Michalopoulos *et al.*, 1984), which has been shown to augment hepatocyte DNA synthesis induced by EGF (Russell and Bucher, 1982). Indeed, both hepatopoetins can stimulate hepatocyte DNA synthesis in the absence or presence of 10 ng/ml EGF, suggesting that each is distinct from, and acts independently of, EGF. However, the two hepatopoetins, A plus B, do appear to interact synergistically with each other to yield hepatotrophic activity comparable to that of rat serum (Michalopoulos *et al.*, 1984).

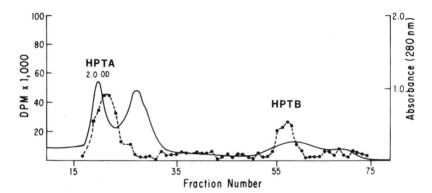

Fig. 3.   Gel filtration of rat serum on Sephadex G-100. The column was equilibrated and eluted with a buffer equivalent to one-tenth the concentration of Earle's salts (without bicarbonate) present in minimum essential medium. Bed volume was 1500 ml, and serum volume was 50 ml. Each 20-ml fraction was assayed at 50% by volume for stimulation of DNA synthesis in the rat hepatocyte bioassay (dpm per culture, dotted curve). Hepatopoietin A (HPTA) corresponds to the peak of activity eluting with the void volume (exclusion limit >150 kDa). Hepatopoetin B (HPTB) is represented by the low-molecular-weight activity peak. Reproduced from Michalopoulos *et al.*, 1984.

Their relative amounts in serum from control vs hepatectomized rats have not yet been measured.

Recent evidence demonstrated that hepatopoetin A is a high-molecular-weight protein (Thaler and Michalopoulos, 1985). Chromatography of rat serum or human plasma on Sephacryl S-200 resolved HPTA activity from the void volume species (>250 kDa). The peak active fractions were pooled and digested for 2 hr at 37°C with tosylphenylalanylchloromethyl ketone (TPCK)-treated trypsin. The original and trypsinized samples of HPTA were then bioassayed for stimulation of DNA synthesis in primary cultures of adult rat hepatocytes. The surprising finding was that trypsin treatment of HPTA substantially increased its hepatotrophic activity. Gel filtration on Sephadex G-50 of HPTA before and after trypsin treatment and subsequent comparison of the activity profiles revealed a shift in the activity from the void volume to low-molecular-weight fractions (Fig. 4a,b) (Thaler and Michalopoulos, 1985). The effect of trypsin on the apparent size and biological activity of hepatopoetin A thus suggest that it could be the precursor protein of a polypeptide growth factor. As discussed earlier, mouse epidermal growth factor has been shown to be encoded within the mRNA of a putative high-molecular-weight precursor protein of about 130,000. The 6 kDa EGF is apparently posttranslationally processed from preproEGF by cleavage at both amino and carboxyl termini behind arginine residues (Gray *et al.*, 1983; Scott *et al.*, 1983). A high-molecular-weight protein fraction (100–200 kDa) with EGF/urogastrone immunoreactivity has been isolated from human serum (Gregory *et al.*, 1979) and platelets (Oka and Orth, 1983). Treatment of this material with trypsin caused an increase in its mitogenic activity on fibroblasts and a decrease in its size estimated by gel filtration. The biological activity and EGF immunoreactivity of the trypsinized serum protein comigrated on Sephadex G-50 with authentic human EGF/urogastrone (Gregory *et al.*, 1979; Oka and Orth, 1983). Consideration of the above information led to the reasonable hypothesis that hepatopoetin A from rat or human serum might be preproEGF. When ^{125}I-labeled mouse EGF and rat HPTA were treated with trypsin and chromatographed on Sephadex G-50, they did not comigrate; the peak of hepatotrophic activity eluted after the peak of radioactivity (Fig. 4b) (Thaler and Michalopoulos, 1985). Furthermore, the bioactivity of native or trypsinized HPTA was not inhibited by polyclonal antibodies raised against murine EGF. Results from these experiments suggest that hepatopoetin A is not identical to the high-molecular-weight precursor of epidermal growth factor. Although the apparent molecular size of HPTB and trypsinized HPTA are similar, there is no data yet available to support the notion that HPTA is the precursor for HPTB. Protease activation or release of growth factors such as HPTA and EGF is an intriguing regulatory mechanism to consider, particularly with regard to liver regeneration. Trypsin-like proteases have been reported to exist at the hepatocyte plasma membrane (Nakamura *et al.*, 1984a) and activate growth hormone

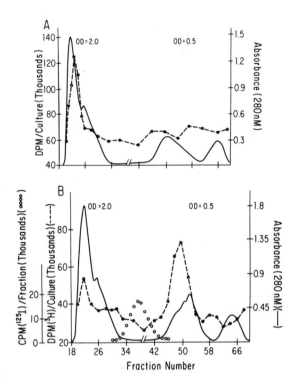

Fig. 4. Activity profiles of untreated and trypsinized Hepatopoetin A (HPTA) chromatographed on Sephadex G-50. Bed volume was 125 ml, and sample volume was 4 ml. The column was equilibrated and eluted with a buffer (pH 7.4) equivalent to Earle's balanced salt solution without bicarbonate. Each 2-ml fraction was assayed at 50% by volume for stimulation of DNA synthesis in the rat hepatocyte bioassay (dashed curve). (A) The activity of untreated HPTA elutes (collected from previous gel filtration on Sephacryl S-200), as expected, in the void volume of Sephadex G-50 (exclusion limit >30 kDa). (B) The same amount of HPTA was mixed with tracer amounts of ^{125}I-labeled EGF and incubated with 20 μg trypsin/mg protein for 2 hr at 37°C. The digestion was stopped with 2μg soybean trypsin inhibitor/μg trypsin, and the samples were subjected to gel filtration as in A. Most of the bioactivity shifts to fractions eluting after the peak of radioactivity (cpm/fraction, open circles), thus representing molecular weights lower than that of trypsinized ^{125}I-labeled EGF. Reproduced from Thaler and Michalopoulos, 1985.

(Schepper *et al.*, 1984). The modification and function of various humoral ligands may be controlled by a balance between proteases exposed at the surface of hepatocytes (or other cells) and protease inhibitors present in the blood.

Protease activation was also suggested in the investigation of a hepatocyte growth factor which shares some, but not all, properties with hepatopoetin A. Hepatotropin (HTP), isolated from rat serum, was described as a 150 kDa anionic, heat ($>55°C$)- and acid-labile protein which stimulated DNA synthesis in primary cultures of adult rat hepatocytes (Nakamura *et al.*, 1984c). The bioactivity of HTP fractions increased over time within 1 week storage at 4°C in phosphate buffered saline. This contrasts with the relative instability of hepatopoetin A activity in neutral, isotonic salt solutions (Thaler and Michalopoulos, 1985). Several serine protease inhibitors prevented the activation of hepatotropin samples during cold storage. Yet the activity of hepatotropin differed remarkably from that of hepatopoetin A in being destroyed by a 2 hr, 37°C incubation with trypsin (Nakamura *et al.*, 1984c). Hepatotropin was further purified by affinity chromatography on heparin–Sepharose. The ability to strongly bind to heparin implies that HTP may be associated with platelets. The activity of hepatotropin in rat serum was found to increase 3- to 5-fold within 24 hr after partial hepatectomy (Nakamura and Ichihara, 1985).

Another liver growth factor has been reported to rise 2- to 5-fold in rat serum and peak 24 hr after partial hepatectomy (Morley and Kingdon, 1973). This "DNA synthesis stimulating factor" was determined to be sensitive to trypsin and acid treatment but resistant to boiling and dialysis. Fractionation on Sephadex G-100 of serum from partially hepatectomized rats gave an M_r approximation of 26,000 for boiled samples and 17,000 for unheated material. In most of these experiments, the activity of serum fractions was assayed by intraperitoneal injection into mice and subsequent isolation and radioactive pulsing of liver cell suspensions. Thus, the hepatocytes were exposed to the growth factor(s) *in vivo*, but DNA synthesis was measured *in vitro* (by liquid scintillation counting or autoradiography of tritiated thymidine incorporation). Whether this serum factor can directly stimulate hepatocyte DNA synthesis in primary culture remains unclear.

VI. FACTORS DERIVED FROM NORMAL LIVER CELLS

A similar combination *in vivo–in vitro* assay was employed in the investigation of growth factors derived from liver tissue. Intraperitoneal injections of cytoplasmic extracts of weanling or regenerating rat livers augmented hepatic DNA synthesis in 34% partially hepatectomized recipients. The tritiated thymidine incorporation was measured *in vivo* or *in vitro* in liver slices prepared from the test animals (LaBrecque and Bachur, 1982). Ethanol-precipitated, aqueous extracts from weanling livers also slightly stimulated DNA synthesis in the livers of unoperated adult rats. However, extracts from normal adult rat liver inhibited DNA synthesis when injected into partially hepatectomized animals. The stimulatory activity in weanling liver extracts, designated hepatic stimulator substance (HSS), was found to be stable to boiling but destroyed by acid. Ultrafiltration and dialysis estimated a molecular weight between 10,000 and 20,000. HSS appeared to be liver specific, but not species specific, and displayed a diurnal rhythm. *In vitro,* HSS stimulated DNA synthesis in serum-supplemented primary cultures of normal adult rat hepatocytes and enhanced the proliferation of all liver-derived cell lines tested (LaBrecque, 1982).

Administration of cytosolic fractions from regenerating rat or porcine liver has been found to increase *in vivo* DNA synthesis measured 24 hr after partial hepatectomy (Makowka *et al.,* 1983). Intraperitoneal injection of regenerating liver cytosol into weanling, but not adult, unoperated rats stimulated the incorporation of tritiated thymidine into hepatocytes isolated from the recipients (L. C. Schwarz *et al.,* 1985). This activity in regenerating liver cytosol, named hepatocyte proliferation factor (HPF), resembled hepatocyte stimulator substance with respect to its heat stability and its hepatotrophic activity and specificity *in vivo.* Ultrafiltration indicated that HPF is larger than 50 kDa. Unfortunately, HPF has not been tested directly on normal adult rat hepatocytes in primary culture. However, other investigators have shown that regenerating liver cytosol (prepared 72 hr posthepatectomy) from several rodent species could stimulate hepatocyte DNA synthesis *in vitro* (Rodgers *et al.,* 1985). Also, media conditioned for 12 hr or more by primary cultures of hepatocytes isolated from regenerating rat liver remnants could induce DNA synthesis in primary cultures of hepatocytes isolated from resting adult rat liver (Demetrious and Levenson, 1977). All of these studies suggest that hepatotrophic substances appear or accumulate within hepatocytes during developmental or regenerative growth. The relationship of these cytoplasmic factors to the serum factors is unknown. Hepatocytes in the liver remnant following partial hepatectomy may synthesize and secrete growth factors into the blood. Alternatively, like EGF, other serum growth factors could be sequestered and concentrated in the remaining liver tissue. Since some of the serum- and liver-derived activities differ in apparent size and physicochemical properties, they are probably distinct growth factors which may interact or integrate in cell cycle control.

VII. FACTORS DERIVED FROM NEOPLASTIC LIVER CELLS

Recent studies indicate that neoplastic liver cells synthesize hepatocyte growth factors. Serum-free medium conditioned for 48 hr by JM1 or JM2 rat hepatoma cells was found to stimulate DNA synthesis in primary cultures of adult rat hepatocytes in a dose-dependent, saturable manner and in the absence of EGF (Fig. 5) (Luetteke and Michalopoulos, 1985). The tumorigenic, metastatic JM1 and JM2 cell lines were clonally established from a hepatocellular carcinoma

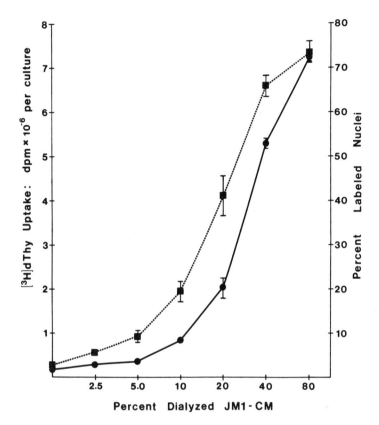

Fig. 5. Dose response of adult rat hepatocytes to JM1-conditioned medium. Conditioned medium was dialyzed (1000 M_r cutoff) overnight against serum-free medium. Volumes ranging from 20 to 800 μl were added with fresh medium (total volume 1.0 ml/dish) to triplicate primary cultures. After 48 hr incubation cultures were processed for measurement of tritiated thymidine incorporation by liquid scintillation counting (dotted curve) or autoradiography (solid curve). DNA synthesis is expressed as dpm per culture (squares) and labeling index (circles). Reproduced from Luetteke and Michalopoulos, 1985.

induced with a diethylnitrosamine initiation and phenobarbitol promotion protocol (Novicki *et al.*, 1983). The hepatotrophic activity secreted by these hepatoma cells was characterized as nondialyzable in 50,000 M_r cutoff membranes, heat and acid stable, and sensitive to trypsin and dithiothreitol treatment (Luetteke and Michalopoulos, 1985). Gel filtration of concentrated JM1- or JM2-conditioned medium on Sephadex G-100 separated the activity into two regions: a major peak migrating with apparent M_r 25,000, and a minor peak with apparent M_r 10,000 (Fig. 6). Both of these proteins, called hepatoma-derived

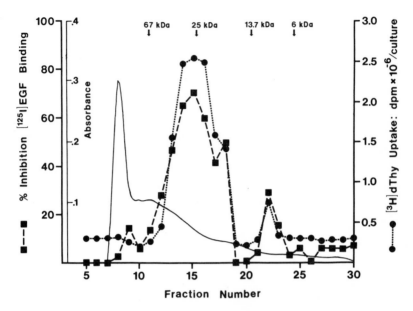

Fig. 6. Sephadex G-100 gel filtration of JM1-conditioned medium. JM1-conditioned medium
was collected after 48 hr, clarified, concentrated, applied to a 1 × 100-cm column, and eluted at 4°C
with Earle's balanced salt solution (without bicarbonate, pH 7.4) at 15–20 ml/hr. Each 3-ml fraction
was tested in duplicate at 20% by volume for stimulation of DNA synthesis in the hepatocyte
bioassay (dotted curve, circles) and at 50% by volume for competition with ^{125}I-labeled EGF in the
radioreceptor assay (dashed curve, squares). Details of the radioreceptor assay are described in
Luetteke and Michalopoulos (1985) from which figure is reproduced. Molecular weight markers are
bovine serum albumin, 67,000; chymotrypsinogen, 25,000; ribonuclease, 13,700; and insulin, 6000.

growth factors (HDGF), can be further purified with ion exchange chromatography and reverse-phase HPLC (Luetteke and Michalopoulos, 1986). Chromatography fractions which stimulated hepatocyte DNA synthesis also com-
petitively inhibited binding of ^{125}I-labeled EGF to hepatocytes (Fig. 6) (Luetteke
and Michalopoulos, 1985). However, antiserum raised against murine EGF did
not block the hepatotrophic activity of HDGF. These results suggest that the JM1
and JM2 hepatoma cells produce proteins which bind to hepatic EGF receptors
but are not immunologically cross reactive with EGF. In addition, specific ^{125}I-
labeled EGF binding to these hepatoma cells was negligible (Luetteke and
Michalopoulos, 1985). Other studies have reported reduced EGF binding in cells
or membranes derived from rat or human hepatomas (Blackshear *et al.*, 1984;
Costrini and Beck, 1983). One possible explanation for this finding is persistent
occupancy and/or down-regulation of EGF receptors by autocrine growth factors
secreted by the hepatoma cells. Although the hepatoma-derived growth factors

stimulate hepatocyte DNA synthesis, it remains to be determined if they stimulate proliferation of hepatoma cells via an autocrine mode.

There have been numerous reports of the production of EGF-like mitogens by neoplastic cells (DeLarco and Todaro, 1978; Todaro *et al.*, 1980). Most of them are heat- and acid-stable, trypsin- and dithiothreitol-sensitive proteins of variable molecular weight (Mr 6000–55,000). They have been classified as type α transforming growth factors (TGF-α), which bind to EGF receptors and interact synergistically with type β transforming growth factors (TGF-β) to stimulate anchorage-independent growth of rodent fibroblasts in soft agar (Roberts *et al.*, 1983). However, in monolayers of many cell types, including isolated rat hepatocytes (Hayashi and Carr, 1985), platelet-derived TGF-β inhibits DNA synthesis induced by EGF. Low-molecular-weight (6000) rat TGF-α has been sequenced and shown to share 33% amino acid homology (Marquardt *et al.*, 1984) but no immunological cross reactivity (DeLarco and Todaro, 1978) with mouse EGF. Cloning and sequence analyses of complementary DNAs encoding rat TGF-α have revealed the probable existence of a precursor protein of at least 159 amino acids (Lee *et al.*, 1985). The TGF-α 4.5-kilobase (kb) mRNA was detected at a low level in normal rat liver. Transforming growth factors have also been extracted from rat fetuses (Matrisian *et al.*, 1982) and human platelets (Assoian *et al.*, 1984). Therefore, they are suspected to function in physiological as well as pathological growth control. Hepatoma-derived growth factor(s) resembles the higher molecular weight TGF-α with respect to EGF receptor binding and physiocochemical properties. Further purification and characterization of HDGF is required for thorough comparison with the precursor proteins of EGF, TGF-α, or other EGF-related proteins, like the recently isolated vaccinia virus growth factor (Stroobant *et al.*, 1985). The ability of HDGF to bind to EGF receptors is particularly interesting in light of previous reports that known liver carcinogens or hepatic tumor promoters can modify EGF binding to liver membranes (Josefsberg *et al.*, 1984; Madhukar *et al.*, 1984). Differential proliferation of normal and genetically altered hepatocytes is an essential feature in most models of hepatocarcinogenesis (Farber, 1984). A growth advantage or autonomy endowed by autocrine secretion and stimulation may selectively favor the development or progression of liver tumors.

Currently, the biochemical and biological characterizations of most of the hepatotrophic factors discussed here are incomplete. The identification and description of these proteins are summarized in Table I. No amino acid sequence data are available for any hepatocyte growth factor except EGF. More precise information is needed concerning the structure and function of these proteins. Future production of specific probes for hepatotrophic factors, such as antibodies or cDNAs, may enable investigators to monitor their expression during liver regeneration and carcinogenesis.

TABLE I

Summary of Hepatocyte Growth Factors

Name	Source	Estimated M_r	Temperature effects	pH effects	Protease effects	Bioactivity or bioassay	Comments	References[a]
Epidermal growth factor (EGF)	Body fluids and tissues	EGF, 6000, pre-proEGF, 130,000	Heat stable	Acid stable	Trypsin sensitive	Stimulates adult rat hepatocyte DNA synthesis and mitoses in vivo and in vitro	Receptor and precursor amino acid sequence known	1,2,3,4,5,6
Hepatoma-derived growth factor (HDGF)	JM1 hepatoma conditioned medium	>25,000	Heat stable	Acid stable	Trypsin sensitive	Stimulates adult rat hepatocyte DNA synthesis in vitro	Competes for binding to EGF receptors	7
Hepatocyte growth factor from rat platelets	Rat serum or platelets	65,000	Heat labile	Acid labile	Trypsin sensitive	Stimulates adult rat hepatocyte DNA synthesis in vitro	Cationic, binds to CM Sephadex	8,9
Hepatopoietin A (HPTA)	Rat and human serum or plasma	>150,000–250,000	Heat labile	Acid labile	Trypsin increases and shifts bioactivity to low M_r	Stimulates adult rat hepatocyte DNA synthesis in vitro	Nondialyzable, very unstable	10,11
Hepatotropin (HTP)	Partially hepatectomized or normal rat serum	150,000	Heat labile	Acid labile	Trypsin sensitive but protease inhibitors prevent activation at 4°C	Stimulates adult rat hepatocyte DNA synthesis in vitro	Anionic, binds to DEAE-cellulose and heparin-Sepharose	12,13

Factor	Source	MW	Heat stability	Acid stability	Protease sensitivity	Activity	Comments	Reference
Hepatopoietin B (HPTB)	Rat and human serum or plasma	<1000	Heat stable	Acid stable		Stimulates adult rat hepatocyte DNA synthesis in vitro	Dialyzable	14
DNA synthesis stimulating factor	Partially hepatectomized rat serum	17,000–26,000	Heat stable	Acid labile	Trypsin sensitive	I.p. injection into adult mice stimulates DNA synthesis in isolated hepatocyte suspensions	Not tested on primary hepatocyte cultures	15
Hepatocyte stimulator substance (HSS)	Weanling or regenerating rat liver	>10,000–<20,000	Heat stable	Acid labile		Stimulates adult rat hepatocyte DNA synthesis in vivo and in vitro	Diurnal rhythm, liver specific	16,17
Hepatocyte proliferation factor (HPF)	Weanling or regenerating rat or porcine liver cytosol	>50,000	Heat stable			I.p. injection into weanling rats or 34% hepatectomized adult rats stimulates DNA synthesis measured in isolated hepatocytes	Not tested on primary hepatocyte cultures	18,19

a Key to references: 1. St. Hilaire and Jones (1982); 2. Carpenter and Cohen (1979); 3. Gray et al. (1983); 4. Scott et al. (1983); 5. Bucher et al. (1978); 6. McGowan et al. (1981); 7. Luetteke and Michalopoulos (1985); 8. Russell et al. (1984a); 9. Paul and Piasecki (1984); 10. Michalopoulos et al. (1984); 11. Thaler and Michalopoulos (1985); 12. Nakamura et al. (1984c); 13. Nakamura and Ichihara (1985); 14. Michalopoulos et al. (1984); 15. Morley and Kingdon (1973); 16. La Brecque and Bachur (1982); 17. LaBreque (1982) 18. Makowka et al. (1983); 19. Schwarz et al. (1985).

REFERENCES

Assoian, R. K., Grotendorst, G. R., Miller, D. M., and Sporn, M. D. (1984). Cellular transformation by coordinated action of three peptide growth factors from human platelets. *Nature (London)* **309,** 804–806.

Benveniste, R., and Carson, S. A. (1985). Binding characteristics of epidermal growth factor receptors in male and female rat liver cell membrane preparations. *Mol. Cell. Endocrinol.* **41,** 147–151.

Berry, M. N., and Friend, D. S. (1969). High yield preparation of isolated rat liver parenchymal cells. A biochemical and fine structural study. *J. Cell Biol.* **43,** 506–520.

Blackshear, P. J., Nemenoff, R. A., and Avruch, J. (1984). Characterization of insulin and epidermal growth factor stimulation of receptor autophosphorylation in detergent extracts of rat liver and transplantable rat hepatomas. *Endocrinology (Baltimore)* **114,** 141–152.

Brown, K. D., Blay, J., Irvine, F. F., Heslop, J. P., and Berridge, M. J. (1984). Reduction of epidermal growth factor receptor affinity by heterologous ligands: Evidence for a mechanism involving the breakdown of phosphoinositides and the activation of protein kinase C. *Biochem. Biophys. Res. Commun.* **123,** 377–384.

Bucher, N. L. R., Patel, U., and Cohen, S. (1978). Hormonal factors and liver growth. *Adv. Enzyme Regul.* **16,** 205–213.

Burwen, S. J., Barker, M. E., Goldman, I. S., Hradek, G. T., Raper, S. E., and Jones, A. L. (1984). Transport of epidermal growth factor by rat liver: Evidence for a nonlysosomal pathway. *J. Cell Biol.* **99,** 1259–1265.

Byyny, R. L., Orth, D. N., Cohen, S., and Doyne, E. S. (1974). Epidermal growth factor: Effects of androgens and adrenergic agents. *Endocrinology (Baltimore)* **95,** 776–782.

Carpenter, G., and Cohen, S. (1979). Epidermal growth factor. *Annu. Rev. Biochem.* **48,** 193–296.

Carpentier, J. L., Gorden, P., Freychet, P., Canivet, B., and Orci, L. (1981). The fate of [^{125}I] iodoepidermal growth factor in isolated hepatocytes: A quantitative electron microscopic autoradiographic study. *Endocrinology (Baltimore)* **109,** 768–775.

Cohen, S., Fava, R. A., and Sawyer, S. T. (1982). Purification and characterization of epidermal growth factor receptor/protein kinase from normal mouse liver. *Proc. Natl. Acad. Sci. U.S.A.* **79,** 6237–6241.

Costrini, N. V., and Beck, R. (1983). Epidermal growth factor-urogastrone receptors in normal human liver and primary hepatoma. *Cancer (Philadelphia)* **51,** 2191–2196.

Cruise, J. L., and Michalopoulos, G. K. (1985). Norepinephrine and epidermal growth factor: Dynamics of their interaction in the stimulation of hepatocyte DNA synthesis. *J. Cell. Physiol.* **125,** 45–50.

Cruise, J. L., and Michalopoulos, G. K. (1987) submitted.

Cruise, J. L., Houck, K. A., and Michalopoulos, G. K. (1985). Induction of DNA synthesis in cultured rat hepatocytes through stimulation of alpha-1 adrenoreceptor by norepinephrine. *Science* **227,** 749–751.

Cruise, J. L., Cotecchia, S., and Michalopoulos, G. K. (1986). Norepinephrine decreases EGF binding in primary rat hepatocyte cultures. *J. Cell. Physiol.* **127,** 39–44.

DeLarco, J. E., and Todaro, G. J. (1978). Growth factors from murine sarcoma virus-transformed cells. *Proc. Natl. Acad. Sci. U.S.A.* **75,** 4001–4005.

Demetrious, A. A., and Levenson, S. M. (1977). Liver origin of hepatotrophic factors. *Surg. Forum* **28,** 383–384.

Diegelmann, R. F., Guzelian, P. S., Gay, R., and Gay, S. (1983). Collagen formation by hepatocytes in primary monolayer culture and *in vivo*. *Science* **219,** 1343–1345.

Downward, J., Parker, P., and Waterfield, M. D. (1984). Autophosphorylation sites on the epidermal growth factor receptor. *Nature (London)* **311**, 483–485.

Draghi, E., Armato, U., Andreis, P. G., and Mengato, L. (1980). The stimulation by epidermal growth factor of the growth of neonatal rat hepatocytes in primary tissue culture and its modulation by serum and associated pancreatic hormones. *J. Cell. Physiol.* **103**, 129–147.

Dunn, W. A., and Hubbard, A. L. (1984). Receptor-mediated endocytosis of epidermal growth factor by hepatocytes in the perfused rat liver: Ligand and receptor dynamics. *J. Cell Biol.* **98**, 2148–2159.

Earp, H. S., and O'Keefe, E. J. (1981). Epidermal growth factor receptor number decreases during rat liver regeneration. *J. Clin. Invest.* **67**, 1580–1583.

Enat, R., Jefferson, D. M., Ruiz-Opazo, N., Gatmaitan, Z., Leinwand, L. A., and Reid, L. M. (1984). Hepatocyte proliferation in vitro: Its dependence on the use of serum-free hormonally defined medium and substrata of extracellular matrix. *Proc. Natl. Acad. Sci. U.S.A.* **81**, 1411–1415.

Farber, E. (1984). The multistep nature of cancer development. *Cancer Res.* **44**, 4217–4223.

Fearn, J. C., and King, A. C. (1985). EGF receptor affinity is regulated by intracellular calcium and protein kinase C. *Cell (Cambridge Mass.)* **40**, 991–1000.

Gray, A., Dull, T. J., and Ullrich, A. (1983). Nucleotide sequence of epidermal growth factor cDNA predicts a 128,000 molecular weight protein precursor. *Nature (London)* **303**, 722–725.

Gregory, H. (1975). The isolation and structure of urogastrone and its relationship to epidermal growth factor. *Nature (London)* **257**, 325–327.

Gregory, H., Walsh, S., and Hopkins, C. R. (1979). The identification of urogastrone in serum, saliva, and gastric juice. *Gastroenterology* **77**, 313–318.

Grisham, J. W., Leong, G. F., and Hold, B. V. (1964). Heterotopic partial autotransplantation of rat liver. Technique and demonstration of structure and function of the graft. *Cancer Res.* **24**, 1474–1501.

Hayashi, I., and Carr, B. I. (1985). DNA synthesis in rat hepatocytes: Inhibition by a platelet factor and stimulation by an endogenous factor. *J. Cell. Physiol.* **125**, 82–90.

Houck, K. A., and Michalopoulos, G. K. (1985). Proline is required for the stimulation of DNA synthesis in hepatocyte cultures by EGF. *In Vitro Cell. Dev. Biol.* **21**, 121–124.

Hunter, T., Ling, N., and Cooper, J. A. (1984). Protein kinase C phosphorylation of the EGF receptor at a threonine residue close to the cytoplasmic face of the membrane. *Nature (London)* **311**, 480–483.

Jirtle, R. L., and Michalopoulos, G. K. (1982). Effects of partial hepatectomy on transplanted hepatocytes. *Cancer Res.* **42**, 3000–3004.

Josefsberg, Z., Carr, B. I., Hwang, D., Barseghian, G., Tomkinson, C., and Lev-Ran, A. (1984). Effect of 2-acetylaminofluorene on the binding of epidermal growth factor to microsomal and Golgi fractions of rat liver cells. *Cancer Res.* **44**, 2754–2757.

LaBrecque, D. R. (1982). *In vitro* stimulation of cell growth by hepatic stimulator substance. *Am. J. Physiol.* **242**, G289–G295.

LaBrecque, D. R., and Bachur, N. R. (1982). Hepatic stimulator substance: Physiochemical characteristics and specificity. *Am. J. Physiol.* **242**, G281–G288.

Lee, D. C., Rose, T. R., Webb, N. R., and Todaro, G. J. (1985). Cloning and sequence analysis of a cDNA for rat transforming growth factor alpha. *Nature (London)* **313**, 489–491.

Lin, Q., Blaisdell, J., O'Keefe, E., and Earp, H. S. (1984). Insulin inhibits the glucocorticoid mediated increase in hepatocyte EGF binding. *J. Cell. Physiol.* **119**, 267–272.

Luetteke, N. C., and Michalopoulos, G. K. (1985). Partial purification and characterization of a hepatocyte growth factor produced by rat hepatocellular carcinoma cells. *Cancer Res.* **45**, 6331–6337.

Luetteke, N. C., and Michalopoulos, G. K. (1986). *Cancer Res.* **27**, 211.

McGowan, J. A., Strain, A. J., and Bucher, N. L. R. (1981). DNA synthesis in primary cultures of adult rat hepatocytes in a defined medium: Effects of epidermal growth factor, insulin, glucagon, and cyclic AMP. *J. Cell. Physiol.* **108,** 353–363.

McGowan, J. C., and Bucher, N. L. R. (1983). Pyruvate promotion of DNA synthesis in serum free primary cultures of adult rat hepatocytes. *In Vitro* **19,** 159–166.

Madhukar, B. V., Brewster, D. W., and Matsumura, F. (1984). Effects of *in vivo* administered 2,3,7,8-tetrachlorodibenzo-*p*-dioxin on receptor binding of epidermal growth factor in the hepatic plasma membrane of rat, guinea pig, mouse, and hamster. *Proc. Natl. Acad. Sci. U.S.A.* **81,** 7407–7411.

Makowka, L., Falk, R. E., Falk, J. A., Teodorczyk-Injeyan, J., Venturi, D., Rotstein, L. E., Falk, W., Langer, B., Blendis, L. M., and Phillips, M. J. (1983). The effect of liver cytosol on hepatic regeneration and tumor growth. *Cancer (Philadelphia)* **51,** 2181–2190.

Marquardt, H., Hunkapiller, M. W., Hood, L. E., and Todaro, G. J. (1984). Rat transforming growth factor type I: Structure and relation to epidermal growth factor. *Science* **223,** 1079–1082.

Matrisian, L. M., Pathak, M., and Magun, B. E. (1982). Identification of an epidermal growth factor-related transforming growth factor from rat fetuses. *Biochem. Biophys. Res. Commun.* **107,** 761–769.

Michalopoulos, D. G., Cianculli, H. D., Novotny, A. R., Kligerman, A. D., Strom, S. C., and Jirtle, R. L. (1982). Liver regeneration studies with rat hepatocytes in primary culture. *Cancer Res.* **42,** 4673–4682.

Michalopoulos, G. K., Houck, K. A., Dolan, M. L., and Luetteke, N. C. (1984). Control of hepatocyte replication by two serum factors. *Cancer Res.* **44,** 4414–4419.

Moolten, C. G. D., and Bucher, N. L. R. (1967). Regeneration of rat liver: Transfer of humoral agent by cross circulation. *Science* **158,** 272–274.

Moriarty, D. M., and Savage, C. R. (1980). Interaction of epidermal growth factor with adult rat liver parenchymal cells in primary culture. *Arch. Biochem. Biophys.* **203,** 506–518.

Morley, C. G. D., and Kingdon, H. S. (1973). The regulation of cell growth: Identification and partial characterization of a DNA synthesis stimulating factor from the serum of partially hepatectomized rats. *Biochim. Biophys. Acta* **308,** 260–275.

Morley, C. G. D., and Royse, V. L. (1981). Adrenergic agents as possible regulators of liver regeneration. *Int. J. Biochem.* **13,** 969–973.

Nakamura, T., and Ichihara, A. (1985). Control of growth and expression of differentiated functions of mature hepatocytes in primary culture. *Cell Struct. Funct.* **10,** 1–16.

Nakamura, T., Asami, O., Tanoka, K., and Ichihara, A. (1984a). Increased survival of rat hepatocytes in serum-free medium by inhibition of a trypsin-like protease associated with their plasma membranes. *Exp. Cell Res.* **154,** 81–91.

Nakamura, T., Teramoto, H., Tomita, Y., and Ichihara, A. (1984b). L-proline is an essential amino acid for hepatocyte growth in culture. *Biochem. Biophys. Res. Commun.* **122,** 884–891.

Nakamura, T., Nawa, K., and Ichihara, A. (1984c). Partial purification and characterization of hepatocyte growth factor from serum of hepatectomized rats. *Biochem. Biophys. Res. Commun.* **122,** 1450–1459.

Novicki, D. L., Jirtle, R. L., and Michalopoulos, G. K. (1983). Establishment of two rat hepatoma cell strains produced by a carcinogen initiation, phenobarbital promotion protocol. *In Vitro* **19,** 191–202.

Oka, Y., and Orth, D. N. (1983). Human plasma epidermal growth factor/urogastrone is associated with blood platelets. *J. Clin. Invest.* **72,** 249–259.

Olsen, P. S., Kirkegaard, P., Poulsen, S. S., and Nexo, E. (1984). Adrenergic effects on exocrine secretion of rat submandibular epidermal growth factor. *Gut* **25,** 1234–1240.

Paul, D., and Piasecki, A. (1984). Rat platelets contain growth factor(s) distinct from PDGF which stimulate DNA synthesis in primary adult rat hepatocyte cultures. *Exp. Cell Res.* **154**, 95–100.

Paul, D., Leffert, H., Sato, G., and Holley, R. W. (1972). Stimulation of DNA and protein synthesis in fetal rat liver cells by serum from partially hepatectomized rats. *Proc. Natl. Acad. Sci. U.S.A.* **69**, 374–377.

Pfeffer, J., and Ullrich, A. (1985). Is the precursor a receptor? *Nature (London)* **313**, 184.

Rall, L. B., Scott, J., Bell, G. I., Crawford, R. J., Penschow, J. D., Niall, H. D., and Coghlan, J. P. (1985). Mouse preproepidermal growth factor synthesis by the kidney and other tissues. *Nature (London)* **313**, 228–231.

Richmond, R. A., Claus, T. H., Pilkis, S. J., and Friedman, D. L. (1976). Hormonal stimulation of DNA synthesis in primary cultures of adult rat hepatocytes. *Proc. Natl. Acad. Sci. U.S.A.* **10**, 3589–3593.

Roberts, A. B., Frolik, C. A., Anzano, M. A., and Sporn, M. B. (1983). Transforming growth factors from neoplastic and non-neoplastic tissues. *Red. Proc., Fed. Am. Soc. Exp. Biol.* **42**, 2621–2626.

Rodgers, G. H., Stephan, R., and Flye, M. W. (1985). Cross species stimulation by regenerating liver and tumor cytosol. *Fed. Proc., Fed. Am. Soc. Exp. Biol.* **44**, 1884.

Rubin, R. A., O'Keefe, E. J., and Earp, H. S. (1982). Alteration of epidermal growth factor-dependent phosphorylation during rat liver regeneration. *Proc. Natl. Acad. Sci. U.S.A.* **79**, 776–780.

Russell, W. E., and Bucher, N. L. R. (1982). Vasopressin augments growth factor stimulated DNA synthesis in primary cultures of adult hepatocytes. *Endocrinology (Baltimore)* **100**, Suppl., p. 162.

Russell, W. E., McGowan, J. A., and Bucher, N. L. R. (1984a). Partial characterization of a hepatocyte growth factor from rat platelets. *J. Cell. Physiol.* **119**, 183–192.

Russell, W. E., McGowan, J. A., and Bucher, N. L. R. (1984b). Biological properties of a hepatocyte growth factor from rat platelets. *J. Cell. Physiol.* **119**, 193–197.

St. Hilaire, R. J., and Jones, A. L. (1982). Epidermal growth factor: Its biologic and metabolic effects with emphasis on the hepatocyte. *Hepatology* **2**, 601–613.

St. Hilaire, R. J., Hradek, G. T., and Jones, A. L. (1983). Hepatic sequestration and biliary secretion of epidermal growth factor: Evidence for a high capacity uptake system. *Proc. Natl. Acad. Sci. U.S.A.* **80**, 3793–3801.

Schepper, J. M., Hughes, E. F., Postel-Vinay, M. C., and Hughes, J. P. (1984). Cleavage of growth hormone by rabbit liver plasmalemma enhances binding. *J. Biol. Chem.* **259**, 12945–12948.

Schwarz, K. R., Lanier, S. M., Carter, E. A. Honey, C. J., and Graham, R. M. (1985). Rapid reciprocal changes in adrenergic receptors in intact isolated hepatocytes during primary culture. *Mol. Pharmacol.* **27**, 200–209.

Schwarz, L. C., Makowka, L., Falk, J. A., and Falk, R. (1985). The characterization and partial purification of hepatocyte proliferation factor. *Ann. Surg.* **20**, 296–302.

Scott, J., Urdea, M., Quiroga, M., Sanchez-Pescador, R., Fong, N., Selby, M., Rutter, W. J., and Bell, G. I. (1983). Structure of a mouse submaxillary messenger RNA encoding epidermal growth factor and seven related proteins. *Science* **221**, 236–240.

Seglen, P. O. (1976). Preparation of isolated rat liver cells. *Methods Cell Biol.* **13**, 29–83.

Simpson, R. J., Smith, J. A., Moritz, R. L., O'Hare, M. J., Rudland, P. S., Morrison, J. R., Lloyd, C. J., Grego, B., Burgess, A. W., and Nice, E. C. (1985). Rat epidermal growth factor: Complete amino acid sequence. *Eur. J. Biochem.* **153**, 629–637.

Skov Olsen, P., Nexo, E., Poulsen, S. S., Hansen, H. F., and Kirkegaard, P. (1984). Renal origin of rat urinary epidermal growth factor. *Regul. Pept.* **10**, 37–45.

Strain, A. J., McGowan, J. A., and Bucher, N. L. R. (1982). Stimulation of DNA synthesis in primary cultures of adult rat hepatocytes by rat platelet-associated substances. *In Vitro* **18,** 108–116.

Stroobant, P., Rice, A. P., Gullick, W. J., Cheng, D. J., Kerr, I. M., and Waterfield, M. D. (1985). Purification and characterization of vaccinia virus growth factor. *Cell* **42,** 383–393.

Thaler, F. J., and Michalopoulos, G. K. (1985). Hepatopoetin A: Partial characterization and trypsin activation of a hepatocyte growth factor. *Cancer Res.* **45,** 2545–2549.

Todaro, G. J., Fryling, C., and DeLarco, J. E. (1980). Transforming growth factors produced by certain human tumor cells: Polypeptides that interact with epidermal growth factor receptors. *Proc. Natl. Acad. Sci. U.S.A.* **77,** 5258–5262.

Tolbert, M. E. M., White, A. C., Aspry, K., Cutts, J., and Fain, J. N. (1980). Stimulation by vasopressin and alphacatecholamines of phosphatidylinositol formation in isolated rat liver parenchymal cells. *J. Biol. Chem.* **255,** 1938–1944.

Tseng, S. C. G., Smuckler, E. A., and Stern, R. (1983). Types of collagen synthesized by normal rat liver hepatocytes in primary culture. *Hepatology* **3,** 955–963.

6

Cytotoxicity Measures: Choices and Methods

CHARLES A. TYSON AND CAROL E. GREEN

Toxicology Laboratory
SRI International
Menlo Park, California 94025

I. INTRODUCTION

Cytotoxicity measurements are made to detect functional or structural perturbations of a cell that impair its capabilities and produce serious injury. In isolated hepatocytes biochemical techniques are convenient, rapid, and easy to apply for detecting and quantitating such perturbations. Commonly employed biochemical markers for a number of different lesions observed *in vivo* are listed in Table I, but the choice *in vitro* is much more diverse. Considering the variety of pathways possible for chemical-induced cytotoxicity that results in necrosis (Zimmerman, 1978; Bridges *et al.*, 1983), a wide array of indicators, biochemical and microscopic, are available and potentially applicable for obtaining information at the cellular, subcellular, and molecular levels.

In designing experiments for evaluating cytotoxic response to chemicals, then, one is clearly faced with the problem of choice and methods to use. Factors that influence indicator choice are as follows: (1) correspondence with the lesion *in vivo* being simulated, (2) the nature of the change anticipated (an early or late event, a reversible or irreversible change in structure or function), (3) the organelle(s) to be monitored, (4) sensitivity, (5) species specificity, (6) reproducibility and reliability, and (7) not the least important, convenience. The focus of this chapter, therefore, will be on reviewing those indicators commonly used for mechanistic studies and cytotoxicity screening and the lessons learned in applying several of them with the above considerations in mind. Other reviews

119

TABLE I

Some Biochemical Indicators for Various Hepatic Lesions

Lesion	Indicator change
Necrosis	Transaminase and dehydrogenase leakage
Steatosis	Fatty acid accumulation
Cholestasis	Alkaline phosphatase elevation
Cirrhosis	Collagen production
Tumorigenesis	Increase or decrease in marker proteins/enzymes
Peroxisome proliferation	Palmitoyl-CoA oxidation (cyanide insensitive) and carnitine acetyltransferase induction

may be consulted for additional information (Krebs *et al.*, 1973; Jeejeebhoy and Phillips, 1976; Grisham, 1979; Bridges *et al.*, 1983).

II. PLASMA MEMBRANE INTEGRITY

A. Cell Viability

Freshly isolated hepatocyte suspensions contain a small percentage of "dead" (~5–15% ordinarily, depending on the assay procedure) as well as "live" cells. What constitutes cell death is still a matter of debate, but operationally it is the point at which the cell loses all capability for conducting useful work functions and cannot recover. That is, the injury has progressed to a severe and irreversible stage, and the cell is no longer a biologically viable entity.

The integrity of the plasma membrane is a key factor in cell function and viability. Indicators commonly used for monitoring the integrity of the plasma membrane in cytotoxicity studies are listed in Table II. These include indicators generally considered to reflect irreversible injury to the membrane under the particular conditions of the assay (dyes or fluorescent agents such as trypan blue, fluorescein diacetate, ethidium bromide) and those that may or may not reflect reversible injury (loss of intracellular K^+ and small molecule constituents to the medium or succinate-stimulated respiration), depending on the nature and length of the alterations produced.

The most commonly used single indicator for assessing cell viability in suspensions or culture is the percentage of hepatocytes that exclude trypan blue. The assay can be conveniently performed at the same time as cell counting of stock solutions. Ethidium bromide is preferred in one or two laboratories; being a

TABLE II

Indicators of Plasma Membrane Integrity

Indicator	Change	Permanency
Cell count	Fewer cells	Irreversible
Trypan blue	Cells stain	Irreversible
Cytosolic enzymes	Leakage to culture medium	Irreversible
K+	Leakage to culture medium	Reversible
Succinate	Stimulated respiration	Reversible

potent mutagen limits its use. An alternate assay for viability in suspensions, also dependent on the leakage of the plasma membrane to small molecules, is NADH penetration (LDH latency test) (Högberg *et al.*, 1975b). This test has the advantage of being quantitatively more precise than cell counting in microscopic fields, although an additional assay involving a little extra time must be performed when characterizing cell preparations for viability.

B. Enzyme Leakage

The incapacity of hepatocytes to retain intracellular enzymes is generally accepted as an indicator for irreversible damage to the plasma membrane. The measurement of enzyme activities is more precise than dye exclusion, and the methods are automated and less time consuming and tedious in large-scale experiments. These advantages often outweigh the disadvantage of the data not being expressed directly as numbers or percentages of altered cells. This disadvantage can be offset to a large degree by expressing enzyme activity in the medium as a percentage of the total (cells + medium), a parameter which should change in close concert with the percentage of cell viability based on dye exclusion.

Several enzymes are used for monitoring the loss of plasma membrane integrity, some indiscriminately. *In vitro,* as *in vivo,* alanine aminotransferase (ALT, GPT) and aspartate aminotransferase (AST, GOT) are, on the surface, logical choices. Changes in AST are more easily detectable because of its larger content in hepatocytes (Story *et al.*, 1983). However, the transaminases are compartmentalized; 80–85% of AST, for example, is localized in the mitochondria (Pappas, 1980). Full release of AST to the medium would presumably entail the compromise of both mitochondrial and plasma membrane integrity by the cytotoxin. It is not surprising, then, that the percentage of cytosolic enzymes such as lactate dehydrogenase (LDH) released to the external medium correlates far better with the percentage of cell viability assessed by trypan blue exclusion for cytotoxins in general (Story *et al.*, 1983). [Argininosuccinate lyase has been reported to be more sensitive than LDH in cultured rat neonate hepatocytes, but from the data shown the marginal increase in sensitivity may not justify the extra effort and time required to execute the assay compared with LDH (Campanini *et al.*, 1970; Acosta *et al.*, 1985).]

In cytotoxicity experiments one may actually take advantage of this distinction in specificity by measuring both AST and LDH release to deduce whether loss of integrity of only the plasma membrane or both membranes occurs at about the same time. If the plasma membrane alone is made leaky to macromolecules as a result of a cytotoxic action, the net percentage of LDH released is roughly four times that of AST; a higher relative percentage for AST would signify disruption of the mitochondrial membrane as well. Net percentage of LDH released is the increase above background release in control (untreated) hepatocytes incubated concurrently (Story *et al.*, 1983). In our experience, leakage of 10–20% of total intracellular LDH during 4- to 6-hr incubations of control hepatocyte suspensions is customary; as a rule of thumb among investigators, ≤25% is acceptable for a valid experiment.

Conclusions as to the relative sensitivity of various enzyme indicators for detecting cytotoxic responses depend in large measure on how the data are expressed. If comparisons are based on the ratio of enzyme activities in the external medium of experimental and control flasks instead of on percentage of enzyme released, AST and LDH, for example, show comparable sensitivity (McQueen and Williams, 1982). In fact, with some cytotoxins a higher ratio is seen for AST than for LDH at the same concentration in their data. The apparent greater sensitivity for AST under these conditions must result from the release of mitochondrial as well as cytosolic isozymes. What one sees as increased or comparable sensitivity for detecting cytotoxicity relative to LDH is counter-balanced by not being able to attribute the change unambiguously to the plasma membrane alone.

There are other situations where enzyme leakage to the medium may not accurately reflect a loss of plasma membrane integrity. Cytotoxins (parent compounds or metabolites) may inhibit or denature one or more indicator enzymes, distorting results (Story *et al.*, 1983; Green *et al.*, 1984; McQueen *et al.*, 1984). Measurement of total enzyme activity in the treated flasks for comparison with total activity in control flasks in preliminary experiments can detect this effect and signal the need to use an alternative indicator for cell death. Also, while it is reasonable to consider that the transaminase in the medium originated in the cytosol as does LDH, this may not always be true. In studies of thioacetamide hepatotoxicity in the rat, it was found from gel electrophoresis on the serum, unexpectedly, that the AST elevation was due to mitochondrial isozyme (Dooley and Masullo, 1982). The results contrast with CCl_4-induced cytotoxicity (Pappas, 1980). The mechanism with thioacetamide remains unexplained, but the observation may dictate caution in interpreting low level release of transaminase to the medium as necessarily involving the cytosolic isozyme initially or exclusively. Further, as is the case with most other cytotoxicity indicators, transaminase release may be increased *or* decreased relative to control cell release (Tyson *et al.*, 1980). A decrease in enzyme release in a cytotoxicity study may reflect a membrane-stabilizing effect of the chemical at those particular concentrations and conditions.

In some studies, questions of relative sensitivity for detecting cytotoxic responses or specificity of enzyme indicators for assessing plasma membrane integrity is of secondary concern, and obtaining a correspondence of *in vitro* with *in vivo* change is dominant. One such case was prompted by an interesting but unexplained observation of a severe depression of ALT in the serum of dogs and rats after short-term exposure to trinitrotoluene (TNT) without a corresponding change in AST (Dilley *et al.*, 1982). This problem was examined *in vitro* with isolated hepatocytes from a beagle. The results in Table III show that after a 4-hr incubation, ALT release was lower in the TNT-treated hepatocytes relative to control hepatocytes, and the total intracellular content was also markedly lower with <0.50 mM TNT in the medium. As in the *in vivo* studies, changes in AST levels were not detected. At 0.50 mM TNT and higher concentrations, both AST and ALT leaked to the medium. These observations allow the following inference. At the doses given *in vivo,* the circulating level of TNT was only high enough to affect ALT. If a higher dose had been tolerated in the animals, overt signs of necrosis—as evidenced by increased levels of both enzymes in the serum—would most likely have been seen. The use of LDH would have been inappropriate for this *in vitro* study. The selective inhibition of ALT relative to

TABLE III

Cytotoxicity of 2,4,6-Trinitrotoluene to Beagle Hepatocyte Suspensions[a]

In vitro TNT concentration (mM)	ALT release (U/liter)	Total ALT[b] (U/liter)	AST release (U/liter)	Total AST[b] (U/liter)
0.0	27	432	50	1075
0.10	11	261	56	1067
0.25	10	284	67	1125
0.50	56	230	223	970
1.0	168	201	465	549
2.5	175	191	443	492

[a]Incubation for 4 hr under carbogen atmosphere (95% O_2:5% CO_2) at 37°C with shaking (70 oscillations/min) in Waymouth's 752/1 hormone-supplemented medium (Green *et al.*, 1983). Hepatocyte concentration, 1.5 × 10[6]/ml. (S. J. Gee, unpublished results).
[b]Medium plus lysed cells.

AST is not without precedent (Meijer *et al.*, 1978), and the hepatocyte system can now be used to investigate the mechanism in detail.

C. Leakage to Ions and Small Molecules

1. Intracellular Components

Increased permeability of cellular membranes to ions and small molecules often precedes total loss of cell viability. Intracellular K^+ may be lost from the cytosol to the medium or Ca^{2+} ion flux altered. The effect may be transient and reversible and may occur early or late relative to other changes. Likewise, small molecules such as pyridine nucleotide and coenzyme A content in hepatocytes treated with a cytotoxin may decrease early or late in the cytotoxic sequelae, but the loss may not always signify a damaged plasma membrane (Thor *et al.*, 1978). When total adenine nucleotide content decreases, irreversible injury is in progress (Ozawa, 1982).

K^+ is usually determined by atomic absorption spectroscopy in the cells after removal of the medium (Baur *et al.*, 1975; Stacey *et al.*, 1980; Thor and Orrenius, 1980). Potentiometric methods theoretically permit on-line, continuous monitoring of K^+ leakage to the medium, though for rapid changes equilibration time may be limiting and there is an obvious need to use K^+-free media to enhance sensitivity. With hepatocyte suspensions centrifugation through a nonaqueous barrier is first required before intracellular ion content can be determined, but measurements on more than one cellular parameter can be made. Cornell (1980) has developed a two-piece separator which has been used with brominated hydrocarbons as the separating phase to avoid losses of cellular water and solutes encountered with the one-piece tubes. The latter appear to be still valid for some solutes including K^+ (Sainsbury *et al.*, 1979).

Ca^{2+} ion content can be determined similarly to K^+ following separation of cells and medium through a silicone oil or phthalate ester barrier (Baur *et al.*, 1975; Krell *et al.*, 1979; Fariss *et al.*, 1985), the latter having been shown to separate viable and nonviable cells at the same time. ^3H_2O and [^{14}C]inulin measurements must be made concurrently to correct for external water adhering to the plasma membrane after passage through the barrier. The use of dyes (murexide, aequorin) for monitoring Ca^{2+} transport is a much more convenient approach but has a disadvantage in requiring low external (nonphysiologic) Ca^{2+} concentrations in the medium which are further lowered through complexation with the spectrophotometric probe.

In contrast to K^+, Ca^{2+} is compartmented in mitochondria and endoplasmic reticulum and complexed with calmodulin and other cell components. Although there is current disagreement over whether mitochondria or endoplasmic reticulum primarily regulate cytosolic free Ca^{2+} concentrations (Joseph et al., 1983; Somlyo et al., 1985), compromise of either organelle can result in toxicity (Moore et al., 1976; Jewell et al., 1982; Bellomo et al., 1982; Thor et al., 1982; Orrenius et al., 1984). Spectrophotometric methods are described in these reports that allow assessments of organelle integrity. The effort to acquire such information is time-consuming, and the interpretations are complex insofar as their relationship to changes in plasma membrane integrity is concerned. For measuring cytosolic Ca^{2+} concentrations, quin2 has use (Tsien et al., 1982). It is preloaded in the cells as the acetoxymethyl ester, where it is hydrolyzed readily to the acid dye which forms a very stable 1:1 complex with free Ca^{2+}. Loading and hydrolysis in hepatocytes are complete within 20 min (Thomas et al., 1984; Berthon et al., 1984).

As with any probe which measures ion concentrations in the range of the dissociation constant for the complex, the chelator lowers the free ion concentration. Ca^{2+} homeostasis is restored by the net entry of Ca^{2+} from the medium (Berthon et al., 1984). The lag time for this reequilibration is negligible in most applications (within 10 sec) (Thomas et al., 1984). However, leakage of the indicator to the medium during incubations ($t_{\frac{1}{2}} = 37$ min) (Berthon et al., 1984) may constitute a more serious limitation to its general use in cytotoxicity studies. Newer probes recently developed may mitigate these and other limitations in using quin2 (Grynkiewicz et al., 1985).

2. Externally Added Solutes

Addition of extraneous ions such as ^{51}Cr or enzyme substrates to the medium have been proposed for assessing plasma membrane integrity (Zawydiwski and Duncan, 1978; Grisham, 1979). In the case of ^{51}Cr, the cells must be preloaded with the radionuclide before experiments begin. The practice has several disadvantages. With hepatocyte suspensions the preloading requirement reduces the incubation time available for studying cytotoxicity. In longer term studies in culture there is concern that the radionuclide may perturb some cell processes. Further, there may be interpretive problems in correlating changes with specific cellular membranes, since ^{51}Cr is taken up by different organelles to different degrees and not all is released on cell lysis (Holme et al., 1982).

A sensitive and more commonly used procedure than ^{51}Cr in hepatocyte studies is to add succinate (1.0 or 10 mM) to the medium and monitor for increased O_2 consumption (Baur et al., 1975). Intracellular accumulation of substrates for amino acid transporters can be monitored for evidence of impaired plasma membrane functional competence, but at least three different systems are involved and, based on studies on one of these with α-(methylamino)isobutyric acid, either inhibition or stimulation of uptake can occur, depending on the mode of action of the chemical (Goethals et al., 1983; Pariza et al., 1976). None of these indicators provides unequivocal evidence of a generalized, irreversible deterioration in plasma membrane integrity.

The same ambiguity in identifying irreversible changes in plasma membrane integrity applies to measurement of LDH latency test in hepatocyte suspensions. In our experience with isolated hepatocytes, NADH penetration has correlated exactly with released LDH activity. However, with isolated kidney tubule fragments NADH penetration is higher than released LDH activity, indicating that there can be circumstances where cell membranes are leaky to NADH or other small molecules and ions but not larger protein molecules. Further, because the assay is performed on the cell suspension directly without prior centrifugation to remove the cells, it must be performed immediately to prevent additional loss of cell viability. This may pose problems when large numbers of samples or indicators are being analyzed in a cytotoxicity study.

III. SUBCELLULAR EFFECTS

A. Indicators of Energy Conservation

1. General Comments

Unless the cytotoxin acts externally on the plasma membrane without penetration into the cells, loss of membrane integrity is almost invariably preceded by subcellular changes. The detection and documentation of these changes provide potentially important information on probable mechanisms of action leading to cell death or on subtle reversible changes that may still impair functional capabilities. The status of energy conservation processes in the mitochondria over time is an overriding consideration (Krebs et al., 1979).

The single indicator most commonly monitored in cytotoxicity studies with isolated hepatocytes that is dependent on the functional integrity of mitochondria is cellular ATP. Measurement of O_2 consumption is also monitored but less frequently. Several processes and components that give an indication of functional integrity besides cellular ATP also depend intimately on the mitochondrial energy status. These include pyridine nucleotide ratios (redox potentials), urea-genesis, gluconeogenesis, protein synthesis, and lactate-to-pyruvate ratios. All of these indicators and processes are compartmentalized (not exclusively localized to the mitochondria). The indicators of mitochondrial integrity and their compartmentalized locations are summarized in Table IV.

2. ATP and Nucleotide Ratios

ATP is most often determined enzymatically with hexokinase/glucose-6-phosphate dehydrogenase and by the luciferin/luciferase assay, which affords high

TABLE IV

Indicators Dependent on Mitochondrial Functional Integrity

Indicator	Compartmentation
ATP (or ATP/ADP)	Mitochondria, cytosol
O_2 consumption	Mitochondria (predominantly), endoplasmic reticulum
Pyridine nucleotide ratios	Mitochondria, endoplasmic reticulum, cytosol
Ureagenesis	Mitochondria, cytosol
Gluconeogenesis	Mitochondria, cytosol
Protein synthesis	Mitochondria, endoplasmic reticulum
Lactate/pyruvate	Mitochondria, cytosol

sensitivity and reliability (Lamprecht and Trautschold, 1974; Strehler, 1974). The cells are deproteinized by acid or ATP released with lysing agents to avoid alterations during workup for assay. With hepatocyte suspensions, aliquots may be taken periodically from the same flask; with cultured hepatocytes, each determination requires a separate dish. Cells cannot be stored frozen for later analysis without prior acidification to stabilize the ATP.

Because ATP is simple to measure periodically in the same flask and is synthesized mainly in the mitochondria, one would ideally like to use this indicator to assess the functional status of the mitochondria over time. ATP content alone, however, is not a good measure of this or, for that matter, of the energy state of the tissue as a whole (Siesjö, 1978). With isolated mitochondria, ATP/ADP, phosphate potentials, or redox potentials give more realistic information on this point. In the cell, however, contributions of cytosolic components to the total are substantial (e.g., more than 60% of the total adenine nucleotides) (Tischler *et al.*, 1977; Soboll *et al.*, 1980; Stier *et al.*, 1980). Evidence has been presented that the two compartments are intimately related and maintained in near equilibrium during energy transfer (Erecinska *et al.*, 1974); changes in mitochondrial energy state should be reflected, therefore, in changes in cellular ATP/ADP with reasonable accuracy. But the ratio is regulated by extramitochondrial factors as well [which factors are important and under what conditions are disputed (Erecinska *et al.*, 1977; Siesjö, 1978; Ozawa, 1982; Lanoue and Schoolwerth, 1984)], which might suggest caution in interpreting changes in this parameter as resulting from direct action of the cytotoxin on the mitochondria.

In spite of these comments, we consider ATP or ATP/ADP are appropriate to use for preliminary indicators of mitochondrial dysfunction. Although we, as others, have relied more on ATP for this purpose in the past (Story *et al.*, 1983), ATP/ADP can have advantages. It is convenient to measure, requiring only the additional determination of ADP, which is straightforward (Jaworek *et al.*, 1974;

Kimmich *et al.*, 1975). Some recent results in our laboratory suggest that the magnitude of the change in ATP/ADP may correlate better with metabolic processes, like urea synthesis, that are energy-dependent rather than ATP. Also, in early, reversible stages of cell injury, ATP/ADP is clearly more sensitive. For example, a 20% depression in cellular ATP will theoretically result in a corresponding change of 55% in ATP/ADP, assuming four molecules of ATP for each ADP at steady-state in the hepatocytes (Soboll *et al.*, 1980). It may be easily verified that ATP/ADP is also a more sensitive indicator of a perturbation in energy supply than energy charge potential, which varies over a small range of values (0.70 to 0.0) in response to loss of mitochondrial function. This theoretical enhancement in sensitivity by measuring both ATP and ADP instead of ATP alone is, in practice, achieved at the expense of the increased variability associated with making two separate measurements instead of one and because of a small increase in AMP as well. The advantages of determining ATP and ADP routinely in a large experiment may be offset by the need and time required to make several other measurements on the same preparation.

3. Oxygen Consumption

O_2 consumption is measured by either polarographic or manometric techniques on hepatocyte suspensions; the results are interchangeable (Krebs *et al.*, 1973). Polarographic techniques (the Clark electrode) are by far the most popular because of their convenience and simplicity. Either specially designed reaction flasks with magnetic stirring or transfer of the cells to a separate, wide-mouth, water-jacketed vessel is required for the measurements. Diffusion of oxygen from the external atmosphere into the solution must be prevented or minimized for linear traces to be obtained. In our experience, stirring (in the Yellow Springs Oxygen Monitor) causes a <10% decrease in cell viability during the minute or two required for kinetic measurements. The manometric technique is preferred by some because it permits continuous monitoring at almost constant O_2 pressure (Krebs *et al.*, 1973); however, it requires specially designed apparatuses, a longer incubation time, and more complex calibration and calculations.

Oxygen consumption rates are modest: ~35 nmol/10^6 cells/min, with hepatocytes suspended in hormone-supplemented culture medium, about one-sixth the dissolved O_2 concentration/min without replenishment from the air. Values in the literature range from 10–45 nmol/10^6 cells/min. The variability is probably mainly due to differences in values for dissolved O_2 concentrations used and in media composition. Reports of O_2 consumption measurement in cultured hepatocytes with polarographic methods are sparse, or nonexistent, because of technical difficulties in excluding atmospheric O_2 from redissolving in the reaction medium.

The direction and magnitude of change seen in various mitochondrial indicators depends on the cytotoxin and its mode of action and the relationship of the indicators to each other. Redox cycling by cytotoxins such as menadione can markedly accelerate O_2 consumption by microsomal enzymes (Thor *et al.*, 1982), adding to the total measured. The microsomal component to overall O_2 uptake can be determined by inhibiting mitochondrial respiration with antimycin. Uncouplers of oxidative phosphorylation may also increase O_2 consumption rates and deplete mitochondrial ATP through ion cycling; inhibitors of cytochrome oxidase or of the electron transport chain, on the other hand, may decrease both in concert. Identical rates of respiration may occur with markedly different ATP/ADP ratios (Letko *et al.*, 1983). Protein synthesis is reportedly highly sensitive to reduced O_2 consumption (van Rossum, 1972), whereas our observations indicate that urea synthesis is somewhat less so. Effects on Na^+/K^+-ATPase activity in plasma membranes by cytotoxins are generally late events (Sayeed, 1982). Only a portion of mitochondrial respiration supports such activities (Hansen, 1985); in rat liver slices a more than 50% reduction in respiration is required for impairment of Na^+ and K^+ ion transport (van Rossum, 1972).

B. Metabolic Competence

1. General Comments

Assessment of cell functional capabilities and metabolic competence often reveals subtle or early changes induced by the cytotoxin, which provide insight into reaction mechanisms that alter cell behavior or are on the pathway to irreversible cell injury (Grisham, 1979; Krebs *et al.*, 1979; Krack *et al.*, 1985). Urea synthesis and protein synthesis are commonly used for this purpose and will be reviewed in some detail. Gluconeogenesis, lipogenesis, glycogen synthesis, and glycolysis are likewise assessed for direct or indirect effects that result in the diversion of carbon from these intermediary metabolic pathways (Thurman and Kaufmann, 1980). These pathways are less frequently monitored in cytotoxicity studies, and, for lack of space, methods for all of them will not be critically reviewed here.

Accurate measurement of gluconeogenesis, ureagenesis, or protein synthesis rates requires from 30 to 60 min, during which time the cytotoxin may have effected substantial additional changes on the cell. The substrates are added directly to the flasks rather than aliquots transferred to assay media to avoid possible reversal of inhibition from dilution. Lactate-to-pyruvate ratios may be monitored as a function of time to estimate when the transition from aerobic to anaerobic pathways in the cells in response to cytotoxins occurs (see, e.g., Santone *et al.*, 1982). Unless extra care is taken, unacceptable variability may be encountered because of the instability of the measures. With cell suspensions, large aliquots must be taken to quantitate lactate and pyruvate, which may necessitate extra flasks in experiments in which several parameters are being monitored.

2. Urea Synthesis

Urea synthesis is a hepatocyte-specific function compartmentalized between mitochondria and cytosol. Its interrelationship with other energy-requiring processes is discussed in the literature (see, e.g., Meijer et al., 1978; Hensgens and Meijer, 1979; Letko et al., 1983). Effects on mitochondrial ATP and the redox potential or on urea synthesis regulation, substrate levels, or the enzymes themselves may result in diminished urea formation. Depression of citrulline as well as urea accumulation can occur in response to a cytotoxin and be detected in the assay (Story et al., 1983). Determination of urea synthesis rates in hepatocyte suspensions with various cytotoxins is sensitive and reproducible (Story et al., 1983). In cultured hepatocyte studies, the urea synthesis activity is lower because fewer cells are present, and the results are, in our experience, more variable. The liver cells may have, therefore, more sufficient reserve capacity to mask small decrements, making the detection of inhibition more difficult. Total urea accumulation in the cells (or in medium) may alternatively be measured (Santone et al., 1982; C. E. Green, unpublished observations).

3. Protein Synthesis

Protein synthesis has been proposed as a sensitive indicator of metabolic competence (Gwynn et al., 1979; Krack et al., 1983; Goethals et al., 1984). The rate of synthesis is calculated from the increase in the specific activity of a radiolabel incorporated into intracellular proteins over a 30- to 60-min incubation period from radiolabeled amino acids added to the external medium. Radiolabeled (^{14}C or ^3H) leucine or valine, which are not readily metabolized, are the most commonly used amino acids for these measurements. ^{14}C-Labeled algal protein hydrolysate has been used for estimates of overall amino acid incorporation (Seglen, 1976a). The disadvantages of its use are stated to include heterogeneity in uptake rates and catabolism of radioactive amino acids, which may result in a general overestimation of the specific activities (Seglen, 1976a, 1978). Isotope dilution can also affect measurements at high cell densities, unless the extracellular amino acid concentration is also made high (Seglen, 1974, 1976a, 1978).

Protein synthesis is usually altered at earlier times or at lower concentrations of the cytotoxin than indicators of cell death (Gwynn et al., 1979; Krack et al., 1983, 1985). A cytotoxin can either induce or inhibit the synthesis of new proteins. Detecting a significant change in itself provides little information on cytotoxic mechanisms without follow-up experiments because the processes are complex and not yet well understood (Seglen *et al.*, 1980). Induction can result from mitogenic stimulation or from increased protein for cell repair or, as in the case of metallothionein, for protection. Inhibition can result from defective amino acid transport, impaired mitochondrial function, polysomal disruption and degeneration of the endoplasmic reticulum, regulation due to altered effector levels, or altered protein catabolism, among other causes. Inhibition at the enzyme level may not reflect potential damage to the rough endoplasmic reticulum exclusively, since protein synthesis also occurs in mitochondria (Gellerfors *et al.*, 1979). Short-term suppression of protein synthesis by inhibitors or cytotoxins can be reversed on their removal from the medium (Mattei *et al.*, 1979; Helinek *et al.*, 1982). Protein synthesis rates are also affected by the nutritional state of the cells and the experimental conditions (Dich and Tønnesen, 1980). Freshly attached hepatocytes in monolayer culture exhibit higher protein synthesis rates than freshly isolated cells in suspension (Le Rumeur *et al.*, 1983), and the rate almost doubles during the next 20 hr in culture (Tyson *et al.*, 1982). Seglen *et al.* (1980) reported contrasting results in observing a slow decline in the rate under their particular conditions.

There is no single indicator for metabolic competence that will be both sensitive and generally applicable. For example, protein synthesis and urea synthesis, two commonly used indicators, apparently proceed independently; cycloheximide inhibits protein synthesis but not urea synthesis, whereas norvaline does the opposite (Hensgens and Meijer, 1979). Aminooxyacetate but not cycloserine inhibits urea synthesis completely; both depress gluconeogenesis from lactate (Meijer *et al.*, 1978).

C. Nonspecific Indicators

1. Commonly Used Indicators

Nonspecific indicators of cytotoxicity are classified here as those that indicate a potential for injury without insight into the particular subcellular site(s). Reduced glutathione (GSH) levels, lipid peroxidation, and covalent binding of the parent or active metabolite to cell macromolecules are commonly measured indicators that fall into this category.

2. Glutathione

GSH protects against cell damage by reacting with cytotoxins such as alkylating agents, oxygen-derived electrophiles, and activated metabolites (although there are instances reported where the complexes themselves are toxic). A reduction in cellular GSH is a reflection of the potential for injury, not of cytotoxicity itself (Grisham, 1979), and usually but not necessarily occurs early in the process of injury. Oxidized glutathione (GSSG), which is approximately 1% of the steady-state GSH level, is measured when information on the fate of GSH is desired, and techniques are also available for measuring protein mixed disulfides and protein sulfhydryl group content (DiMonte et al., 1984).

Colorimetric, spectrofluorometric, and enzymatic methods are available for the quantitation of GSH in isolated hepatocytes (Cohn and Lyle, 1966; Hissin and Hilf, 1976; Akerboom and Sies, 1981; Saville, 1958). The colorimetric method requires little analytical time but actually measures total nonprotein sulfhydryl content in the cells and not GSH specifically. Because GSH constitutes approximately 90% of the total (Reed and Beatty, 1980), colorimetric and fluorometric techniques give similar results (Högberg and Kristoferson, 1977). The spectrofluorometric methods, which are used more often, are highly sensitive and reproducible and also require minimal workup time. The general procedure involves homogenization of the cells, acidification and centrifugation to remove proteins, and reaction with a fluorophore (o-phthalaldehyde). Fluorescence intensity at pH 8 is linear over a 500-fold concentration range with detection limits of 0.02 μg GSH/ml. Accurate determination of GSSG using the same fluorophore at pH 12, however, has been questioned (Beutler and West, 1977), and a revised procedure using N-ethylmaleimide as a trapping agent for GSH has been proposed to minimize autoxidation of GSH to GSSG which may occur during workup (Akerboom and Sies, 1981).

Fluorometric and enzymatic methods for GSH determination have produced excellent agreement in studies on the relationship of formaldehyde metabolism to cytotoxicity (Ku and Billings, 1984). However, the methods do not discriminate between free GSH and the formaldehyde–GSH adduct, since the latter readily dissociates. Ku and Billings (1984) used the enzymatic method to quantitate extracellular GSSG and the HPLC method of Reed et al. (1980) to confirm the presence of both GSH and GSSG in the medium. They did not comment on the relative merits of the methods. In our opinion, the latter technique offers distinct advantages for the detailed analysis of the thiol content of isolated hepatocytes; possible disadvantages are the relatively low (49%) derivatization of GSSG and the unavailability of suitable commercial columns, requiring preparation by the investigators themselves.

Many investigators report changes in GSH levels as percentages rather than in actual units. GSH levels in untreated hepatocyte suspensions usually increase during incubations, although in culture medium containing methionine and cysteine this increase was not statistically significant during a 6-hr incubation period (Green et al., 1983). In cultured hepatocytes, an initial and notable decrease and then an increase to or above initial levels have been reported (Morrison et al., 1985; Hayes et al., 1986). Also, GSH is compartmentalized, about 15% being present in the mitochondria and the remainder in the cytosol (Meredith and Reed, 1982). When a decrease in GSH plateaus at a level above 15% of the total intracellular content as the cytotoxin concentration is increased, it may be that only the cytosolic component is involved, and a combination of mild detergent action and centrifugation to quantitate mitochondrial levels may confirm this suspicion (Meredith and Reed, 1982). GSH depression is usually reversible, unless irreversible cell injury or enzyme inactivation has occurred, after removal of the chemical insult. GSH may also be elevated by response to cytotoxins.

3. Lipid Peroxidation

Lipid peroxidation results from the interaction of unsaturated lipid components (primarily membrane phospholipids) with oxygen free radicals or excess H_2O_2 generated by cytotoxins during metabolism. Unlike with GSH, however, many pathways and many different products may result from the peroxidation process. A variety of analytical methods to measure lipid peroxidation products are available based on these differences, and most have been reviewed elsewhere (Recknagel et al., 1982; Kappus, 1985). The techniques used in hepatocyte studies include measurement of malondialdehyde (MDA) formation using the thiobarbituric acid (TBA) reaction, spectrophotometric detection of lipid-conjugated dienes, measurement of evolved gaseous hydrocarbons (ethane and pentane, primarily), quantitation of fluorescent pigments, and chemiluminescence. Methods (e.g., iodometric) for quantitating lipid peroxides also exist and appear to be reliable, providing that the peroxides are first extracted into organic solvents to avoid complications from the presence of interfering substances in unreacted biological tissue and uncontrolled variables (Recknagel et al., 1982).

A summary of the principal advantages and disadvantages of the various techniques, based in large part of these reports, is given in Table V. Some data are available comparing these techniques for sensitivity and reliability in detecting lipid peroxidation (Högberg et al., 1975a,b; Stacey et al., 1980; Gee and Tappel, 1981; Stacey and Klaassen, 1981; Koster et al., 1982; Recknagel et al., 1982; Stacey and Kappus, 1982; Stacey et al., 1982; Smith et al., 1982; Kappus, 1985). These papers may be consulted for experimental methods, results, and conclusions. Some observations, ours as well as the authors', are pertinent here.

Sensitivity and reliability of different methods for detecting lipid peroxidation are dependent on many factors, among the more important being the experimental conditions, possible interferences, background, and mechanism of action (Stacey et al., 1980). As an example, special conditions (phenobarbital pretreatment, nonphysiologic media, etc.) are required to detect CCl_4-induced lipid peroxidation using the TBA reaction; sensitivity of detection is also influenced by O_2 content in the atmosphere, hepatocyte content in the medium, and the

TABLE V

Advantages and Disadvantages of Techniques for Lipid Peroxidation (LP)

Technique	Advantages	Disadvantages
TBA reactants	Fast, simple	Often understates extent of LP because of (1) metabolism and reactivity of MDA, (2) components in medium catalyzed (e.g., Fe^{3+} contamination) Possible artifacts: autoxidation of unsaturated lipids during workup; reactions of TBA with aldehydes and other cell components; MDA from oxidation of nonlipid cell constituents
Hydrocarbon gas evolution	Sensitive[a]	Air-tight conditions required. Separate flasks required for additional end points if flasks are stored for later analysis. Shorter incubation times to avoid anoxia and increase detection levels Possible artifacts: perturbation of normal metabolism when small changes observed; a parent compound metabolism (e.g., triethyl Pb)
Chemiluminescence	Sensitive[b]	Results expressed in arbitrary units. Potential problem with drug quenching
Fluorescent pigments	Stable products, sensitive[c]	Potential interferences necessitate more workup time. Causal relation to LP not yet shown. Small yield of products/O_2 consumed
Conjugated dienes	Unequivocal evidence for LP with O_2 present	Possible artifact from lipid peroxide contaminants. More workup time involved. Not very sensitive

[a]Ethane appears to be the best single indicator to measure in general. Pentane can be metabolized, and ethylene is the least sensitive of the three (Gee and Tappel, 1981; Smith et al., 1982; Kappus, 1985).

[b]Smith et al. (1982) rank the sensitivity of five techniques as follows: ethane evolution ≥ chemiluminescence ≥≥ TBA reactants ≃ pentane evolution ≃ fluorescent lipids.

[c]Recknagel et al. (1982).

shaking rate (Stacey et al., 1982). Those investigators who have compared hydrocarbon gas evolution with TBA reactants or other techniques tend to conclude that, monitoring ethane in particular, it is the most reliable indication of lipid peroxidation (Gee and Tappel, 1981; Smith et al., 1982; Kappus, 1985). On-line analysis is not necessarily required; Gee and Tappel (1981) obtained reasonably good correlations of ethane and pentane with TBA reactants in experiments in which the flasks were stored frozen before headspace analysis. With pentane in contrast to ethane, a correction factor must be applied for gas remaining dissolved in the aqueous phase, and liver can metabolize it (Smith *et al.*, 1982; Wendel and Dumelin, 1981). Smith *et al.* (1982) concluded from their studies that MDA, pentane, and fluorescent-product measurements underestimate the extent of lipid peroxidation in the cells, attributed in part to further metabolism of MDA and pentane. Metabolism of ethane by liver has not been detected, in contrast to other hydrocarbons (Wendel and Dumelin, 1981).

Stacey *et al.* (1982) noted that whereas ethane evolution was low in normal, untreated hepatocytes incubated in Eagle's minimal essential medium and Tris–phosphate buffer under different atmospheres, TBA reactants were increased under a carbogen atmosphere relative to corresponding control flasks incubated under air. This probably contributes to the relatively lower sensitivity of the method under carbogen. With metal ion cytotoxins, measuring TBA reactants was slightly more sensitive initially, but reactant formation leveled off and ethane evolution was more sensitive as the incubation was continued. The authors applied the method of Buege and Aust (1978) to whole-cell suspensions rather than analyzing the supernatant after acid precipitation of cell macromolecules and lipids, as is more often done. The Buege–Aust method measures both MDA and lipid peroxides, which Stacey and colleagues suggest may be more representative of the extent of lipid peroxidation and thereby more comparable in sensitivity to ethane formation. Butylated hydroxytoluene was added during workup to prevent artifactual production of lipid peroxides during heating (Buege and Aust, 1978; Asakawa and Matsushita, 1981). Heating under N_2 has recently been recommended to minimize autoxidation in lieu of adding antioxidants (Kirkpatrick *et al.*, 1986). Significant numbers of damaged cells in the medium may also enhance lipid peroxidation (Chiarpotto *et al.*, 1981).

Overall, these results indicate the most reliable indicator of lipid peroxidation in general is probably ethane evolution. Whether one considers it the most sensitive depends on the perspective, since ethane evolution on a molar basis is 100- to 500-fold lower than MDA, and MDA formation itself is estimated to relate to only 10% of the total lipids oxidized (Kappus, 1985). One serious constraint in using ethane is the need to decrease headspace volume in the flask to lower detection levels for GC analysis (Smith *et al.*, 1982). This limits available O_2, and carbogen atmospheres are required to extend the lifetime of cells. This tactic may have undesirable effects on the extent of peroxidation with some cytotoxins or be of questionable relevance to *in vivo* conditions. Thus, until these technical problems are overcome, MDA analysis will continue to be preferred, at least in initial experiments, because of its simplicity and ease of use. Other indicators of lipid peroxidation also have utility, but conjugated-diene formation as a measure, although apparently the most definitive, with rare exceptions, is reportedly the least sensitive. A suggestion that the sensitivity problem in biological systems can probably be overcome by using second derivative spectra at 233 nm (Corongiu and Milia, 1983; Kappus, 1985) needs experimental verification.

4. Covalent Binding

The covalent binding of activated metabolites to cell macromolecules has long been thought to be a principal mechanism of hepatotoxicity by xenobiotics (Gillette, 1974; Mitchell and Gillette, 1975). Binding of activated molecules to cell proteins occurs early, either plateauing before or continuing to increase up to the point of cell death. It thus may serve as a trigger for a sequence of events leading to cell death. Analogous to GSH and to lipid peroxidation, cellular proteins are thought to bind activated metabolites up to some threshold above which proteins critical to the structural or functional integrity of the cell are altered (Mitchell *et al.*, 1982; Black, 1984). Although the critical-site hypothesis is reasonable and attractive, no such proteins or their locations in the cell have yet been identified in cytotoxicity studies; the exact relationship of covalent binding to cell death, therefore, remains obscure.

Despite these uncertainties as to the mechanism by which covalent binding of chemicals produces cytotoxicity, there are examples in which the formation of activated metabolites that bind to protein have been correlated to hepatotoxicity (Gillette, 1974). Isolated hepatocytes have been used frequently as the test system for investigating adduct formation, since exposure conditions can be so readily altered. The method of choice for quantitating covalently bound adducts is to use a radiolabeled substrate and measure the amount of radiolabel that is strongly associated with protein. Macromolecules are usually precipitated with acid and repeatedly extracted with solvents to remove weakly bound chemical. Samples may also be repeatedly dissolved in NaOH and then reprecipitated with acid or digested to demonstrate that the radiolabel remains associated with amino acids and small peptides, confirming that the bond between chemical and protein is covalent (Thorgeirsson and Wirth, 1977).

In the conduct of these studies, it is important to be aware that a number of factors can cause misleading results in covalent binding studies. These include the presence of radiolabeled impurities in the test compound, nonspecific binding, chemical breakdown in the *in vitro* system that would not occur *in vivo*, and biosynthetic incorporation of radiolabeled metabolites. To correct for nonspecific binding that is not removed by the precipitate wash procedure, the radioactivity associated with heat- or acid-denatured cells is usually subtracted from the radioactivity associated with live cells. However, it has been reported that in hepatocytes treated with acetaminophen or *N*-hydroxyacetaminophen, the radioactivity associated with denatured cell material is about twice the level of binding measured in viable cells (Holme *et al.*, 1982; Green *et al.*, 1984). Presumably this is because additional sites for nonspecific binding, which are not normally available, are exposed in the denatured protein. In this case, a flask or culture of liver cells, to which the radiolabeled substrate is added and the reaction immediately stopped (zero time sample), is used as the control specimen providing data on the level of nonspecific binding that is subtracted from the results obtained from live hepatocytes. Other potential artifacts encountered in the determination of covalent adduct formation and possible approaches for minimizing these have been reviewed (Thorgeirsson and Wirth, 1977; Pohl and Branchflower, 1981; Lutz, 1979).

D. Organelle-Specific Indicators

The effects on organelles beside mitochondria or the plasma membrane may also be important in the pathologic process. Examples are numerous. Deterioration of the endoplasmic reticulum may affect Ca^{2+} homeostasis, inducing surface blebbing and an inability to maintain ion gradients critical for cell survival (Jewell et al., 1982; Hayes and Pickering, 1985). Metal ions and many noxious substances may concentrate in lysosomes, causing release of hydrolytic enzymes and damage to the cells (Davies and Allison, 1972). Inhibition of lipoprotein synthesis and secretion by CCl_4 in Golgi complexes may lead to fatty acid accumulation and cell dysfunction (Chiarpotto et al., 1981). Peroxisome proliferation can be induced by hypolipidemic agents (Reddy and Lalwani, 1983). In some cases the nuclear capsule may be selectively dissolved (Stammati et al., 1981).

Ideally, when initiating investigations with a cytotoxin for which the mechanism of action is unknown, the availability of a limited set of organelle-specific biochemical indicators for detecting subcellular changes and their temporal relationship would be useful. The indicators discussed so far are, for the most part, not organelle specific. Some others, which may be more so, may be sought from the literature. Glucose-6-phosphatase release or inactivation is taken as an indicator of smooth and rough endoplasmic reticulum integrity (Feuer et al., 1965; Platt and Cockrill, 1967), but failure to observe a change in this parameter does not signify freedom from injury to enzymes localized there (Poli et al., 1981). Alternatively, the Ca^{2+}-sequestering capability of microsomes prepared from hepatocytes has been reported to be a sensitive measure of injury to that membrane (Pencil et al., 1982). Acid phosphatase or β-glucuronidase can be determined for indications of a loss of lysosomal membrane integrity (Dujovne, 1978). Cells should be centrifuged to remove external medium and freeze/ thawed first to assay for free activity in the cytosol, to which loss of plasma membrane integrity may be related. Cyanide-insensitive palmitoyl-CoA oxidation has been used as a specific marker for peroxisomal proliferation (Gray et al., 1983).

Use of purely biochemical approaches for surveying organelle integrity is effort intensive and may require several experiments with different indicators for thoroughness. Another complimentary approach for use in this regard is electron microscopy. With transmission electron microscopy, all organelles can be monitored simultaneously for subcellular changes during incubation. Scanning electron microscopy allows visualization of the surface of the plasma membrane for perturbations that signal loss of integrity. Neither of these techniques, however, is sufficient without determination of the functional alterations that precede or occur concurrently with the structural. The most comprehensive approach is one that combines electron microscopic with cytochemical and biochemical measures, but it is seldom applied because of the expense of electron microscopic examination and the semiquantitative nature of cytochemical results.

IV. DATA PRESENTATION

A. Reference Data

Various investigators have used different ways of expressing experimental results in isolated hepatocyte studies. The most common way is in terms of 10^6 cells, but milligrams of protein or of DNA and grams of wet or dry weight of liver tissue are also used (Krebs *et al.*, 1973). Expressions based on protein or DNA content in the flask may be quantitatively more precise than counting hepatocytes in microscopic fields because of nonuniform cell dispersion. However, 10^6 cells is the preferred reference in cytotoxicity studies because it facilitates conceptualization of changes in integrity at the cellular level, the primary focus. Interspecies differences in response are also easier to visualize when results are compared on the basis of cell number rather than protein and DNA content, which vary considerably with species (Table VI). In addition, the cell content of protein may change during monolayer culture (Table VI). Use of 10^6 cells also eliminates a source of variability when some assays are standardized to protein or DNA content, which is subject to individual animal variation, pretreatments, or the nutritional status at sacrifice (Pfaff *et al.*, 1980).

The importance of presenting pertinent reference data in study reports cannot be understated. Most researchers express the change in an indicator produced over time in incubation or in response to a cytotoxin as the percentage of some initial or control cell value. Unfortunately, the actual values often are omitted. Some laboratories no longer report basic information such as viability data on their preparations. On occasion, the same viability number may appear in a series of papers from the same laboratory, suggesting that the reported viability is an

TABLE VI

Protein and DNA Content of Isolated Hepatocytes

	Freshly isolated		Monolayer cultures[b]	
Species[a]	DNA (μg/10^6 cells)	Protein (mg/10^6 cells)	DNA (μg/10^6 cells)	Protein (mg/10^6 cells)
Rats	12.8 ± 3.4[c]	1.19 ± 0.23	12.4 ± 1.1	1.73 ± 0.21
Hamster	10.1 ± 0.5	1.27 ± 0.25	13.8 ± 0.78	1.34 ± 0.37
Rabbit	10.7 ± 2.7	1.31 ± 0.26	11.1 ± 0.87	1.60 ± 0.36
Dog	8.74 ± 1.5	0.76	7.68 ± 1.5	0.86 ± 0.16
Squirrel monkey	12.5 ± 1.5	1.11	ND[d]	ND
Human	16.2 ± 4.6	1.14 ± 0.32	ND	ND

[a]Male Sprague–Dawley rats; male Syrian hamsters; male New Zealand rabbits; male beagle dogs; female squirrel monkeys; male and female human donors of organs for transplantation.
[b]DNA and protein contents were analyzed 27 hr after hepatocytes were plated. Data are from Green et al. (1984).
[c]Means ± SD.
[d]ND, Not determined.

assumed one based on historical experience and not actually determined before use of the cells for study. Likewise, no initial viability data at all may be reported, raising the question of whether this key benchmark of cell quality was ever assessed and detracting from the creditability of an otherwise good study. In this regard, simple measurements (trypan blue exclusion or LDH latency, e.g.) are better than none. Analyzing complex anabolic processes, however, to characterize cell preparations on a routine basis is time consuming, counterproductive, and not useful because of the greater variability possible in such data (Pfaff et al., 1980). Each laboratory, however, should have evaluated its preparative procedures for both cell viability and selected functional properties before embarking on investigative studies with the cells.

B. Data Expression and Analysis

Presentation of cytotoxicity data may take different forms. The data are usually expressed as a percentage change relative to either control values or total cellular content. The former is sensitive in detecting small changes in a parameter but gives no information on the magnitude of the change relative to the maximum possible. For example, an observed increase of 150% in LDH activity in the incubation medium relative to control cell release can easily indicate a cytotoxic response but cannot be used to infer the percentage of cells that have died in treated or control flasks. Control cell leakage during the incubation period should be specified for assurance that high background enzyme release may not have also occurred concurrently.

When data are expressed as a percentage of total intracellular content, the data should first be corrected for normal (control) cell leakage occurring concurrently during the incubation to allow focusing on the effects of the cytotoxin. One can do this by simply subtracting control cell leakage both from enzyme content measured in the medium of the treated flask and from the total enzyme content in that flask before calculating the percentage change as an approximation, or more precisely by correcting for differences in total cell content in treated and control flasks (Story *et al.*, 1983). Underlying any of the above treatments are the assumptions that background leakage of enzymes to the medium is the same in treated and control flasks and that enzyme release from control cells is proportional to the concentration of cells present in the flasks.

Various other treatments of the data can be attempted to obtain information on the possible relationships of different cytotoxic indicators to each other in an investigative study. Log probit plots of a series of structurally related chemicals can be examined for overlap and indications of common steps in the cytotoxic pathways. Log probit plots of various indicators as a function of time in the same concentration range may similarly overlap, data consistent with their being related to a common cytotoxic pathway. The success of these treatments will be influenced substantially by variability, particularly in end points for cell death, as reflected in standard deviations for trypan blue exclusion and LDH release measurements with hepatocyte suspensions and various cytotoxins (Story *et al.*, 1983; Goethals *et al.*, 1984). Animal-to-animal variation will contribute, but some reduction in variability may be realized by conducting more replicate experiments (Inmon *et al.*, 1981), using cleaner preparations [e.g., Percoll gradients during purification (Kreamer *et al.*, 1986)], or simply by pooling in order to improve the chances of correlating different indicator changes.

V. *IN VIVO/IN VITRO* CORRESPONDENCE

Of fundamental interest in cytotoxicity investigations is the correspondence of *in vitro* and *in vivo* events (Klaassen and Stacy, 1982). The choice of indicators may be dictated in part by such considerations. One way to investigate correspondence of *in vitro* with *in vivo* response for structurally or functionally related chemicals is to rank them quantitatively using appropriate end points for the response. Despite pharmacokinetic, extrahepatic, and other factors that can thwart such ranking correlations, some modestly successful results have been reported (Tyson *et al.*, 1983a; Cantilena *et al.*, 1983; Gray *et al.*, 1983). Such successes are based on the use of cytotoxins with widely differing reactivity, clearly defined and relevant end points of toxicity, and large, easily detectable and quantifiable changes in those end points.

Figure 1 compares *in vitro* and *in vivo* response to eight halogenated aliphatics to illustrate some of these points. Here dose *in vivo* (since circulating levels of the halogenated hydrocarbons were unavailable), LDH rather than AST or ALT *in vitro* [since LDH is more precisely correlated with cell death (necrosis) (Story *et al.*, 1983)], and air : water partition coefficients to partly reconcile the pharmacokinetic differences of comparing results in open (whole-animal) and closed (reaction flask *in vitro*) systems with chemicals of different volatility (Tyson *et al.*, 1983a) are used. Presumably, dose in this case must be closely related to the peak serum concentrations of the hydrocarbons, and other factors must be of secondary importance for a good correlation to be obtained (r = 0.92 by linear regression analysis of the data) (Plaa and Hewitt, 1982). One might expect further improvement by appropriate manipulation of the experimental condi-

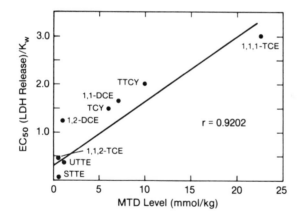

Fig. 1. EC_{50}/K_w vs. maximum tolerated dose in Osborne–Mendel rats for eight chlorinated hydrocarbon solvents. EC_{50} values are the dissolved haloalkane concentrations releasing 50% of the intracellular LDH, after subtracting out background release, from young adult rat (Osborne–Mendel) hepatocytes in a cytotoxicity screen during a 2-hr incubation. K_w values are published water–air partition coefficients for the chlorinated aliphatics in the plot (Sato and Nakajima, 1979). Maximally tolerated dose (MTD) levels are from NCI bioassays with the chemicals in the same rat strain or, in the case of UTTE, from SRI studies. 1,1-DCE, 1,1-Dichloroethane; 1,2-DCE, 1,2-dichloroethane; 1,1,1-TCE, 1,1,1-trichloroethane; 1,1,2-TCE; 1,1,2-trichloroethane; TCY, trichloroethylene; TTCY, tetrachloroethylene; UTTE, 1,1,1,2-tetrachloroethane; STE, 1,1,2,2-tetrachloroethane. (D. L. Story and K. Hawk-Prather, unpublished results.)

tions, such as using an air:CO_2 rather than a carbogen atmosphere, as has been suggested (Stacey et al., 1982; Bridges et al., 1983). The O_2 tension in liver corresponds to from 5 to 20% O_2 content in the atmosphere, depending on proximity to arterial and portal blood supplies (Derrick and Russell, 1964; Richter et al., 1972); at 5% atmosphere oxygen concentration, hepatocytes do not survive short (30-min) incubations when suspended in Hanks' Balanced Salt Solution (Bond and Rickert, 1981).

There are many other tests that one can use to establish in vitro/in vivo correspondence, such as the use of inducers of metabolism or inhibitors of processes thought to be related to the chemical's cytotoxicity. When one is interested in the early events in cytotoxic processes, the sensitivity and accuracy of biochemical measures are compounded by the large amount of unaffected tissue in the intact organism, and a further perturbation in the measurement if the tissue must be harvested for analyses. The use of microscopic techniques instead of biochemical ones can often circumvent these difficulties. An example is the use of transmission electron microscopy to better define the concentration range in vitro to be used in CCl_4 studies that produce time-course pathologic changes similar to those in vivo (Tyson et al., 1983b).

Establishing in vitro/in vivo correspondence for selected members of structurally related chemical series whenever possible gives confidence in the use of hepatocytes as a screening system for hepatotoxic potency for untested chemicals in the series. When data are available, correlations with peak serum levels or area under the curve are more accurate than dose. Also, instead of the EC_{50} values in vitro, rankings based on LECT, the lowest effective concentration tested, when the dose–response curve differs markedly for different cytotoxins (Inmon et al., 1981), may give better results. Since we are studying cells from target tissues that undergo similar pathologic changes in vivo, a reasonable correlation or correspondence between potency and events in vivo and in vitro is more to be expected than not. Failure of such correlation attempts may be due not to the failure of hepatocytes themselves to be useful for such purposes, but to inappropriate choice of experimental conditions or parameters measured, relative insensitivity, or other deficiencies that might be remedied by intelligent modification of the variables under experimental control. An example might be to attempt correlations of lipid peroxidation using ethane evolution for a series of analogs in vitro and in vivo without taking into account potential variability from the impact of gut metabolism or other tissues on the expired level (Recknagel et al., 1982; Gelmont et al., 1981; Kappus, 1985). Another example seen is to compare results with hepatocytes isolated from enzyme-induced animals with those in vivo from uninduced animals. Thus, even with rational optimization of controllable variables, the results obtained from the use of hepatocytes in a validated cytotoxicity screen are only approximations, though still useful for guidance (Zbinden et al., 1984).

VI. METHODS

The methods described in the following sections are for the indicators of cytotoxicity most commonly used in isolated hepatocytes. In general, the techniques are applicable to liver cells both in suspension and in monolayer culture. Differences between these two systems will be noted when they are significant. The detailed methods have been written assuming that the experimental conditions most frequently used in our laboratory are being employed (Tyson *et al.*, 1983a,b; Story *et al.*, 1983; Green *et al.*, 1983, 1984).

A. Incubation Conditions

Hepatocytes in suspension are incubated at a density of $1.0-1.5 \times 10^6$ viable cells/ml, at 37°C, 60–70 oscillations/min, in serum-free complex culture medium containing either 0.2% or 2.0% bovine serum albumin (BSA). The cells are incubated under an atmosphere of 5% CO_2:95% air or O_2 in gas-tight, all-glass flasks that have a side arm to allow serial sampling and a center well for the addition of volatile test compounds. The flasks are approximately the volume of a 25-ml Erlenmeyer flask, and each contains 4 ml of incubation mixture. Primary monolayer cultures of hepatocytes are routinely plated on collagen-coated 60-mm culture dishes, 2.5×10^6 viable cells/4 ml hormone-supplemented culture medium, and incubated under 5% CO_2; 95% air.

A variety of different incubation media have been used ranging from simple buffered salt solutions to complex culture media. Although simple media (most commonly, Krebs-Henseleit buffer or Hanks BSS) are effective in short-term studies (e.g., Holme *et al.*, 1982; Jewell *et al.*, 1982; Dankovic and Billings, 1985), the viability of control cells is better maintained in more complex, physiologically-based media (Green *et al.*, 1983). The medium we use includes several factors known from the literature to be advantageous. Since the absence of Ca^{2+} is harmful to hepatocytes (Smith *et al.*, 1981; Jewell *et al.*, 1982; Fariss and Reed, 1985; C. A. Tyson and C. E. Green, personal observations), it is present. To best mimic *in vivo* response to cytotoxins, the normal extracellular concentration should be used (1–2.5 mM). Cysteine and methionine, precursor amino acids for glutathione synthesis, are necessary for maintaining glutathione levels of isolated hepatocytes (Reed and Orrenius, 1977). Hormone-supplementation is also important, particularly in monolayer cultured hepatocytes, for sustaining cytochrome P-450 levels and activities (Decad *et al.*, 1977; Dickins and Peterson, 1980). Most hormones have a finite lifetime in the medium, and others may selectively induce certain P-450 isozymes or cause other perturbations that can be of potential concern for long-term incubations (Schuetz *et al.*, 1984). Other factors have been included in hepatocyte incubation media such as serum, bovine serum albumin, fatty acids, or trace elements, but the effect of these variables on cytotoxicity results have not been well-documented. Many different incubation media appear to be suitable for hepatocyte incubations and cytotoxicity assessments, and the final choice of media components must be based on the specific aims of the studies. At present, there is no single medium that can be claimed to be superior for all purposes.

B. Plasma Membrane Integrity

1. Dye Exclusion

The trypan blue assay is based on the premise that viable cells have an intact plasma membrane that is impermeable to the dye. Nonviable cells absorb the dye, and the nucleus is stained blue as a result. Cells sometimes exhibit varying degrees of staining; however, any blue color indicates that the cell is nonviable. As an aid to identification of nonviable cells, it should be noted that these cells also have a nonrefractile appearance. Although other dyes are sometimes recommended for the determination of liver cell viability, most problems associated with the use of trypan blue can be avoided if a standardized technique and source of dye are used.

To determine the viability of isolated cells, add 0.10 ml of cells (approximately 10×10^6 cells/ml of protein-free medium) to 0.85 ml of buffer (also protein free) and 0.050 ml of 0.4% trypan blue. Hold at room temperature for 5 min and then load into an improved Neubauer counting chamber. Count the numbers of viable and nonviable cells and calculate the viability.

To determine the viability of suspensions of hepatocytes exposed to chemicals, remove 0.10 ml of cells (about $1.0–1.5 \times 10^6$ cells/ml) and mix with 0.10 ml of protein-free buffer and 0.050 ml of 0.4% trypan blue (Story et al., 1983). A higher concentration of dye is used in this case because the incubation medium contains BSA with which the dye will complex. Continue exactly as described above. The viability of monolayer cultures can be measured similarly, by adding 0.95 ml of protein-free buffer and 0.05 ml of 0.4% trypan blue to the culture dish and counting attached cells in five different areas, 0.0025 mm² each, using an eyepiece reticle (Green et al., 1984). If trypan blue will interfere with other assays that are planned for these cultures, determining the number of cells per culture by counting the attached cells with cuboidal flattened morphology correlates nearly exactly with trypan blue exclusion.

2. Enzyme Release

To measure enzyme release (LDH or AST) in liver cells in suspension, two aliquots (about 0.30 ml each) are removed and held on ice. An equal volume of

1% Triton X-100 is added to one aliquot with gentle shaking for 10 min to lyse the cells so that total enzyme content can be determined. Both aliquots are centrifuged for 3 min at 1200 g to pellet the cells and debris, and the supernate is aspirated (Tyson *et al.*, 1983a). Aliquots for both LDH and AST determinations may be stored at 4°C for as long as 24 hr before analysis. To measure enzyme release from hepatocytes in primary culture, the enzyme content of both the medium and attached cells is determined. Released enzyme is measured in the medium. The total enzyme content is measured by scraping the cells into 1.0 ml of 0.5% Triton X-100, and the enzyme activity of the medium and cell lysate is summed to obtain total activity. The rest of the procedure is the same as for cells in suspension.

LDH is measured by the method of Wroblewski and LaDue (1955). This technique follows the formation of NADH by measuring the increase in the absorbance at 340 nm. The assay can be performed manually on a spectrophotometer; or, to efficiently and conveniently analyze large numbers of samples from cytotoxicity studies, the assay can be performed with an automated analyzer, such as the GEMENI Minicentrifugal Analyzer (Electro-Nucleonics, Inc., Fairfield, NJ). AST is determined similarly using the method optimized by Rodgerson and Osberg (1974), which measures the decrease in absorbance at 340 nm as NAD^+ is formed.

The data are expressed as the percentage of total cellular activity released by the chemical after adjusting for control leakage during the same period. This can be done by subtracting the percentage of enzyme release in control cells from both the numerator and denominator for treated cells; or, if the total enzyme varies among the samples, the following formula should be used:

$$\text{Net percentage enzyme release} = \frac{E_{RTr} - (E_{RC}/E_{TC})}{E_{TTr} - (E_{RC}/E_{TC})} \times 100$$

where E_{RTr} = released enzyme in cells treated with test chemical; E_{TTr} = total enzyme in cells treated without chemical; E_{RC} = released enzyme in control cells; E_{TC} = total enzyme in control cells.

Alternatively, NADH penetration (LDH latency) may be measured (Moldeus *et al.*, 1978). The assay is performed as described above for LDH release except that the cells in suspension are not pelleted by centrifugation but are left in the reaction mixture.

To perform the assay, mix 0.10 ml of hepatocytes or cell lysate (cell suspensions in 0.5% Triton X-100) with 2.80 ml of 0.10 M potassium phosphate buffer, pH 7.5, and 0.20 mg NADH. Add 0.10 ml 0.02 M pyruvate to start the reaction, and measure the change in absorbance at 340 nm. The percentage of penetration is calculated by dividing the LDH activity of intact cells by the LDH activity in the lysed cells and multiplying the result by 100.

3. Cellular K^+ Content

K^+ content of hepatocytes is believed to be a more sensitive indicator of the intactness of the plasma membrane than enzyme leakage or dye exclusion (Baur et al., 1975). K^+ is assayed by flame photometry after rapid separation of the liver cells from the incubation medium by centrifugal filtration (Stacey et al., 1980). An aliquot of hepatocytes (0.10 ml) is layered onto the top of 0.050 ml of silicone oil (density, 1.02) floating on 0.10 ml of 0.7 M HClO$_4$ in a 0.4-ml microfuge tube. The samples are centrifuged at 9400 g for 10 sec, and the bottom layer containing precipitated cells is analyzed for K^+.

C. Mitochondrial Function

ATP analysis is performed using the Packard Luminometer (Downers Grove, IL) which detects light produced by luciferase using luciferin and ATP as substrates. This assay is extremely sensitive and the ATP content of about 1000 hepatocytes can be detected. As a result, precise aliquots and dilutions are necessary to obtain reproducible data.

With hepatocyte suspensions, 0.25 ml of cell suspension is added to 0.25 ml 6% perchloric acid (PCA). The samples can then be analyzed immediately or stored at $-70°C$ for 24 hr before analysis. Prior to assay, acid extracts are neutralized with approximately 0.04 ml of 7.5 N KOH, 50 mM K$_2$HPO$_4$, centrifuged and diluted, 0.10 ml in 2.9 ml of 0.05 M Tris buffer, pH 7.7, containing 4.0 mM MgSO$_4$. After completion of these steps, samples can be stored for 2–3 weeks at $-135°C$ with no changes in ATP or ADP levels.

To determine ATP content, an aliquot (0.010 ml) is removed, added to the assay test tube (6 × 50 mm) and mixed with 0.020 ml of buffer. The reaction is started by the addition of 0.040 ml of luciferin–luciferase complex (Immunex Radiochemical Diagnostics, Carson City, NV), which is prepared by resuspending the lyophilized reagent in 10 ml 25 mM HEPES, pH 7.5, and subjecting it to freeze–thaw five times.

Monolayer cultures are assayed similarly, except that the attached cells are scraped into 1.0 ml 3% PCA and the acid extracts are neutralized with approximately 0.080 ml 7.5 N KOH, 50 mM K$_2$HPO$_4$. Subsequent dilutions are performed as described for the hepatocyte suspensions. The amount of ATP is quantified by extrapolation from a standard curve of 10^{-8} to 10^{-5} M ATP.

When the ATP to ADP ratio is desired, ADP is measured by incubating an aliquot (0.05 ml) of the diluted sample with 0.05 ml of 50 mM phosphoenolpyruvate, 100 mM Tris-HCl 35 mM KCl, 6 mM MgCl$_2$, and 10 U pyruvate kinase for 20 min to convert ADP to ATP (Kimmich et al., 1975). The quantities of ATP and ADP are calculated from the ATP standard curve and the ratio of the two determined.

D. Metabolic Competence

1. Protein Synthesis

Either radiolabeled valine or leucine can be used to determine the protein-synthetic capability of isolated hepatocytes (Seglen, 1976b; Seglen and Solheim, 1978; Green *et al.*, 1983). To minimize the effect of intracellular protein degradation and prevent gradual isotope dilution, the incubation medium should be supplemented with the substrate amino acid at a level several-fold higher than the intracellular level (Seglen and Solheim, 1978).

To perform the assay, 1 μCi [U-^{14}C]valine/20 μmol valine in 0.10 ml of incubation medium is added to the flask of cells in suspension or to the monolayer culture. The hepatocytes are incubated for 30 or 60 min longer; then 0.50 ml is removed from the cell suspension and added to 0.10 ml of 10% PCA. Samples can be stored for later analysis at this point.

After thawing, the samples are centrifuged at 1200 g for 5 min. The precipitated protein is washed three times with 3 ml of 2% PCA by vortexing the suspension and pelleting the precipitate by centrifugation at 1200 g for 5 min. The precipitate is then dissolved in 0.50 ml of 0.3 N NaOH, transferred to scintillation vials with 2 ml of distilled water, mixed with 10 ml of Neutralizer scintillation fluid (Research Products Intl., Mount Prospect, IL), and the radioactivity determined by liquid scintillation counting. Attached cultures are assayed essentially like suspensions of hepatocytes, except that cultures are frozen on dry ice at the end of the incubation period and scraped in 1.0 ml phosphate buffered saline after thawing. Protein is precipitated by the addition of 0.20 ml of 10% PCA. Data are expressed as micromoles of valine incorporated/min.

2. Urea Synthesis

The capability of the hepatocytes to synthesize urea can be determined either by the addition of substrates to sustain the maximum rate of synthesis (Story *et al.*, 1983) or by the measurement of the amount of urea that is formed by the liver cells using the incubation medium constituents as substrates (Santone *et al.*, 1982).

To determine the maximum rate of urea synthesis, a 0.50-ml aliquot is removed from each flask for measurement of initial urea content and the substrate mixture (10 mM NH$_4$Cl + 10 mM ornithine, final concentrations, delivered in 0.050 ml of incubation medium) is added. At the end of the 30-min incubation period, another 0.50-ml aliquot is removed. The reaction in each 0.50-ml aliquot is stopped by precipitating the proteins with 0.05 ml of 60% PCA. The precipitate is pelleted by centrifugation at 1200 g for 3 min, and 0.40 ml of the supernate is neutralized with 0.10 ml of 3 M K$_2$CO$_3$ + 0.5 M triethanolamine. Samples are again centrifuged at 1200 g for 3 min. Because citrulline is an endogenous interference to the urea assay, 0.25 ml of each sample is incubated with 0.10 ml of urease (1 mg/ml) for 1 hr at 37°C to destroy urea so that citrulline can be quantitated. These samples as well as those containing both citrulline and urea (0.050 ml sample + 0.30 ml incubation medium to dilute) are mixed with 5.0 ml urea assay reagent (Foster and Hochholzer, 1971). The capped samples are then heated at 90°C for 15 min, cooled, and the absorbance read at 460 nm. The amounts of citrulline and urea are determined by extrapolation from a concentration vs A460 curve. Urea is determined by difference, and the rate of urea synthesis expressed as micromoles of urea synthesized per min.

3. Gluconeogenesis

Measurement of the synthesis of glucose by hepatocytes can be used as an indicator of the capability of the cells to perform a complex multistep function. Although glucose can be measured readily by the glucose oxidase method, this technique is not appropriate for use with isolated hepatocytes under most conditions because of the high levels of glycogen and glycogenolysis present in liver cells prepared from fed animals. An additional disadvantage in the measurement of gluconeogenesis as a cytotoxicity indicator is that the assay should be performed in the absence of glucose, a common substituent of the complex incubation media that are most compatible with hepatocyte viability.

To perform the assay, the hepatocytes are incubated for 30 min with 10 mM [3-^{14}C]lactate:pyruvate (9:1), using [^{14}C]lactate with a specific activity of 0.025 mCi/mmol in Krebs–Ringer buffer (Green et al., 1983). After 30 min, the cell suspension (4.0 ml) is acidified with 0.40 ml of 60% perchloric acid and centrifuged at 1200 g to remove the precipitate. The supernatant is neutralized with 1.1 ml of 20% KOH (or more as needed). The entire sample is then applied to a Dowex 50 column (1.2 × 5.0 cm; H$^+$ form, AG 50W-S8) placed above a Dowex 1 column (1.2 × 5.0 cm; acetate, Dowex 1-X8). Glucose is eluted from the column with 30 ml distilled water and the radioactivity in an aliquot is determined by liquid scintillation counting (Dong and Freedland, 1980) and micromoles of glucose formed per minute calculated.

4. Lactate-to-Pyruvate Ratio

Lactate is assayed by the method of Hohorst (1963), and pyruvate is assayed by the method described by Bücher et al., (1957). At the desired time point after addition of the cytotoxin, an aliquot of the cell suspension is removed and immediately centrifuged at 1200 g for 3 min. The test sample (cells or supernate) is then added to an equal volume of ice-cold HPO$_3$ (5 g/100 ml). The samples can then be analyzed immediately or stored at -70°C for later analysis the following day.

To measure lactate, 100 μl of the sample supernate is mixed with 2.9 ml of the test kit reagent (kit available from Behring Diagnostics, La Jolla, CA). The reaction mix is incubated at 37°C for 45 min and then the absorbance is read at 340 nm.

To measure pyruvate, the following are added to a 1-cm light path cuvette: 2.0 ml of sample supernate, 0.50 ml TRIXMA base solution, and 0.50 ml NADH solution −A (from Sigma kit procedure no. 726-ar). The absorbance at 340 nm is read, then 0.05 ml of LDH is added. The samples are maintained at 25°C and the absorbance is read again after 2–5 min.

E. Nonspecific Indicators

1. Glutathione

The GSH assay developed by Hissin and Hilf (1976) is readily adaptable to measurements with isolated hepatocytes. Cell samples (approximately 1.5–4 × 10^6 hepatocytes) are rapidly frozen on dry ice and stored at −70°C until analysis. After thawing, 1.5 ml of buffer (0.1 M sodium phosphate, 0.005 M EDTA, pH 8.0) is added to the samples and the cell material is dispersed by repeated pipeting. To quantify the amount of cells, 0.25 ml is removed for measurement of DNA or protein content, if desired. To the remaining 1.25 ml, 0.33 ml of 25% HPO_3 is added to precipitate protein. The samples are centrifuged at 100,000 g for 30 min and 0.50 ml of the supernate is diluted with 1.5 ml of phosphate–EDTA buffer. The pH of each sample is individually checked and adjusted if necessary to pH 8.0 with 0.1 M NaOH. An aliquot (0.10 ml) is then mixed with 1.80 ml of phosphate–EDTA buffer and with 0.10 ml of o-phthalaldehyde (1 mg/ml) in methanol. The fluorescence of the samples is determined at an emission wavelength of 420 nm using an excitation wavelength of 350 nm. The quantity of GSH is determined using a standard curve of 0.10–2.0 μg/ml.

2. Lipid Peroxidation

Although a variety of end points can be used, measurement of malondialdehyde and of ethane have been the most frequently and successfully used with isolated hepatocytes.

To measure MDA (or thiobarbituric acid reactants), a 0.25-ml aliquot is removed from the incubation flask and mixed immediately with 0.5 ml 15% trichloroacetic acid (TCA) in a screw-top test tube. The samples may be frozen at this point for later analysis. The TBA reagent (0.67%) is prepared by mixing 2 parts of 1% TBA (neutralized) with 1 part of 0.5 M sodium citrate buffer. The thawed samples are each mixed with 1 ml of this reagent, capped, and immediately heated at 92°C for 30 min. The samples are then cooled to room temperature and centrifuged at 1200 g for 10 min. The supernate is transferred to another test tube and the absorbance of the samples read at 530 nm. The amount of TBA reactants is quantitated against a standard curve of MDA (1,1,3,3-tetramethoxypropane) ranging from 0.50 to 10 μM. The same method can be used for hepatocytes both in suspension and in monolayer culture.

Ethane production is measured by gas chromatographic analysis of the headspace gas using a Porapak column (Smith et al., 1982). To detect ethane, it is necessary to modify the hepatocyte incubation conditions, increasing the ratio of cells to headspace to increase the sensitivity of the measurement. A ratio of 1.5 headspace to solution permits detection of ~50 ng ethane/ml of headspace with cumene hydroperoxide as the cytotoxin, 10-fold higher than the GC limit of detection. Cell survival time in control flasks is reduced about 50% because of the smaller amount of oxygen in the flask than available under usual conditions (4.0 ml of solution in a 25-ml flask). Therefore, this assay is limited to use with rapidly acting chemicals.

3. Covalent Adduct Formation

The determination of the amount of a toxic chemical that binds covalently to cellular macromolecules is, in general, an indication of the amount of activated metabolites that are formed. The easiest assay to perform is the determination of the total quantity of covalent adducts to protein and nucleic acids. However, if more specific information is required, techniques are available for the measurement of covalent adducts to specific macromolecules (DNA, RNA, protein, lipids), which can be readily adapted to cells in vitro. By using either cesium chloride density gradient centrifugation or hydroxylapatite chromatography, DNA, RNA, and protein can be separated and covalent adducts to these fractions quantitated (Shaw et al., 1975; Beland et al., 1979).

To determine total covalent binding, the radiolabeled test chemical is added to hepatocytes, both live cells and heat- or acid-denatured cells, and the incubation continued. At the desired time point, the reaction is stopped either by fast freezing the cell suspension in a dry ice–acetone bath or by adding TCA to give a final concentration of 10%. Monolayer cultures of hepatocytes are usually frozen to stop the reaction and later scraped into 1.0 ml phosphate buffered saline prior to precipitation with TCA (Green et al., 1984). In both cases, cellular macromolecules (primarily protein) are collected by vacuum filtration on glass fiber filters or by centrifugation as described for the protein synthesis assay. The filters or pellets are then washed with ice-cold 10% TCA and with 5–10 volumes of organic solvents to remove noncovalently bound radioactive substrate and metabolites. The wash procedure is continued until wash solvents contain only background levels of radioactivity. The particular solvents used depend on the chemical characteristics of the test compound but generally methanol, methanol:ether (3:1), and ethanol:ether (3:1) (Pohl and Branchflower, 1981; Green *et al.*, 1984) are appropriate. Precipitated protein is dissolved in tissue solubilizer such as Protosol (New England Nuclear, Boston, MA) by heating at 50°C for 2 hr. Dissolved protein is assayed for radioactivity by liquid scintillation counting and for protein content by spectrophotometric assay (Lowry *et al.*, 1951; Bradford, 1976).

REFERENCES

Acosta, D., Anufuro, D., Mitchell, D. B., Santone, K. S., and Nelson, K. F. (1985). A primary cell culture system for hepatotoxic assessment. *Lab. Anim.* **14**(5), 31–36.

Akerboom, T. P. M., and Sies, H. (1981). Assay of glutathione, glutathione disulfide, and glutathione mixed disulfides in biological samples, *Methods Enzymol.* **77**, 373–382.

Asakawa, T., and Matsushita, S. (1981). Thiobarbituric acid test for detecting lipid hydroperoxides under anaerobic conditions. *Agric. Biol. Chem.* **45**, 453–457.

Baur, H., Kasperek, S., and Pfaff, E. (1975). Criteria of viability of isolated liver cells. *Hoppe-Seyler's Z. Physiol. Chem.* **356**, 827–838.

Beland, F. A., Dooley, K. L., and Casciano, D. A. (1979). Rapid isolation of carcinogen-bound DNA and RNA by hydroxyapatite chromatography. *J. Chromatogr.* **174**, 177–186.

Bellomo, G., Jewell, S. A., Thor, H., and Orrenius, S. (1982). Regulation of intracellular calcium compartmentation: Studies with isolated hepatocytes and *t*-butyl hydroperoxide. *Proc. Natl. Acad. Sci. U.S.A.* **79**, 6842–6846.

Berthon, B., Binet, A., Mauger, J.-P., and Claret, M. (1984). Cytosolic free Ca^{2+} in isolated rat hepatocytes as measured by quin2. *FEBS Lett.* **167**, 19–24.

Beutler, E., and West, C. (1977). Comment concerning a fluorometric assay for glutathione. *Anal. Biochem.* **81**, 458–460.

Black, M. (1984). Acetaminophen hepatotoxicity. *Annu. Rev. Med.* **35**, 577–593.

Bond, J. A., and Rickert, D. E. (1981). Metabolism of 2,4-dinitro[^{14}C]toluene by freshly isolated Fischer-344 rat primary hepatocytes. *Drug Metab. Dispos.* **9**, 10–14.

Bradford, M. M. (1976). A rapid and sensitive method for the quantitation of microgram quantities of protein utilizing the principle of protein-dye binding. *Anal. Biochem.* **72**, 248–254.

Bridges, J. W., Benford, D. J., and Hubbard, S. A. (1983). Mechanisms of toxic injury. *Ann. N.Y. Acad. Sci.* **407**, 42–63.

Bücher, T., Czok, R., Lamprecht, W., and Latzko, E. (1963). Pyruvate. *In* "Methods of Enzymatic Analysis" (H.-U. Bergmeyer, ed.), pp. 253–259. Academic Press, New York.

Buege, A., and Aust, S. D. (1978). Microsomal lipid peroxidation. *Methods Enzymol.* **52**, 302–310.

Campanini, R. Z., Tapia, R. A., Sarnat, W., and Natelson, S. (1970). Evaluation of serum argininosuccinate lyase (ASAL) concentrations as an index to parenchymal liver disease. *Clin. Chem.* **16**, 44–53.

Cantilena, L. R., Stacey, N. H., and Klaassen, C. D. (1983). Isolated rat hepatocytes as a model system for screening chelators for use in cadmium intoxication. *Toxicol. Appl. Pharmacol.* **67**, 257–263.

Chiarpotto, E., Olivero, E., Albano, E., Poli, G., Gravela, E., and Dianzani, M. U. (1981). Studies on lipid peroxidation using whole liver cells: Influence of damaged cells on the prooxidant effect of ADP-Fe^{3+} and CCl_4. *Experientia* **37**, 396–397.

Cohn, V. H., and Lyle, J. (1966). A fluorometric assay for glutathione. *Anal. Biochem.* **14**, 434–440.

Cornell, N. W. (1980). Rapid fractionation of cell suspensions with the use of brominated hydrocarbons. *Anal. Biochem.* **102**, 326–331.

Corongiu, F. P., and Milia, A. (1983). An improved and simple method for determining diene conjugation in autoxidized polyunsaturated fatty acids. *Chem.-Biol. Interact.* **44**, 289–297.

Dankovic, D. A., and Billings, R. E. (1985). The role of 4-bromophenol and 4-bromocatechol in bromobenzene covalent binding and toxicity in isolated rat hepatocytes. *Toxicol. Appl. Pharmacol.* **79**, 323–331.

Davies, P., and Allison, A. C. (1972). The significance of the lysosome in toxicology. *CRC Crit. Rev. Toxicol.* **1**, 283–323.

Decad, G. M., Hsieh, D. P. H., and Byard, J. L. (1977). Maintenance of cytochrome *P*-450 and metabolism of aflatoxin B1 in primary hepatocyte cultures. *Biochem. Biophys. Res. Commun.* **78**, 279–287.

Derrick, J. R., and Russell, D. (1964). Oxygen tensions in tissues. *Arch. Surg. (Chicago)* **88**, 1059–1062.

Dich, J., and Tønnesen, I. C. (1980). Effects of alcohol, nutritional status, and composition of the incubation medium on protein synthesis in isolated rat liver parenchymal cells. *Arch. Biochem. Biophys.* **204**, 640–647.

Dickens, M., and Peterson, R. E. (1980). Effects of a hormone-supplemented medium on cytochrome *P*-450 and mono-oxygenase activities of rat hepatocytes in primary culture. *Biochem. Pharmacol.* **29**, 1231–1238.

Dilley, J. V., Tyson, C. A., Spanggord, R. J., Sasmore, D. P., Newell, G. W., and Dacre, J. C. (1982). Short-term oral toxicity of 2,4,6-trinitrotoluene in mice, rats, and dogs. *J. Toxicol. Environ. Health* **9**, 565–585.

DiMonte, D., Ross, D., Bellomo, G., Eklöw, L., and Orrenius, S. (1984). Alterations in intracellular thiol homeostasis during the metabolism of menadione by isolated rat hepatocytes. *Arch. Biochem. Biophys.* **235**, 334–342.

Dong, F. M., and Freedland, F. A. (1980). Effects of alanine on gluconeogenesis in isolated rat hepatocytes. *J. Nutr.* **110**, 2341–2349.

Dooley, J. F., and Masullo, K. (1982). Activation of serum aspartate aminotransferase (AST) in rats after acute thioacetamide hepatotoxicity by use of pyridoxal 5′-phosphate (PLP). *Clin. Chem.* **28**, 1640.

Dujovne, C. A. (1978). Hepatotoxic and cellular uptake interactions among surface active components of erythromycin preparations. *Biochem. Pharmacol.* **27**, 1925–1930.

Erecinska, M., Wilson, D. F., and Nishiki, K. (1974). Homeostatic regulation of cellular energy metabolism: Experimental characterization *in vivo* and fit to a model. *Am. J. Physiol.* **234**, C82–C89.

Erecinska, M., Stubbs, M., Miyata, Y., Ditre, C. M., and Wilson, D. F. (1977). Regulation of cellular metabolism by intracellular phosphate. *Biochim. Biophys. Acta* **462**, 20–35.

Fariss, M. W., and Reed, D. J. (1985). Mechanism of chemical-induced toxicity. II. Role of extracellular calcium. *Toxicol. Appl. Pharmacol.* **79**, 296–306.

Fariss, M. W., Brown, M. K., Schmitz, J. A., and Reed, D. J. (1985). Mechanism of chemical-induced toxicity. I. Use of a rapid centrifugation technique for the separation of viable and nonviable hepatocytes. *Toxicol. Appl. Pharmacol.* **79**, 283–295.

Feuer, G., Gaunt, I. F., Golberg, L., and Fairweather, F. A. (1965). Liver response tests. VI. Applications to a comparative study of food antioxidants and hepatotoxic agents. *Food Cosmet. Toxicol.* **3**, 457–469.

Foster, L. B., and Hochholzer, J. M. (1971). A single reagent manual method for directly determining urea nitrogen in serum. *Clin. Chem.* **17**, 921–925.

Gee, D. L., and Tappel, A. L. (1981). Production of volatile hydrocarbons by isolated hepatocytes: An *in vitro* model for lipid peroxidation studies. *Toxicol. Appl. Pharmacol.* **60**, 112–120.

Gellerfors, P., Wielburski, A., and Nelson, B. D. (1979). Synthesis of mitochondrial proteins in isolated rat hepatocytes. *FEBS Lett.* **108**, 167–170.

Gelmont, D., Stein, R. A., and Mead, J. F. (1981). The bacterial origin of rat breath pentane. *Biochem. Biophys. Res. Commun.* **102**, 932–936.

Gillette, J. R. (1974). A perspective on the role of chemically reactive metabolites of foreign compounds in toxicity. I. Correlation of changes in covalent binding of reactive metabolites with changes in the incidence and severity of toxicity. *Biochem. Pharmacol.* **23**, 2927–2938.

Goethals, F., Krack, G., and Roberfroid, M. (1983). Effects of diethyl maleate on protein synthesis in isolated hepatocytes. *Toxicology* **26**, 47–54.

Goethals, F., Krack, G., Deboyser, D., Vossen, P., and Roberfroid, M. (1984). Critical biochemical functions of isolated hepatocytes as sensitive indicators of chemical toxicity. *Fundam. Appl. Toxicol.* **4**, 441–450.

Gray, T. J., Lake, B. G., Beamand, J. A., Foster, J. R., and Gangolli, S. D. (1983). Peroxisome proliferation in primary cultures of rat hepatocytes. *Toxicol. Appl. Pharmacol.* **67**, 15–25.

Green, C. E., Dabbs, J. E., and Tyson, C. A. (1983). Functional integrity of isolated rat hepatocytes prepared by whole liver vs. biopsy perfusion. *Anal. Biochem.* **129**, 269–276.

Green, C. E., Dabbs, J. E., and Tyson, C. A. (1984). Metabolism and cytotoxicity of acetaminophen in hepatocytes isolated from resistant and susceptible species. *Toxicol. Appl. Pharmacol.* **76**, 139–149.

Grisham, J. W. (1979). Use of hepatic cell cultures to detect and evaluate the mechanisms of action of toxic chemicals. *Int. Rev. Exp. Pathol.* **20**, 123–209.

Grynkiewicz, G., Poenie, M., and Tsien, R. Y. (1985). A new generation of Ca^{2+} indicators with greatly improved fluorescence properties. *J. Biol. Chem.* **260**, 3440–3450.

Gwynn, J., Fry, J. R., and Bridges, J. W. (1979). The effect of paracetamol and other foreign compounds on protein synthesis in isolated adult rat hepatocytes. *Biochem. Soc. Trans.* **7**, 117–118.

Hansen, A. J. (1985). Effect of anoxia on ion distribution in the brain. *Physiol. Rev.* **65**, 101–148.

Hayes, M. A., and Pickering, D. B. (1985). Comparative cytopathology of primary rat hepatocyte cultures exposed to aflatoxin B_1, acetaminophen, and other hepatotoxins. *Toxicol. Appl. Pharmacol.* **80**, 345–356.

Hayes, M. A., Murray, C. A., and Rushmore, T. H. (1986). Influences of glutathione status on different cytocidal responses of monolayer rat hepatocytes exposed to aflatoxin B_1 or acetaminophen. *Toxicol. Appl. Pharmacol.* **85**, 1–10.

Helinek, T. G., Devlin, T. M., and Ch'ih, J. J. (1982). Initial inhibition and recovery of protein synthesis in cycloheximide-treated hepatocytes. *Biochem. Pharmacol.* **31**, 1219–1225.

Hensgens, H. E. S. J., and Meijer, A. J. (1979). The interrelationship between ureogenesis and protein synthesis in isolated rat-liver cells. *Biochim. Biophys. Acta* **582**, 525–532.

Hissin, P. J., and Hilf, R. (1976). A fluorometric method for determination of oxidized and reduced glutathione in tissues. *Anal. Biochem.* **74**, 214–226.

Högberg, J., and Kristoferson, A. (1977). A correlation between glutathione levels and cellular damage in isolated hepatocytes. *Eur. J. Biochem.* **74**, 77–82.

Högberg, J., Orrenius, S., and Larson, R. (1975a). Lipid peroxidation in isolated hepatocytes. *Eur. J. Biochem.* **50**, 595–602.

Högberg, J., Moldeus, P., Arbourgh, B., O'Brien, P., and Orrenius, S. (1975b). The consequences of lipid peroxidation in isolated hepatocytes. *Eur. J. Biochem.* **59**, 457–462.

Hohorst, H. J. (1963). L-(+)-lactate. Determination with lactate dehydrogenase and DPN. *In* "Methods of Enzymatic Analysis" (H.-U. Bergmeyer, ed.), pp. 266–270. Academic Press, New York.

Holme, J. A., Wirth, P. J., Dybing, E., and Thorgeirsson, S. S. (1982). Cytotoxic effects of N-hydroxyparacetamol in suspensions of isolated rat hepatocytes. *Acta Pharmacol. Toxicol.* **51,** 87–95.

Inmon, J., Stead, A., Waters, M. D., and Lewtas, J. (1981). Development of a toxicity test system using primary rat liver cells. *In Vitro* **17,** 1004–1010.

Jaworek, D., Gruber, W., and Bergmeyer, H. U. (1974). Adenosine-5'-diphosphate and adenosine-5'-monophosphate. *In* "Methods of Enzymatic Analysis" (H.-U. Bergmeyer, ed.), pp. 2127–2131. Academic Press, New York.

Jeejeebhoy, K. H., and Phillips, M. T. (1976). Isolated mammalian hepatocytes in culture. *Gastroenterology* **71,** 1086–1096.

Jewell, S. A., Bellomo, G., Thor, H., Orrenius, S., and Smith, M. T. (1982). Bleb formation in hepatocytes during drug metabolism is caused by disturbances in thiol and calcium ion homeostasis. *Science* **217,** 1257–1258.

Joseph, S. K., Coll, K. E., Cooper, R. H., Marks, J. S., and Williamson, J. R. (1983). Mechanisms underlying calcium homeostasis in isolated hepatocytes. *J. Biol. Chem.* **256,** 731–741.

Kappus, H. (1985). Lipid peroxidation: Mechanisms, analysis, enzymology and biological relevance. *In* "Oxidative Stress" (H. Sies, ed.), pp. 273–310. Academic Press, London.

Kimmich, G. A., Randles, J., and Brand, J. S. (1975). Assay of picomole amounts of ATP, ADP, and AMP using the luciferase enzyme assay. *Anal. Biochem.* **89,** 187–206.

Kirkpatrick, D. T., Guth, D. J., and Mavis, R. D. (1986). Detection of *in vivo* lipid peroxidation using the thiobarbituric acid assay for lipid hydroperoxides. *J. Biochem. Toxicol.* **1,** 93–104.

Klaassen, C. D., and Stacey, N. H. (1982). Use of isolated hepatocytes in toxicity assessments. *In* "Toxicology of the Liver" (G. L. Plaa and W. R. Hewitt, eds.), pp. 117–179. Raven, New York.

Koster, J. F., Slee, R. G., and Van Berkel, T. J. C. (1982). On the lipid peroxidation of rat liver hepatocytes, the formation of fluorescent chromolipids and high molecular weight protein. *Biochim. Biophys, Acta* **710,** 230–235.

Krack, G., Goethals, F., Deboyser, D., and Roberfroid, M. (1983). Metabolic competence of isolated hepatocytes in suspension: A new tool for *in vitro* toxicological evaluation. *In* "Isolation, Characterization, and Use of Hepatocytes" (R. A. Harris and N. W. Cornell, eds.), pp. 391–398. Elsevier, New York.

Krack, G., Deboyser, D., Goethals, F., Vossen, P., and Roberfroid, M. (1985). An *in vitro* model for acute toxicity testing using hepatocytes freshly isolated from adult mammals. *In* "Safety Evaluation and Regulation of Chemicals. II: Impact of Regulations: Improvement of Methods" (F. Homburger, ed.), pp. 286–294. Karger, Basel.

Kreamer, B. L., Staecker, J. L., Sawada, N., Sattler, G. L., Hsia, M. T. S., and Pitot, H. C. (1986). Use of a low-speed iso-density percoll centrifugation method to increase the viability of isolated rat hepatocyte preparations. *In Vitro* **22,** 201–211.

Krebs, H. A., Cornell, N. W., Lund, P., and Hems, E. (1973). Isolated liver cells as experimental material. *In* "Regulation of Hepatic Metabolism" (F. Lundquist and N. Tygstrup, eds.), pp. 726–755. Munksgaard, Copenhagen.

Krebs, H., Lund, P., and Edwards, M. (1979). Criteria of metabolic competence of isolated hepatocytes. *In* "Methodological Surveys B, Biochemistry" (E. Reid, ed.), Vol. 8, pp. 1–6. Horwood, Chichester, England.

Krell, H., Baur, H., and Pfaff, E. (1979). Transient ^{45}Ca uptake and release in isolated rat-liver cells during recovery from deenergized states. *Eur. J. Biochem.* **101,** 349–364.

Ku, R. H., and Billings, R. E. (1984). Relationships between formaldehyde metabolism and toxicity and glutathione concentrations in isolated rat hepatocytes. *Chem.-Biol. Interact.* **51,** 25–36.

Lamprecht, W., and Trautschold, I. (1974). Determination with hexokinase and glucose-6-phosphate dehydrogenase. *In* "Methods of Enzymatic Analysis" (H.-U. Bergmeyer, ed.), pp. 2101–2110. Academic Press, New York.

Lanoue, K. F., and Schoolwerth, A. C. (1984). Metabolite transport in mammalian mitochondria. *In* "Bioenergetics" (L. Ernster, ed.), pp. 250–254. Elsevier, Amsterdam.

Le Rumeur, E., Guguen-Guillouzo, C., Beaumont, C., Saunier, A., and Guillouzo, A. (1983). Albumin secretion and protein synthesis by cultured diploid and tetraploid rat hepatocytes separated by elutriation. *Exp. Cell Res.* **147**, 247–254.

Letko, G., Küster, U., and Pohl, K. (1983). Influence of precursors of biosynthesis on the energy metabolism of the liver cells. *Biomed. Biochim. Acta* **42**, 323–333.

Lowry, O. H., Rosebrough, N. J., Farr, A. L., and Randall, R. J. (1951). Protein measurement with the Folin phenol reagent. *J. Biol. Chem.* **193**, 256–275.

Lutz, W. K. (1979). *In vivo* covalent binding of organic chemicals to DNA as a quantitative indicator in the process of chemical carcinogenesis. *Mutat. Res.* **65**, 289–356.

McQueen, C. A., and Williams, G. M. (1982). Cytotoxicity of xenobiotics in adult rat hepatocytes in primary culture. *Fundam. Appl. Toxicol.* **2**, 139–144.

McQueen, C. A., Merrill, B. M., and Williams, G. M. (1984). Comparison of several indicators of cytotoxicity in rat hepatocytes in primary culture. *Toxicologist* **4**, 134.

Mattei, E., Delpino, A., and Ferrini, U. (1979). Protein synthesis inhibition induced by dimethylnitrosamine and diethylnitrosamine on isolated rat hepatocytes. *Experientia* **35**, 1213–1215.

Meijer, A. J., Gimpel, J. A., Deleeuw, G., Tischler, M. E., Tager, J. M., and Williamson, J. R. (1978). Interrelationships between gluconeogenesis and ureogenesis in isolated hepatocytes. *J. Biol. Chem.* **253**, 2308–2320.

Meredith, M. J., and Reed, D. J. (1982). Status of the mitochondrial pool of glutathione in the isolated hepatocyte. *J. Biol. Chem.* **257**, 3747–3753.

Mitchell, J. R., and Gillette, J. R. (1975). Metabolic activation of drugs to toxic substances. *Gastroenterology* **68**, 392–410.

Mitchell, J. R., Corcoran, G. B., Smith, C. V., Hughes, H., and Lauterberg, B. H. (1982). Alkylation and peroxidation injury from chemically reactive metabolites. *Adv. Exp. Med. Biol.* **136A**, 199–223.

Moldeus, P., Hogberg, J., and Orrenius, S. (1978). Isolation and use of liver cells. *Methods Enzymol.* **52**, 60–71.

Moore, L., Davenport, G. R., and Landon, E. J. (1976). Calcium uptake of a rat liver microsomal subcellular fraction in response to *in vivo* administration of carbon tetrachloride. *J. Biol. Chem.* **251**, 1197–1201.

Morrison, M. H., DiMonte, D., and Jernstrom, B. (1985). Glutathione status in primary cultures of rat hepatocytes and its role in cell attachment to collagen. *Chem.-Biol. Interact.* **53**, 3–12.

Orrenius, S., Thor, H., and Bellomo, G. (1984). Alterations in thiol and calcium-ion homeostasis during hydroperoxide and drug metabolism in hepatocytes. *Biochem. Soc. Trans.* **12**, 23–28.

Ozawa, K. (1982). Energy metabolism. *In* "Pathophysiology of Shock, Anoxia, and Ischemia" (R. A. Cowley and B. F. Trump, eds.), pp. 74–83. Williams & Wilkins, Baltimore, Maryland.

Pappas, N. J., Jr. (1980). Increased rat liver homogenate, mitochondrial, and cytosolic aspartate aminotransferase activity in acute carbon tetrachloride poisoning. *Clin. Chim. Acta* **106**, 223–229.

Pariza, M. W., Butcher, F. R., Kletzien, R. F., Becker, J. E., and Potter, V. R. (1976). Induction and decay of glucagon-induced amino acid transport in primary cultures of adult rat liver cells: Paradoxical effects of cycloheximide and puromycin. *Proc. Natl. Acad. Sci. U.S.A.* **73**, 4511–4515.

Pencil, S. D., Glende, E. A., Jr., and Recknagel, R. O. (1982). Loss of calcium sequestration capacity in endoplasmic reticulum of isolated hepatocytes treated with carbon tetrachloride. *Res. Commun. Chem. Pathol. Pharmacol.* **36**, 413–428.

Pfaff, E., Schuler, B., Krell, and Höke, H. (1980). Viability control and special properties of isolated rat hepatocytes. *Arch. Toxicol.* **44**, 3–21.

Plaa, G. L., and Hewitt, W. R. (1982). Quantitative evaluation of indices of hepatotoxicity. *In* "Toxicology of the Liver" (G. L. Plaa and W. R. Hewitt, eds.), pp. 103–120. Raven, New York.

Platt, D. S., and Cockrill, B. L. (1967). Liver enlargement and hepatotoxicity: An investigation into the effects of several agents on rat liver enzyme activities. *Biochem. Pharmacol.* **16**, 2257–2270.

Pohl, L. R., and Branchflower, R. V. (1981). Covalent binding of electrophilic metabolites to macromolecules. *Methods Enzymol.* **77**, 43–50.

Poli, G., Cheeseman, K., Slator, T. F., and Dianzani, M U. (1981). The role of lipid peroxidation in CCl_4-induced damge to liver microsomal enz .   Comparative studies *in vitro* using microsomes and isolated liver cells. *Chem.-Biol. Interact.* **37**, 13–24.

Recknagel, R. O., Glende, E. A., Jr., Waller, R. L., and Lowrey, K. (1982). Lipid peroxidation: Biochemistry, measurement, and significance in liver cell injury. *In* "Toxicology of the Liver" (G. L. Plaa and W. R. Hewitt, eds.), pp. 213–241. Raven, New York.

Reddy, J. K., and Lalwani, N. D. (1983). Carcinogenesis by hepatic peroxisome proliferators: Evaluation of the risk of hypolipidemic drugs and industrial plasticizers to humans. *CRC Crit. Rev. Toxicol.* **12**, 1–58.

Reed, D. J., and Beatty, P. W. (1980). Biosynthesis and regulation of glutathione: Toxicological implications. *In* "Reviews in Biochemical Toxicology" (E. Hodgson, J. Bend, and B. Philpot, eds.), pp. 213–241. Elsevier/North-Holland, New York.

Reed, D. J., and Orrenius, S. (1977). The role of methionine in glutathione biosynthesis by isolated hepatocytes. *Biochem. Biophys. Res. Commun.* **77**, 1257–1264.

Reed, D. J., Babson, J. R., Beatty, P. W., Brodie, A. E., Ellis, W. W., and Potter, D. W. (1980). High-performance liquid chromatography analysis of nanomole levels of glutathione, glutathione disulfide, and related thiols and disulfides. *Anal. Biochem.* **106**, 55–62.

Richter, A., Sanford, K. K., and Evans, V. J. (1972). Influence of oxygen and culture media on plating efficiency of some mammalian tissue cells. *J. Natl. Cancer Inst.* **49**, 1705–1712.

Rodgerson, D. O., and Osberg, I. M. (1974). Sources of error in spectrophotometric measurement of aspartate aminotransferase and alanine aminotransferase activities in serum. *Clin. Chem.* **20**, 43–50.

Sainsbury, G. M., Stubbs, M., Hems, R., and Krebs, H. A. (1979). Loss of cell constituents from hepatocytes on centrifugation. *Biochem. J.* **180**, 685–688.

Santone, K. S., Acosta, D., and Bruckner, J. V. (1982). Cadmium toxicity in primary cultures of rat hepatocytes. *J. Toxicol. Environ. Health* **10**, 169–177.

Sato, A., and Nakajima, T. (1979). A structure-activity relationship of some chlorinated hydrocarbons. *Arch. Environ. Health* **34**, 69–75.

Saville, B. (1958). A scheme for the colorimetric determination of microgram amounts of thiols. *Analyst (London)* **83**, 670–672.

Sayeed, M. M. (1982). Membrane Na^+-K^+ transport and ancillary phenomena in circulatory shock. *In* "Pathophysiology of Shock, Anoxia, and Ichemia" (R. A. Cowley and B. F. Trump, eds.), pp. 112–132. Williams & Wilkins, Baltimore, Maryland.

Schuetz, E. G., Wrighton, S. A., Varwick, J. L., and Guzelian, P. S. (1984). Induction of cytochrome *P*-450 by glucocorticoids in rat liver. I. Evidence that glucocorticoids and pregnenolone-16α-carbonitrile regulate *de novo* synthesis of a common form of cytochrome *P*-450 in cultures of adult rat hepatocytes and in the liver *in vivo*. *J. Biol. Chem.* **259**, 1999–2006.

Seglen, P. O. (1974). Autoregulation of glycolysis, respiration, gluconeogenesis and glycogen synthesis in isolated parenchymal rat liver cells under aerobic and anaerobic conditions. *Biochim. Biophys. Acta* **338**, 317–336.

Seglen, P. O. (1976a). Incorporation of radioactive amino acids into protein in isolated rat hepatocytes. *Biochim. Biophys. Acta* **442**, 391–404.

Seglen, P. O. (1976b). Preparation of isolated rat liver cells. *Methods Cell Biol.* **13**, 31–83.

Seglen, P. O. (1978). Effects of amino acids, ammonia and leupeptin on protein synthesis and degradation in isolated rat hepatocytes. *Biochem. J.* **174**, 469–474.

Seglen, P. O., and Solheim, A. E. (1978). Valine uptake and incorporation into protein in isolated rat hepatocytes. *Eur. J. Biochem.* **85**, 15–25.

Seglen, P. O., Solheim, A. E., Grinde, B., Gordon, P. B., Schwarze, P. E., Gjessing, R., and Poli, A. (1980). Amino acid control of protein synthesis and degradation in isolated rat hepatocytes. *Ann. N.Y. Acad. Sci.* **349**, 1–17.

Shaw, J. L., Blanco, J., and Mueller, G. C. (1975). A simple procedure for isolation of DNA, RNA, and protein fractions from cultured animal cells. *Anal. Biochem.* **65**, 125–131.

Siesjö, B. K. (1978). "Brain Energy Metabolism," pp. 179–186. Wiley, New York.

Smith, M. T., Thor, H., and Orrenius, S. (1981). Toxic injury to isolated hepatocytes is not dependent on extracellular calcium. *Science* **213**, 1257–1259.

Smith, M. T., Thor, H., Hartzell, P., and Orrenius, S. (1982). The measurement of lipid peroxidation in isolated hepatocytes. *Biochem. Pharmacol.* **31**, 19–26.

Soboll, S., Akerboom, P. M., Schwenke, W.-D., Haase, R., and Sies, H. (1980). Mitochondrial and cytosolic ATP/ADP ratios in isolated hepatocytes. *Biochem. J.* **192**, 951–954.

Somlyo, A. P., Bond, M., and Somlyo, A. V. (1985). Calcium content of mitochondria and endoplasmic reticulum in liver frozen rapidly *in vivo*. *Nature* **314**, 622–625.

Stacey, N. H., and Kappus, H. (1982). Comparison of methods of assessment of metal-induced lipid peroxidation in isolated rat hepatocytes. *J. Toxicol. Environ. Health* **9**, 277–285.

Stacey, N. H., and Klaassen, C. D. (1981). Inhibition of lipid peroxidation without prevention of cellular injury in isolated rat hepatocytes. *Toxicol. Appl. Pharmacol.* **58**, 8–18.

Stacey, N. H., Cantilena, L. R., and Klaassen, C. D. (1980). Cadmium toxicity and lipid peroxidation in isolated rat hepatocytes. *Toxicol. Appl. Pharmacol.* **53**, 470–480.

Stacey, N. H., Ottenwälder, H., and Kappus, H. (1982). CCl_4-induced lipid peroxidation in isolated rat hepatocytes with different oxygen concentrations. *Toxicol. Appl. Pharmacol.* **62**, 421–427.

Stammati, A. P., Silano, V., and Zucco, F. (1981). Toxicology investigations with cell culture systems. *Toxicology* **20**, 91–153.

Stier, A., Nolte, K. H., Schlenker, A., Schumann, W., and Zuretti, F. M. (1980). Toxicological studies of liver cells by microspectrofluorometry, *Arch. Toxicol.* **44**, 45–54.

Story, D. L., Gee, S. J., Tyson, C. A., and Gould, D. H. (1983). Response of isolated hepatocytes to organic and inorganic cytotoxins. *J. Toxicol. Environ. Health* **11**, 483–501.

Strehler, B. L. (1974). Adenosine-5′-triphosphate and creatine phosphate determination with luciferase. *In* "Methods of Enzymatic Analysis" (H.-U. Bergmeyer, ed.), pp. 2112–2121. Academic Press, New York.

Thomas, A. P., Alexander, J., and Williamson, J. R. (1984). Relationship between inositol polyphosphate production and the increase of cytosolic free Ca^{2+} induced by vasopressin in isolated hepatocytes. *J. Biol. Chem.* **259**, 5574–5584.

Thor, H., and Orrenius, S. (1980). The mechanism of bromobenzene-induced cytotoxicity studied with isolated hepatocytes. *Arch. Toxicol.* **44**, 31–43.

Thor, H., Moldeus, P., Hermanson, R., Högberg, J., Reed, D. J., and Orrenius, S. (1978). Metabolic activation and hepatotoxicity. Toxicity of bromobenzene in hepatocytes isolated from phenobarbital- and diethylmaleate-treated rats. *Arch. Biochem. Biophys.* **188**, 122–129.

158   Charles A. Tyson and Carol E. Green

Thor, H., Smith, M. T., Hartzell, P., Bellomo, G., Jewell, S. A., and Orrenius, S. (1982). The metabolism of menadione (2-methyl-1,4-naphthoquinone) by isolated hepatocytes. *J. Biol. Chem.* **257**, 12419–12425.

Thorgeirsson, S. S., and Wirth, P. J. (1977). Covalent binding of foreign chemicals to tissue macromolecules. *J. Toxicol. Environ. Health* **2**, 873–881.

Thurman, R. G., and Kaufmann, F. C. (1980). Factors regulating drug metabolism in intact hepatocytes. *Pharmacol Rev.* **31**, 229–251.

Tischler, M. E., Hecht, P., and Williamson, J. R. (1977). Determination of mitochondrial/cytosolic metabolite gradients in isolated rat liver cells by cell disruption. *Arch. Biochem. Biophys.* **181**, 278–292.

Tsien, R. Y., Pozzan, T., and Rink, T. J. (1982). Calcium homeostasis in intact lymphocytes: Cytoplasmic free calcium monitored with a new, intracellularly trapped fluorescent indicator. *J. Cell Biol.* **94**, 325–334.

Tyson, C. A., Mitoma, C., and Kalivoda, J. (1980). Evaluation of hepatocytes isolated by a nonperfusion technique in a prescreen for cytotoxicity. *J. Toxicol. Environ. Health* **6**, 197–205.

Tyson, C. A., Green, C. E., LeValley, S. E., and Stephens, R. J. (1982). Characterization of isolated Fe-loaded rat hepatocytes prepared by collagenase perfusion. *In Vitro* **18**, 945–951.

Tyson, C. A., Hawk-Prather, K., Story, D. L., and Gould, D. H. (1983a). Correlations of *in vitro* and *in vivo* hepatotoxicity for five haloalkanes. *Toxicol. Appl. Pharmacol.* **70**, 289–302.

Tyson, C. A., Story, D. L., and Stephens, R. J. (1983b). Ultrastructural changes in isolated rat hepatocytes exposed to different CCl_4 concentrations. *Biochem. Biophys. Res. Commun.* **114**, 511–517.

van Rossum, G. D. V. (1972). The relation of sodium and potassium ion transport to the respiration and adenine nucleotide content of liver slices treated with inhibitors of respiration. *Biochem. J.* **129**, 427–438.

Wendel, A., and Dumelin, E. E. (1981). Hydrocarbon exhalation. *Methods Enzymol.* **77**, 10–15.

Wroblewski, F., and LaDue, J. S. (1955). Lactic dehydrogenase activity in blood. *Proc. Soc. Exp. Biol. Med.* **90**, 210–213.

Zawydiwski, R., and Duncan, R. (1978). Spontaneous ^{51}Cr release by isolated rat hepatocytes: An indicator of membrane damage. *In Vitro* **14**, 707–714.

Zbinden, G., Elsner, J., and Boelsterli, U. A. (1984). Toxicological screening. *Regul. Toxicol. Pharmacol.* **4**, 275–286.

Zimmerman, H. J. (1978). "Hepatotoxicity—the Adverse Effects of Drugs and Other Chemicals on the Liver." Appleton, New York.

7

Metals, Hepatocytes, and Toxicology

CURTIS D. KLAASSEN* AND NEILL H. STACEY†

*Department of Pharmacology, Toxicology and Therapeutics
School of Medicine
The University of Kansas Medical Center
Kansas City, Kansas 66103

†National Occupational Health and Safety Commission
The University of Sydney
Sydney, NSW, 2006, Australia

I. INTRODUCTION

Hepatocytes, both freshly isolated suspensions and primary cultures, have become recognized as useful tools in studies pertaining to many aspects of liver function or dysfunction. The suitability of such preparations for investigations of a toxicological nature has recently been reviewed (Grisham, 1979; Schwarz and Greim, 1981; Klaassen and Stacey, 1982). Only data obtained from either of the above tissue preparations will be discussed in this review, although it is acknowledged that many other cellular systems have been used in studies on metal toxicity, including cell lines originally derived from liver (see, e.g., Potter and Matrone, 1977; Jacobs *et al.*, 1978; Rudd and Herschman, 1978; Alexander *et al.*, 1979; Skilleter and Paine, 1979; Walum and Marchner, 1983).

Some metals, such as cadmium and mercury, are generally considered only from a toxicological point of view as they are not essential for the well being of the organism. However, other metals, such as iron and copper, are essential for a viable organism but present significant health problems when in excess. Still other metals, for example calcium and zinc, although rarely considered as presenting a direct toxicological threat to an organism, may be intimately involved,

159

TABLE I

Studies of Metals with Hepatocytes in Suspension or Primary Culture

Metal	Suspension (S) or culture (C)	Primary aspect investigated	Reference
Arsenic	C	Toxicity	Inmon et al. (1981)
	C	Uptake	Lerman et al. (1983)
	S	Toxicity	Story et al. (1983)
Cadmium	C	Toxicity	Acosta and Sorensen (1983)
	S	Uptake	Cain and Skilleter (1980)
	S	Chelator	Cantilena et al. (1983)
	C	Uptake	Caperna and Failla (1984)
	S	Toxicity	Din and Frazier (1985)
	C	Uptake	Failla et al. (1979)
	C	Uptake	Gerson and Shaikh (1982)
	C	Uptake	Gerson and Shaikh (1984)
	S	Mechanism	Hidalgo et al. (1978)
	C	Toxicity	Inmon et al. (1981)
	C	Toxicity	Morselt et al. (1983a)
	C	Mechanism	Morselt et al. (1983b)
	S	Mechanism	Müller (1983, 1986a,b)
	S	Mechanism	Müller and Ohnesorge (1982)
	S	Mechanism	Müller and Ohnesorge (1984)
	C	Toxicity	Puvion and Lange (1980)
	C	Toxicity	Puvion and Dutilleul and Puvion (1981)
	C	Toxicity	Santone et al. (1982)
	S	Mechanism	Sciortino et al. (1982)
	S	Mechanism	Stacey (1986a,b)
	S	Mechanism	Stacey and Kappus (1982a)
	S	Uptake	Stacey and Klaassen (1980)
	S	Mechanism	Stacey and Klaassen (1981a)
	S	Mechanism	Stacey and Klaassen (1981b)
	S	Mechanism	Stacey et al. (1980)
	S	Toxicity	Story et al. (1983)
Calcium	C	Mechanism	Acosta and Sorensen (1983)
	S	Mechanism	Albano et al. (1985)
	S	Mechanism	Bellomo et al. (1982)
	C	Mechanism	Chenery et al. (1981)
	S	Mechanism	Di Monte et al. (1984)
	C	Mechanism	Farber (1981)
	S	Mechanism	Fariss and Reed (1985)
	S	Mechanism	Fariss et al. (1985)
	S	Mechanism	Jewell et al. (1982)
	C	Mechanism	Long and Moore (1986)
	S	Mechanism	Maridonneau-Parini et al. (1986)
	S	Mechanism	Moore et al. (1985)

(continued)

TABLE I (*Continued*)

Metal	Suspension (S) or culture (C)	Primary aspect investigated	Reference
	S	Mechanism	Okuno et al. (1983)
	S	Mechanism	Richelmi et al. (1984)
	C	Mechanism	Schanne et al. (1979)
	S	Mechanism	Smith and Sandy (1985)
	S	Mechanism	Smith et al. (1981)
	S	Mechanism	Stacey and Klaassen (1982)
	C	Mechanism	Starke et al. (1986)
	S	Mechanism	Thor et al. (1984)
Chromium	C	Toxicity	Inmon et al. (1981)
	S	Interaction	Stacey and Klaassen (1981a)
	S	Mechanism	Stacey and Klaassen (1981b)
	S	Interaction	Stacey and Klaassen (1981e)
Cobalt	C	Toxicity	Inmon et al. (1981)
Copper	S	Mechanism	Stacey and Klaassen (1981b)
	S	Mechanism	Stacey and Klaassen (1981e)
	C	Uptake	Weiner and Cousins (1980)
Iron	C	Chelator	Baker et al. (1984)
	S	Mechanism	Cheeseman et al. (1984)
	S	Mechanism	Chiarpotto et al. (1981)
	Both	Mechanism	Doolittle and Richter (1981)
	S	Uptake	Grohlich et al. (1977)
	S	Uptake	Grohlich et al. (1979)
	S	Mechanism	Hultcrantz et al. (1983)
	S	Mechanism	Koster and van Berkel (1983)
	C	Chelator	Octave et al. (1983)
	C	Uptake	Page et al. (1984)
	S	Mechanism	Poli et al. (1979)
	C	Uptake	Rama et al. (1981)
	C	Uptake	Sibille et al. (1981)
	S	Uptake	Smith and Morgan (1981)
	S	Mechanism	Stacey and Kappus (1982a)
	S	Mechanism	Stacey and Klaassen (1981a)
	S	Mechanism	Stacey and Klaassen (1981b)
	S	Mechanism	Stacey and Klaassen (1981e)
	S	Mechanism	Stacey and Priestly (1978)
	C	Mechanism	Starke and Farber (1985)
	C	Uptake	Tyson et al. (1982)
	S	Uptake	Young and Aisen (1980)
	S	Uptake	Young and Aisen (1981)
	C	Uptake	Zaman and Verwilghen (1981)

(*continued*)

TABLE I *(Continued)*

Metal	Suspension (S) or culture (C)	Primary aspect investigated	Reference
Lead	S	Toxicity	Guttas *et al.* (1983)
	C	Toxicity	Pounds *et al.* (1982a)
	C	Toxicity	Pound *et al.* (1982b)
	S	Interaction	Stacey and Klaassen (1981a)
	S	Toxicity	Stacey and Klaassen (1981b)
Manganese	S	Mechanism	Stacey and Kappus (1982a)
	S	Interaction	Stacey and Klaassen (1981a)
	S	Toxicity	Stacey and Klaassen (1981b)
	S	Interaction	Stacey and Klaassen (1981e)
Mercury	S	Chelator	Aaseth *et al.* (1981)
	S	Uptake	Alexander and Aaseth (1982)
	C	Uptake	Gerson and Shaikh (1982)
	C	Uptake	Gerson and Shaikh (1984)
	C	Toxicity	Inmon *et al.* (1981)
	S	Mechanism	Stacey and Kappus (1982b)
	S	Mechanism	Stacey and Klaassen (1981b)
	S	Toxicity	Story *et al.* (1983)
Nickel	C	Toxicity	Inmon *et al.* (1981)
	S	Interaction	Stacey and Klaassen (1981a)
	S	Toxicity	Stacey and Klaassen (1981b)
Selenium	S	Mechanism	Anundi *et al* (1982)
	S	Mechanism	Hill and Burk (1982)
	S	Mechanism	Hogberg and Kristoferson (1979)
	C	Toxicity	Inmon *et al.* (1981)
	S	Interaction	Stacey and Klaassen (1981a)
	S	Toxicity	Stacey and Klaassen (1981b)
	S	Mechanism	Stahl *et al.* (1984)
	S	Toxicity	Story *et al.* (1983)
Tin	S	Toxicity	Wiebkin *et al.* (1982)
Vanadium	S	Mechanism	Agius and Vaartjes (1982)
	C	Toxicity	Inmon *et al.* (1981)
	S	Toxicity	Seglen and Gordon (1981)
	S	Mechanism	Stacey and Kappus (1982a)
	S	Mechanism	Stacey and Klaassen (1981b)
	S	Mechanism	Stacey and Klaassen (1981d)
Zinc	C	Uptake	Failla and Cousins (1978a,b)
	C	Uptake	Failla and Cousins (1978a)
	S	Mechanism	Sciortino *et al.* (1982)
	S	Interaction	Stacey and Klaassen (1981a)
	S	Toxicity	Stacey and Klaassen (1981b)
	S	Uptake	Stacey and Klaassen (1981c)
	S	Interaction	Stacey and Klaassen (1981e)
	C	Interaction	Zaman and Verwilghen (1981)

TABLE II

Studies Using Hepatocyte Suspensions or Cultures in Which Toxic Responses Have Been Demonstrated

Metal	Exposure conditions (concentration/time)	Parameter(s) of toxicity	Reference
Arsenic	EC_{50} 0.044 mM/20 hr LECT 0.030 mM/20 hr	Trypan blue	Inmon et al. (1981)
	1–10 mM/2–5 hr	Trypan blue GOT(AST) release LDH release Urea	Story et al. (1983)
Cadmium	50–200 μM/2–4 hr	Trypan blue LDH release Urea	Acosta and Sorensen (1983)
	5–40 μM/6 hr	Inhibition of protein synthesis	Din and Frazier (1985)
	EC_{50} 0.013 mM/20 hr LECT 0.001 mM/20 hr	Trypan blue	Inmon et al. (1981)
	1 μg/ml $CdCl_2$ (5 μM)/1 hr	Chromatin condensation Trypan blue	Morselt et al. (1983a)
	50–100 μM/30–60 min	Trypan blue	Müller and Ohnesorge (1982); Müller (1983)
	10–100 μM/15–60 min	Trypan blue Oxygen consumption ATP/ADP Lactate/pyruvate	Müller and Ohnesorge (1984); Müller (1986a)
	50 μM/60 min	Trypan blue LDH release NADH oxidation Nitrophenylphosphate uptake Fluorescence of 8-anili-nonaphthalenesulfon-ate-1	Müller (1984)
	1 μ/ml $CdCl_2$ (5 μM/30 min–4 hr	Perichromatin granules	Puvion and Lange (1980)
	50–400 μM/60 min	Trypan blue LDH release Lactate/pyruvate Urea	Santone et al. (1982)
	50–400 μM/10–75 min	Intracellular K^+ AST release Lactate/pyruvate	Stacey (1986a,b); Stacey et al. (1980); Stacey and Klaassen (1981a,1981b, 1982); Stacey and Kappus (1982a)

(continued)

TABLE II (*Continued*)

Metal	Exposure conditions (concentration/time)	Parameter(s) of toxicity	Reference
	0.01–1 mM/2 hr	Trypan blue GOT(AST) release LDH release Urea ATP	Story *et al.* (1983)
Chromium (3+)	EC$_{50}$ 2.196 mM/20 hr LECT 2.000 mM/20 hr	Trypan blue	Inmon *et al.* (1981)
Cobalt	EC$_{50}$ 2.512 mM/20 hr LECT 0.500 mM/20 hr	Trypan blue	Inmon *et al.* (1981)
Copper	25–200 μM/15–75 min	Intracellular K$^+$ AST release	Stacey and Klaassen (1981b, 1982)
Iron (+3)	1 mM/45 min	Intracellular K$^+$	Stacey and Klaassen (1981b)
Mercury	ED$_{50}$ 0.159 mM/20 hr LECT 0.040 mM/20 hr	Trypan blue	Inmon *et al.* (1981)
	25–100 μM/20–60 min 50–200 μM/20–45 min	LDH release Intracellular K$^+$ AST release	Stacey and Kappus (1982b) Stacey and Klaassen (1981b)
	0.01–0.2 mM/2–5 hr	GOT(AST) release	Story *et al.* (1983)
Nickel	EC$_{50}$ 3.578 mM/20 hr LECT 2.000 mM/20 hr	Trypan blue	Inmon *et al.* (1981)
Selenium	50–250 μM/1–4 hr	Latency of LDH (NADH permeability)	Anundi *et al.* (1982; 1984)
	EC$_{50}$ 0.072 mM/20 hr LECT 0.001 mM/20 hr	Trypan blue	Inmon *et al.* (1981)
	1–10 mM/2–5 hr	Trypan blue GOT(AST) release LDH release Urea	Story *et al.* (1983)
Tin (organo)	10–100 μM/1 hr	Trypan blue Latency of LDH	Wiebkin *et al.* (1982)
Vanadium	EC$_{50}$ 0.042 mM/20 hr LECT 0.0049 mM/20 hr	Trypan blue	Inmon *et al.* (1981)
	20 mM/2 hr 400 μM/1–6 hr	Trypan blue Intracellular K$^+$ AST release	Seglen and Gordon (1981) Stacey and Klaassen (1981d)

through their presence or absence, in cellular dysfunction. Each of the above situations will be considered in this review.

This chapter is divided into five major sections: uptake and efflux of metals from hepatocytes, use of hepatocytes to determine the effectiveness of chelators, toxicity of metals using hepatocytes, toxic interactions among metals in hepatocytes, and use of hepatocytes to determine the mechanism of toxicity of metals. For easy reference to studies on particular metals, the reader is referred to Table I. Similarly, those studies which have reported positive toxic effects for metals are listed in Table II, with concentrations, time of exposure, and the parameter(s) of toxicity assayed.

II. UPTAKE AND EFFLUX OF METALS

Isolated hepatocytes have been widely used to investigate the cellular uptake of a variety of compounds (Klaassen and Stacey, 1982; Schwenk, 1980). The hepatocyte suspensions have two major advantages over alternative systems. First, they are functionally very similar to *in vivo* liver which gives them an advantage over cultured cell lines and liver slices. Second, isolated hepatocyte preparations allow rapid and repeated sampling at a frequency not as readily feasible with the intact organ or attached cells.

Apart from defining the characteristics of uptake per se, uptake–efflux studies are of importance from other aspects. For example, as approximately half of an intravenous or absorbed oral dose of cadmium is rapidly accumulated by the liver, alterations in uptake may cause changes in the toxicity to the liver, as well as other target organs such as testes and kidney, due to different concentrations of cadmium in the particular organ.

Efflux of chemicals from isolated hepatocytes has not been widely investigated, but some reports indicate that excretion can be studied (Horne *et al.*, 1979; Craik and Elliott, 1979; Le Cam *et al.*, 1979; Schwenk *et al.*, 1981) and that it probably occurs in a manner similar to *in vivo* excretion (Vonk *et al.*, 1978; Schwarz *et al.*, 1979; Blom *et al.*, 1981; Gewirtz *et al.*, 1981, 1982; Tarao *et al.*, 1982).

Freshly prepared isolated hepatocyte suspensions have some disadvantages in this respect. For example, isolated hepatocytes lack the polarity of *in vivo* hepatocytes (Schwenk, 1980; Gebhardt and Jung, 1982). This contrasts with primary cultures of hepatocytes which re-form bile canaliculi similar to the intact organ while retaining many of their differentiated functions. Indeed, Gebhardt and Jung (1982) found that sodium fluorescein was taken up and subsequently accumulated in bile canaliculi-like structures in hepatocyte cultures. Thus, there seems to be a good correlation, especially qualitatively, between uptake studies with hepatocyte suspensions and perfused liver where comparative data are available (Schwenk, 1980).

A. Method for Rapid Separation of Hepatocyte

For studying uptake of chemicals into hepatocytes, a method for rapid separation of cells from incubation media is needed. To achieve this, a method for separating fluid layers with silicone oil has been useful. The technique consists of layering an aliquot (200 μl) of cell suspension containing radiolabeled chemical onto 50 μl silicone oil (density 1.02, from Wacker-Chemie GmbH, Munich, or density 1.05 from Aldrich Chemical Co., Milwaukee, Wisconsin) which has been previously layered over 50 μl 3 M KOH in a 400 μl polyethylene microfuge tube. The samples are then centrifuged at 11,400 rpm (9400 g) for 10 sec in a tabletop microfuge (Beckman Instruments, Fullerton, California). The cells spin through the oil layer in 1–2 sec. The amount of chemical taken up by the hepatocytes can then be determined by cutting the tube at the oil interface and counting the bottom portion in a scintillation spectrometer and relating it to protein concentration. Relevant kinetic assessment of the movement of the chemical can then be made. This would be much more difficult to achieve using whole-organ techniques.

B. Uptake of Specific Metals

The uptake of several metals by hepatocytes in suspension or primary culture has been examined. Cadmium has been shown to be taken up in a biphasic manner with a rapid initial phase and a slower second phase (Stacey and Klaassen, 1980) (Fig. 1). The first phase shows some characteristics of being carrier mediated while there is no indication of carrier involvement for the second phase. These observations are consistent with the findings of Frazier and Kingsley (1976, 1977) who used isolated perfused rat livers.

Zinc was found to interfere with the first phase of cadmium uptake (Stacey and Klaassen, 1980), demonstrating another way in which hepatocytes can be used to study aspects of metal toxicity. While this seems to be in agreement with the data of Kingsley and Frazier (1979) from isolated perfused rat livers, there are inconsistencies at long exposure times. Explanation of such differences may well increase our understanding of the overall uptake process for cadmium.

Uptake of cadmium in primary cultures of hepatocytes has also been evaluated (Failla et al., 1979). This study examined uptake over hours rather than minutes, as well as subcellular distribution and induction of metallothionein synthesis. If the experimental hypothesis requires incubation times in excess of about 6 hr, the use of attached cultures rather than suspensions is necessary. Gerson and Shaikh (1984) employed hepatocytes in primary culture to compare the uptake of cad-

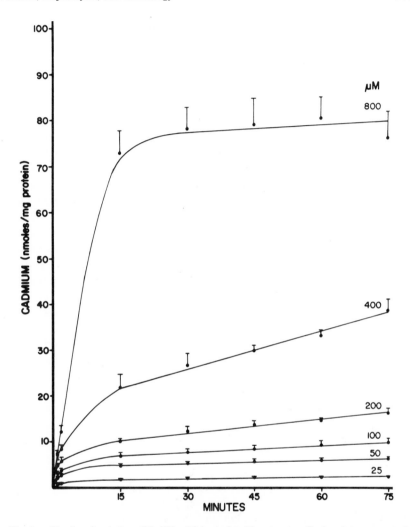

Fig. 1. Uptake of cadmium (25–800 μ*M*) by isolated hepatocytes. Each point represents the mean (±SE) of four or five separate hepatocyte preparations. [Reproduced with permission from Stacey and Klaassen (1980).]

mium and mercury. Their data indicate that sulfhydryl groups are involved in uptake of cadmium but not mercury. Another approach to the study of cadmium uptake by liver cells has been documented by Cain and Skilleter (1980) who pretreated rats with radiolabeled cadmium and examined its distribution in subsequently isolated parenchymal and nonparenchymal cells. They reported that the parenchymal cells accumulated cadmium more readily than nonparenchymal cells. It is not clear, however, if volume differences in cells were taken into account when ascribing a selective uptake to the parenchymal cells. In a further study, these authors compared the effects of monothiols and dithiols on cadmium accumulation by rat liver parenchymal and Kupffer cells (Skilleter and Cain, 1981). Their data indicate that monothiols reduce the availability of cadmium for uptake by both cell types but that dithiols stimulate the rate of cadmium accumulation by both cell types.

Isolated hepatocytes have been used to gain further insight into the dynamics of methyl mercury disposition (Aaseth et al., 1981; Alexander and Aaseth, 1982). Complexation with glutathione (GSH) was found to decrease methyl mercury uptake by hepatocytes which corresponds with and extends the in vivo data. Lead, too, has received some attention, but only with primary cultures (Pounds et al., 1982b). The uptake of lead was found to be biphasic over a period of hours. These authors also studied the subcellular distribution under various conditions to gain further insight into the kinetic compartments. They concluded that the kinetics of the cellular disposition of lead is similar to that for calcium in hepatocytes. A subsequent study, however, cautions that cell fractionation techniques may provide data that reflect metal redistribution after cell disruption (Mittelstaedt and Pounds, 1984). Primary cultures have also been used to investigate the uptake and metabolism of arsenic (Lerman et al., 1983). This study showed that arsenic in the 3^+ rather than 5^+ oxidation state is taken up by parenchymal cells.

In conjunction with studies on metabolism and toxicity of selenium compounds, amino acid uptake in the presence of selenite has been found to be inhibited (Hogberg and Kristoferson, 1979), and the uptake of selenite itself has been examined (Stahl et al., 1984). Uptake of selenite occurred quite rapidly in the initial stages but cellular levels then decreased, reflecting formation of volatile selenium metabolites.

The accumulation and metabolism of copper has also been investigated with the aid of primary cultures (Weiner and Cousins, 1980). Whereas some similarities to the handling of zinc and cadmium were shown, some differences also became apparent, suggesting unique regulatory mechanisms that differentiate among these metals (Weiner and Cousins, 1983a,b). Collectively, the results suggest glucocorticoids increase hepatic zinc accumulation through augmented metallothionein synthesis, whereas they increase copper output by increasing secretion as ceruloplasmin. The kinetics of copper uptake by hepatocytes have been investigated by Schmitt et al. (1983), who reported that the process was saturable and that zinc could inhibit the process. In a subsequent study it was shown that albumin decreased the uptake of copper because it decreased the free concentration of copper in the media, whereas histidine increased copper uptake because it mobilized copper from albumin (Darwish et al., 1984a). In brindled mice (a model of Menkes disease), copper uptake and efflux were similar to control mice but the steady-state accumulation of copper was found to be less in hepatocytes isolated from brindled mice as compared to normals (Darwish et al., 1983). This correlates well with in vivo observations and strengthens the existence of the primary defect being at the cellular level in the liver of the mouse mutants.

Zinc uptake and metabolism by hepatocytes in primary culture have also been investigated (Failla and Cousins, 1978a,b). Zinc was shown to be accumulated by a process that was temperature, energy, and sulfhydryl group dependent, saturable and subject to regulation by glucocorticoids. Zinc uptake has also been examined in isolated hepatocytes, with uptake being quantitated at much earlier exposure times (Stacey and Klaassen, 1981c). The uptake was found to be biphasic, the first of these showing characteristics of being carrier mediated. No energy dependence of either phase could be demonstrated in this system.

Iron, which is both an essential metal and hepatotoxic in high concentrations, has been the subject of a number of studies relating to its uptake and metabolism by liver cells. This may reflect the complexity of the disposition of this essential metal in the liver. Ferrous ions were found to be taken up by a simple diffusion process, whereas ferric ion uptake appeared to be saturable (Grohlich et al., 1977, 1979). These authors also investigated uptake of iron from the transferrin-bound form, showing the involvement of a carrier in at least part of this process. Other studies have supported the role of receptors in transferrin–iron uptake (Young and Aisen, 1980, 1981; Cole and Glass, 1983), but this has not gone undisputed (Sibille et al., 1981). A more recent study has indicated that only a small part of the total iron uptake by hepatocytes in primary cultures is due to transferrin receptors (Page et al., 1984). A study of the interaction of zinc with iron metabolism showed that, at high concentrations, zinc could depress the uptake of iron from the transferrin complex (Zaman and Verwilghen, 1981). Pretreatment of animals with iron and subsequent liver cell isolation has shown that 95% of the iron is found in parenchymal and about 5% in nonparenchymal cells (Tyson et al., 1982; Hultcrantz et al., 1983).

C. Efflux of Metals

Efflux of metals from hepatocytes has not been as intensively studied as uptake. Iron again has attracted the majority of attention with some emphasis being placed on aspects of mechanism, such as the role of apotransferrin in the release process (Young and Aisen, 1980, 1981; Rama et al., 1981; Baker et al., 1981; Octave et al., 1983). The efflux of other metals from hepatocytes has received only minor attention, with some aspects of lead (Pounds et al., 1982b), cadmium (Failla et al., 1979), and copper (Weiner and Cousins, 1980; Darwish et al., 1984b) release being investigated. This apparent lack of attention to efflux mechanisms is perhaps surprising considering that the biliary route is a relatively common method of excretion for metals (Klaassen, 1976; Klaassen and Watkins, 1984).

III. USE OF HEPATOCYTES TO STUDY CHELATORS

A potentially important application of *in vitro* hepatocyte techniques in the study of metal toxicity concerns the investigation of chelators to remove metals from preloaded cells. Concurrently, any effects on cellular viability can be ascertained. A few studies have addressed this topic of chelators with hepatocytes. The ability of several chelators to remove cadmium from hepatocytes was found to compare to *in vivo* results reasonably well (Cantilena *et al.*, 1983). The chelators, diethylenetriaminepentaacetic acid, ethylenediaminetetraacetic acid, 2,3-dimercaptosuccinic acid, diethyldithiocarbamate, and 2,3-dimercaptopropanol, removed cadmium from the hepatocytes *in vitro* similar to that in the intact animal. This was evident in hepatocytes isolated from untreated and metallothionein-induced rats. Three other chelators, nitrilotriacetic acid, *dl*-penicillamine, and 2,3-dimercaptosuccinic acid did not correlate as well but the differences could be explained on a concentration basis.

The ability of chelators to ameliorate copper-induced toxicity in isolated hepatocytes has also been examined (Stacey and Klaassen, 1981e). When added to the cell suspension just before copper chloride, diethyldithiocarbamate, and penicillamine effectively blocked the toxic response, reflecting the chelating properties of these agents. Further studies with copper-loaded hepatocytes would provide additional useful information.

Iron is the only other metal to have received significant attention in this regard. Baker *et al.* (1981) compared the ability of three known chelators to extract iron from hepatocytes isolated from preloaded rats, concluding that hepatocytes provide a useful model for such studies. The potential for this type of study using *in vivo* loading of tissues has also been recognized by Tyson *et al.* (1982). Enhanced removal of iron from *in vitro* preloaded cells was found in response to desferrioxamine by Octave *et al.* (1983). Tyson *et al.* (1984) used iron-loaded hepatocytes to investigate ionophore–polymeric chelator combinations for their ability to remove intracellular iron. Iron removal from preloaded cells was also studied by Baker *et al.* (1984), who showed that lipophilic chelators were more effective than their hydrophilic counterparts in enhancing iron release, which led the authors to the conclusion that the lipophilic chelators may provide a clinical treatment to replace desferrioxamine. Part of the experimental design included an examination of potential chelators on a structure–activity relationship basis. This demonstrates a general approach where such *in vitro* systems may provide a useful tool in developmental studies.

In view of such potential applications, it is surprising that more emphasis has not been placed on screening chelators in these *in vitro* systems. The obvious advantage is the number of chelators and concentrations that can be investigated in a given time as compared to *in vivo* studies. Confirmatory *in vivo* studies of the most promising chelators would still be required, however. It would also seem prudent to investigate chelator removal of metals after pretreatment of rats on a subchronic basis as well as acute administration. This would have significant clinical application, particularly for a metal such as cadmium.

Metallothionein is an inducible low-molecular-weight protein that can be viewed as a chelator in that it has the ability to sequester some metal ions inside the cell. It appears to play a protective role in the toxicity of cadmium. Hepatocytes isolated from metallothionein-induced rats were less sensitive to the inhibitory effects of cadmium on synthesis of cellular proteins (Din and Frazier, 1985). The induction of this protein in response to cadmium has been investigated at the cellular level using hepatocyte suspensions and cultures (Hidalgo *et al.*, 1978; Gerson and Shaikh, 1982; Bracken and Klaassen, 1985). The role of nonparenchymal vs parenchymal cells in metallothionein metabolism has been facilitated by the application of the isolated liver cell techniques (Sciortino *et al.*, 1982; Cain and Skilleter, 1983; Caperna and Failla, 1984). Due to the important role of this sulfhydryl-rich protein in cadmium toxicity, studies aimed at understanding its cellular functions are of obvious importance.

IV. TOXICITY OF METALS

Hepatocytes in suspension or in primary monolayer culture are being increasingly used to study metal-induced toxicity. These *in vitro* systems are particularly useful as a tool in screening metals for various effects relevant to toxicity. Furthermore, mechanisms of the toxic response to metals have been extensively studied with hepatocytes *in vitro*. As the system is by definition cellular, it lends itself particularly well to the examination of the pathways by which cellular injury is produced. As well as metals which are primarily considered for their toxicity, this section will also deal with essential metals which have deleterious effects when in excess and are of significant importance to the human organism. However, only investigations pertaining to toxicity will be considered.

A. Parameters of Toxicity

Generally, two types of parameters have been used in assessing the toxicity of chemicals in hepatocytes *in vitro*. Membrane integrity, as measured by enzyme or potassium ion leakage and uptake of vital dyes, has been the method receiving the most frequent use. However, albeit somewhat less frequently, parameters of metabolic integrity such as urea production and lactate-to-pyruvate ratio have also been employed. Both parameters are important in assessing the toxic response. Leakage of cytoplasmic enzymes indicates a severe disruption of the cell membrane and is generally taken to indicate cell death. Interference with metabolic parameters may not necessarily lead to death of the cell but may be involved in a toxic effect on the organism as a whole. Of course, interference with intermediary metabolism may eventually lead to death of the cell as well. It is considered prudent that both structural and metabolic parameters be included in a protocol to examine toxicity, at least initially. (See also Chapter 6.)

With regard to choice of parameters for assessment of cytotoxicity, leakage of cytoplasmic enzymes and loss of potassium ions from the cell suspensions provide a relatively simple and yet effective screening procedure. For cultures, counting of attached viable cells, in conjunction with the above parameters, may also be of some value.

The silicone–oil layer technique is again useful, as perchloric acid ($0.7\,M$) can be used as the bottom layer into which only the cells gravitate. This bottom layer can then be quantitated for potassium ion concentration by flame photometry, while the fraction above the oil layer can be used to determine enzyme leakage. An experimenter needs to be vigilant to ensure that the chosen assessment procedure is not being inhibited directly by the metal under test. For example, we found that alanine aminotransferase was inhibited by cadmium. One should also be aware that enzymes in physiological media may not be as stable as in plasma under frozen storage (Stacey *et al.*, 1978).

In using metabolic parameters as a general procedure for assessing toxicity, it is the authors' experience that urea synthesis and lactate-to-pyruvate ratio determinations provide parameters somewhat different in nature, but both seem to be sensitive to various toxicants. In conjunction with the use of two parameters of structural integrity, estimation of at least one parameter of metabolic function should be carried out. This may be particularly wise in the early stages of a new project. If little extra information is being obtained, the assay need not necessarily be continued through the entire project.

B. Cytotoxicity Studies

Studies have been undertaken which have compared the cytotoxic effects of various metals on hepatocytes *in vitro*. Within the same experimental protocol, copper, mercury, and cadmium were found to consistently cause a cytotoxic response, while chromium, manganese, zinc, nickel, lead, selenium, vanadium, and iron were without consistent effect (Stacey and Klaassen, 1981b). It should be noted that this serves only as a comparison among metals. Some of those without toxicity in this study have elicited cell damage under other conditions. In assessing attached hepatocytes as a system for toxicity testing, Inmon *et al.* (1981) compared several metals. Interestingly, they found, with a 20-hr exposure period that cadmium, on an EC_{50} basis, was the most cytotoxic metal followed by vanadium, arsenic, selenium, mercury, chromium, cobalt, and nickel. Among a variety of chemicals, Story *et al.* (1983) investigated four metals for their ability to invoke a toxic response in hepatocytes. Cadmium, mercury, arsenite, and selenate were found to result in loss of cytoplasmic enzymes from the cells.

The cytotoxicity of cadmium in primary cultures has been comprehensively investigated by Santone *et al.* (1982) who showed that concentrations of 50–400 μM produced evidence of toxicity measured by parameters of both structural and metabolic integrity. The same laboratory has reported similar toxicity using morphometric analysis with concentrations of 50–200 μM cadmium (Sorensen and Acosta, 1984). It should be noted that the hepatocytes were from young rather than adult rats, which may make comparison difficult as there are known differences between livers of young and adult animals. Indeed, ouabain uptake in cells isolated from 12-day-old rats was so low that it could not be determined, whereas it is readily taken up by hepatocytes from adult rats (Stacey and Klaassen, 1979). Nevertheless, on a quantitative basis, cadmium cytotoxicity in cultured cells (Santone *et al.*, 1982) and freshly prepared suspensions (Stacey *et al.*, 1980) was similar. A number of tests that quantitate membrane integrity have been compared, and it has been found that cadmium alters them at similar concentrations (Müller, 1984). The appearance of perichromatin granules in hepatocyte cultures has been found in response to cadmium (Puvion and Lange, 1980). After exposure of the hepatocyte cultures to cadmium, condensation of chromatin was found to be an early event preceding membrane leakage (Morselt *et al.*, 1983a,b).

Several studies have investigated effects of cadmium on metabolic parameters, either alone or in conjunction with an estimation of toxicity on a structural basis. Cadmium was found to increase lactate-to-pyruvate ratio in hepatocyte suspensions before evidence of structural damage (Stacey *et al.*, 1980). Similar observations were also made by Santone *et al.* (1982), who also found urea synthesis to be sensitive to the presence of cadmium. Müller and Ohnesorge (1984) examined the effects of cadmium on metabolic parameters of isolated hepatocytes and indicated that interference with mitochondrial function was an early event in cadmium-induced toxicity in isolated rat hepatocytes. In studies on chromatin, cadmium was found to induce a preferential inhibition of nucleolar RNA synthesis, and it was suggested that perichromatin granules may develop during transcription by coiling of nascent RNA fibrils (Puvion and Lange, 1980; Puvion-Dutilleul and Puvion, 1981).

In vivo hepatotoxic effects in response to cadmium have been mentioned in previous studies (Stowe *et al.*, 1972; Cook *et al.*, 1974; Singhal *et al.*, 1974; Hoffmann *et al.*, 1975), but only recently has the question been comprehensively examined. Dudley *et al.* (1982) showed that a high acute dose of cadmium causes massive hepatic necrosis in the rat. An approximate comparison of dose *in vivo* to concentration *in vitro* indicates that cell death occurs at similar levels. This demonstrates a good quantitative as well as qualitative correlation of the toxicity studies using hepatocyte suspensions with those in the whole animal.

Lead was examined for its ability to produce cytotoxicity in hepatocytes in culture over a 24-hr incubation period, but none was detected (Pounds *et al.*, 1982b). Lead has been reported to interfere with the homeostasis of cellular calcium in cultures of hepatocytes, particularly affecting calcium in the deep intracellular compartment, which includes mitochondria (Pounds *et al.*, 1982a). The intact membrane of isolated hepatocytes has also been used to put into perspective alterations in respiratory rate induced by lead in tissue homogenates, since intact cells were relatively resistant to respiratory inhibition (Guttas *et al.*, 1983).

The toxic potential of organotin compounds has been compared in hepatocyte suspensions. Using parameters of structural integrity, triethyltin bromide was more toxic than diethyltin dichloride (Wiebkin *et al.*, 1982). The triethyl- and diethyltin salts were found to reduce oxygen consumption, decrease ATP levels, and inhibit xenobiotic biotransformation in isolated hepatocytes (Wiebkin *et al.*, 1982).

Vanadium has also been investigated in hepatocytes with regard to interference in metabolic functions. It was found to inhibit protein synthesis at high concentrations (Seglen and Gordon, 1981) and to stimulate fatty acid synthesis (Agius and Vaartjes, 1982). One study has reported that chromium (VI) inhibits mitochondrial respiration of isolated hepatocytes and may be important in producing cytotoxicity (Ryberg and Alexander, 1984). Cobalt has been shown to inhibit heme synthesis in hepatocyte cultures, which is consistent with *in vivo* data (Lodola, 1981).

No convincing evidence exists that iron is toxic to isolated hepatocytes (Stacey and Priestly, 1978; Stacey and Klaassen, 1981b; Stacey and Kappus, 1982a). This metal has been examined in hepatocytes at rather high concentrations but generally for only short exposure periods, which may account for a general lack of toxicity. Cheeseman *et al.* (1984) reported that iron caused a decline in cytochrome *P*-450 levels in hepatocytes but did not reduce cell viability. Hepatocytes isolated from iron-loaded rats have also been used to study lysosomal enzymes in response to treatment (Hultcrantz *et al.*, 1983). No effect could be demonstrated. Other aspects relating to iron metabolism, with possible bearing on toxic manifestations of the metal, have also been reported (Doolittle and Richter, 1981; Smith and Morgan, 1981). A role for endogenous iron in the killing of hepatocytes by hydrogen peroxide has been documented (Starke and Farber, 1985). This raises the possibility that iron may be involved in cell death induced by other chemicals.

Copper, which has hepatotoxic properties in overexposure and in Wilson's disease, was shown to be toxic to isolated hepatocytes (Stacey and Klaassen, 1981b,e). Whereas on a quantitative basis it is difficult to compare *in vitro* to *in vivo* hepatotoxicity in response to copper, the existing studies certainly demonstrate that hepatocellular death can occur in both systems. Further studies are warranted in cultures with longer duration of exposure in order to further delineate the relevance of such studies to the *in vivo* situation.

At short exposure times, we did not observe a toxic response with sodium selenite (Stacey and Klaassen, 1981b). However, in a comprehensive study of the cytotoxic effects of selenite, Anundi *et al.* (1982; 1984) demonstrated a significant toxic response. Their data indicate that the toxicity is linked to selenite metabolism and that an insufficient NADPH supply could be a critical factor. Hepatocytes have also been used to investigate some aspects by which selenium can indirectly affect toxic responses (Hill and Burk, 1982).

It would seem that there is a very great scope for investigations in hepatocytes with essential metals that can either directly cause a cytotoxic response or can be indirectly involved through their absence. Overall, those metals which have been associated with a hepatotoxic response have shown a propensity to cause a toxic response in *in vitro* hepatocellular preparations. The data available for cadmium allow a quantitative as well as qualitative comparison and show a good correlation between *in vitro* and *in vivo* studies.

V. INTERACTIONS AMONG METALS

Many reports have documented interactions among metals, and these can be toxicologically relevant. Hepatocytes *in vitro* provide a valuable tool for studying such interactions at the cellular level. Two main aspects have already been addressed: the ability of one metal to interfere with the kinetics of a second metal and interactions having a direct effect on cytotoxicity. Zinc and cadmium have been found to inhibit each other's transport into hepatocytes both in culture and in suspension (Failla and Cousins, 1978a; Stacey and Klaassen, 1980, 1981c), whereas the uptake of cadmium was shown to enhance zinc efflux (Failla *et al.*, 1979). Zinc has also been reported to inhibit deposition of iron into cellular ferritin, whereas it inhibits actual uptake into cells only at high zinc concentrations (Zaman and Verwilghen, 1981). In contrast, zinc was without effect on copper accumulation by hepatocyte cultures (Weiner and Cousins, 1980).

The ability of other metals to protect against cadmium toxicity *in vivo* has been demonstrated in a variety of studies. The occurrence of such interactions at the cellular level has been investigated to a limited extent. Cadmium-induced cytotoxicity was found to be consistently ameliorated by salts of chromium, manganese, zinc, lead, and iron (Stacey and Klaassen, 1981a) (Fig. 2). The protection was not due to a decrease in uptake of cadmium by the cells. The protective effect of zinc was of interest because zinc protects against toxic effects of cadmium *in vivo*. Thus similar protection was found at the cellular level. In contrast, chromium, which is highly protective to the cells, had not been documented to have protective properties *in vivo*. We found that cadmium-induced toxicity could be inhibited by pretreatment of the animal with chromic chloride (Stacey *et al.*, 1983). Thus, an *in vivo* interaction of two metals was accurately predicted by *in vitro* observations. A

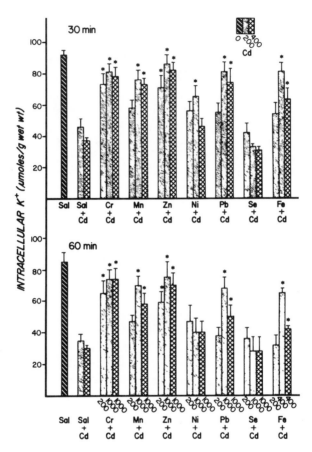

Fig. 2. Effects of Cd (200 or 400 μM) and Cd (200 or 400 μM) plus Cr, Mn, Zn, Ni, Pb, Se, or Fe (200, 400, or 1000 μM) on intracellular K^+ concentration of isolated rat hepatocytes after 30 or 60 min of incubation. Sal represents hepatocytes incubated with saline vehicle as the only addition. Columns represent means of 4–6 separate hepatocyte preparations and bars indicate SE. *, Significantly different from appropriate incubation with Cd only, $p < 0.05$. [Reproduced with permission from Stacey and Klaassen (1981a), copyright Hemisphere Publishing.]

critical role for calcium in the cytotoxic properties of cadmium and copper could not be demonstrated in hepatocyte suspensions (Stacey and Klaassen, 1982) or for cadmium with hepatocyte cultures (Acosta and Sorensen, 1983). This aspect of calcium and cytotoxicity will be dealt with in more detail in a following section.

Chromium was also found to protect against the toxic effects of copper in hepatocyte suspensions. However, zinc showed no evidence of the protective actions it displayed for cadmium-induced cytotoxicity (Stacey and Klaassen, 1981e). It would be most interesting to determine if similar observations could be made for the *in vivo* toxic actions of copper.

VI. MECHANISMS OF TOXICITY

Various mechanisms have been suggested as being responsible for the cellular toxicity of metals. These include interaction with sulfhydryl groups (Rothstein, 1959; Vallee and Ulmer, 1972; Chvapil, 1973; Webb, 1977), inhibition of mitochondrial function (Chvapil, 1973; Webb, 1977), association with lipid peroxidation (Chvapil, 1973; Willson, 1977; Kinter and Pritchard, 1977), and inhibition of adenosine triphosphatases (ATPases) (Vallee and Ulmer, 1972; Webb, 1977).

Several studies using hepatocytes have concentrated on the evaluation of lipid peroxidation as the mechanism of toxic response. In this process, free radicals attack the lipids in the cellular membranes and, if severe enough, cause cell death. With respect to assays for lipid peroxidation, it is considered that thiobarbituric acid (TBA) reactants provide a satisfactory reflection of the peroxidative events of the hepatocytes. However, it is wise to confirm the presence of peroxidized lipids with a second assay such as diene conjugation or ethane evolution.

Lipid peroxides are characteristically formed in biological membranes in response to iron, and this has been suggested to be involved in its toxic manifestations. Several studies have used iron, often in conjunction with ADP, to investigate lipid peroxidation in hepatocytes. While evidence for peroxidation has been found, a cytotoxic response has not been an accompanying feature (Stacey and Priestly, 1978; Stacey and Klaassen, 1981b; Cheeseman *et al.*, 1984). Iron-induced decrease in glucose-6-phosphatase and cytochrome *P*-450 is still produced in hepatocytes when lipid peroxidation has been inhibited (Poli *et al.*, 1979; Cheeseman *et al.*, 1984). As with cadmium, experiments with mercury (Stacey and Kappus, 1982b) and copper (Stacey and Klaassen, 1981e) have not supported a role for lipid peroxidation in the damaging effects of these metals to isolated hepatocytes. Results with vanadium, however, are less definitive and require further investigation to delineate the role that lipid peroxidation has in the cytotoxic effects of this metal (Stacey and Klaassen, 1981d).

Cadmium has been shown to elicit lipid peroxidation in isolated hepatocytes as determined by assay for thiobarbituric acid reacting substances and diene conjugation (Stacey *et al.*, 1980). However, concurrent treatment with antioxidants showed an inhibition of the lipid peroxidation without effect on the accompanying cytotoxicity (Fig. 3). This is strong evidence that lipid peroxidation is not the mechanism by which cadmium exerts its toxic effects in hepatocytes. A comparison of lipid peroxidation and cell injury in response to metals has shown a

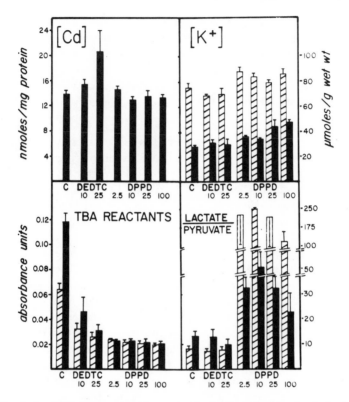

Fig. 3. Isolated hepatocytes exposed to antioxidant compounds alone (hatched columns) or antioxidant plus cadmium (200 μm) (solid columns) for a 45-min incubation. C, Hepatocytes incubated without chelator. Concentrations are ×10⁻⁶ M. Columns represent the mean of 4–6 separate preparations and the bars indicate the SE. Upper left, Uptake of cadmium into the hepatocytes and effects of the antioxidants. Bottom left, Formation of TBA reactants in hepatocyte suspensions and the effects of antioxidants in the absence and presence of cadmium. Upper right, Intracellular [K⁺] of isolated hepatocytes and the effects of the antioxidants in the absence and presence of cadmium. Bottom right, Lactate-to-pyruvate ratio of hepatocyte suspensions and the effects of antioxidants in the absence and presence of cadmium. [Reproduced with permission from Stacey *et al.* (1980).]

dichotomy between the two events, which also suggests that the lipid peroxidation in response to cadmium is not causally related to the cytotoxicity (Stacey and Klaassen, 1981a,b; Stacey and Kappus, 1982a). A similar conclusion using antioxidants and dietary status of liver donors has been reached by another laboratory (Müller and Ohnesorge, 1982; Müller, 1983). The effects of metals on lipid peroxidation in cell membranes in general have recently been reviewed by Christie and Costa (1984), who also expressed reservations about the importance of this process.

The role of ATPase inhibition in cell injury has not been comprehensively investigated. It is of interest to note, however, that in a comparison of metals there was no indication that those with a known inhibitory effect on this enzyme caused cell damage while those without this property did not (Stacey and Klaassen, 1981b). Similarly, inhibition of mitochondrial function has not been extensively examined in isolated hepatocytes. The recent studies of Müller and Ohnesorge (1984) and Müller (1986a) suggest that this may be deserving of further investigation.

Other more recent studies have attempted to define the role of thiol groups in the expression of cadmium-induced toxicity. Dithiothreitol, a known thiol-reducing agent, was found to protect against the cytotoxicity in response to this metal (Stacey, 1986a). However, it wasn't clear whether the protection was due to maintenance of thiol groups in the reduced state or an intracellular chelation of Cd. A similar pattern of response was found with exogenously added reduced glutathione (GSH), which remains extracellular (Stacey, 1986b). Both studies demonstrate that Cd-induced toxicity to hepatocytes is not due to a simple, single irreversible interaction with cellular thiol groups. The latter study also found that cellular protein-bound sulfhydryl groups exhibited a small decrease on exposure to Cd, but that this occurred only after cell injury had been initiated. Interaction of Cd with components of the biochemical pathways involving GSH has also been recently investigated. Müller (1986b) found that glutathione peroxidase was unaffected while glucose-6-phosphate dehydrogenase and gluthathione reductase were inhibited. This suggested to the author that these enzymes may well be involved in a cadmium-induced decrease in cellular SH groups. It is unlikely that this is directly related to cell viability however, as decreases in GSH were only seen after cytotoxicity had been initiated (Stacey and Kappus, 1982a).

A report by Morselt et al. (1983b) has suggested that a disturbance in ribosomal RNA synthesis due to cadmium may be the main cause leading to cell lysis. This remains to be critically evaluated, however.

There has been considerable research over the past few years on the role of calcium in chemically induced cell death. A report by Schanne et al. (1979) suggested that calcium is a mediator in a final common pathway in cell death in response to a variety of chemical agents. Several subsequent reports from this group, most of which have been reviewed (Farber, 1981), substantiate this claim. The overall theory suggests that, after injury to plasma membranes, there is an influx of calcium into the cell along a steep concentration gradient. This converts potentially reversible alterations into the irreversible response of cell death. Chenery et al. (1981) also suggested that calcium was involved in cell death in response to carbon tetrachloride, although scrutiny of their data fails to consistently support this claim. Investigations with isolated hepatocyte suspensions, however, were unable to provide support for the proposal that calcium was a ubiquitous mediator of cell death (Smith et al., 1981; Stacey and Klaassen, 1982; Smith and Sandy, 1985; Fariss and Reed, 1985). As these studies used suspensions rather than primary cultures, this difference in methodology may be responsible for the opposite findings. In support of this, calcium was found to play a role in anoxic liver cell injury in isolated perfused liver but *not* in suspensions of isolated hepatocytes (Okuno *et al.*, 1983). However, experiments using primary cultures of hepatocytes from very young rats were unable to find any evidence of protection against cadmium-induced cytototoxicity by omission of calcium from the incubation medium (Acosta and Sorensen, 1983; Sorensen *et al.*, 1984). Further studies by a Swedish group have indicated that, while influx of extracellular calcium may not be the critical event in cell death, alteration in intracellular calcium homeostasis may be of considerable importance (Jewell *et al.*, 1982; Bellomo *et al.*, 1982; Orrenius *et al.*, 1984; Thor *et al.*, 1984). This contention is supported by the recent study of Long and Moore (1986). The possible involvement of subcellular organelles has been highlighted in the work of Richelmi *et al.* (1984) and Albano *et al.* (1985). An apparent link has also been drawn between alterations in intracellular calcium homeostasis and changes in cellular thiol status (protein- and nonprotein-bound) (Di Monte *et al.*, 1984; Moore *et al.*, 1985; Maridonneau-Parini *et al.*, 1986). This may prove to be of importance with regard to a mechanism of hepatocellular toxicity.

The reason for the major discrepancy among laboratories with regard to the role of calcium has recently been identified as the difference in vitamin E content of the different incubation media employed (Fariss *et al.*, 1985). Furthermore, Farber's group has also recently been able to dissociate alteration in intracellular calcium ion from cell injury in response to oxidative stress (Starke *et al.*, 1986).

VII. CONCLUSION

Hepatocytes in suspension and in primary culture have been used to examine several aspects of metal metabolism pertaining to toxic events. These include mechanisms of uptake of metals, the ability of chelators to reduce cellular metal load, toxicity of metals at the cellular level, interactions among metals, mechanisms by which metal toxicity occurs, and the role of the endogenous metal, calcium, in cell death in general. Considerable information has been gained from these studies, but all areas require further investigation. In particular, efflux of metals from hepatocytes has been studied very little, the mechanisms by which toxicity occur are still unclear, and the role of calcium in cell death is still an area of controversy. Both suspensions and primary cultures of hepatocytes have their advantages and disadvantages. It seems apparent that both preparations have a future role to play in extending the studies reviewed here. The choice of either preparation should be made according to the information that is required. Further additional investigations with hepatocytes will enhance our understanding of the toxicology of metals.

REFERENCES

Aaseth, J., Alexander J., and Deverill, J. (1981). Evaluation of methyl mercury chelating agents using red blood cells and isolated hepatocytes. *Chem.-Biol. Interact.* **36**, 287–297.

Acosta, D., and Sorensen, E. M. B. (1983). Role of calcium in cytotoxic injury of cultured hepatocytes. *Ann. N.Y. Acad. Sci.* **407**, 78–92.

Agius, L., and Vaartjes, W. (1982). The effects of orthovanadate on fatty acid synthesis in isolated rat hepatocytes. *Biochem. J.* **202**, 791–794.

Albano, E., Bellomo, G., Carini, R., Biasi, F., Poli, G., and Dianzani, M. U. (1985). Mechanisms responsible for carbon tetrachloride-induced perturbation of mitochondrial calcium homeostasis. *FEBS Lett.* **192**, 184–188.

Alexander, J., and Aaseth, J. (1982). Organ distribution and cellular uptake of methyl mercury in the rat as influenced by the intra- and extracellular glutathione concentration. *Biochem. Pharmacol.* **31**, 685–690.

Alexander, J., Hostmark, A. T., Forre, O., and Von Kraemer, B. M. (1979). The influence of selenium on methyl mercury toxicity in rat hepatoma cells, human embryonic fibroblasts and human lymphocytes in culture. *Acta Pharmacol. Toxicol.* **45**, 379–386.

Anundi, I., Hogberg, J., and Stahl, A. (1982). Involvement of glutathione reductase in selenite metabolism and toxicity, studied in isolated rat hepatocytes. *Arch. Toxicol.* **50**, 113–123.

Anundi, I., Stahl, A., and Hogberg, J. (1984). Effects of selenite on O_2 consumption, glutathione oxidation and NADPH levels in isolated hepatocytes and the role of redox changes in selenite toxicity. *Chem. Biol. Interact.* **50**, 277–288.

Baker, E., Vicary, F. R., and Huehns, E. R. (1981). Iron release from isolated hepatocytes. *Br. J. Haematol.* **47**, 493–504.

Baker, E., Torrance, J., and Grady, R. (1984). The effect of iron chelators on hepatocyte iron metabolism. *Proc. Aust. Soc. Med. Res.* **17**, 4.

Bellomo, G., Jewell, S. A., Thor, H., and Orrenius, S. (1982). Regulation of intracellular calcium compartmentation: Studies with isolated hepatocytes and *t*-butyl hydroperoxide. *Proc. Natl. Acad. Sci. U.S.A.* **79**, 6842–6846.

Blom, A., Keulemans, K., and Meijer, D. K. F. (1981). Transport of dibromosulphthalein by isolated rat hepatocytes. *Biochem. Pharmacol.* **30**, 1809–1816.

Bracken, W. M., and Klaassen, C. D. (1985). Metal-induced metallothionein synthesis in primary rat hepatocyte cultures. *Toxicologist* **5**, 530.

Cain, K., and Skilleter, D. N. (1980). Selective uptake of cadmium by the parenchymal cells of liver. *Biochem. J.* **188**, 285–288.

Cain, K., and Skilleter, D. N. (1983). Comparison of cadmium-metallothionein synthesis in parenchymal and non-parenchymal rat liver cells. *Biochem. J.* **210**, 769–773.

Cantilena, L. R., Jr., Stacey, N. H., and Klaassen, C. D. (1983). Isolated rat hepatocytes as a model system for screening chelators for use in cadmium intoxication. *Toxicol. Appl. Pharmacol.* **67**, 257–263.

Caperna, T. J., and Failla, M. L. (1984). Cadmium metabolism by rat liver endothelial and Kupffer cells. *Biochem. J.* **221**, 631–636.

Cheeseman, K. H., Milia, A., Chiarpotto, E., Biasi, F., Albano, E., Poli, G., Dianzani, M. U., and Slater, T. F. (1984). Iron overload, lipid peroxidation and loss of cytochrome P-450: Comparative studies in vitro using microsomes and isolated hepatocytes. *IRCS Med. Sci.* **12**, 61–62.

Chenery, R., George, M., and Krishna, G. (1981). The effect of ionophore A23187 and calcium on carbon tetrachloride-induced toxicity in cultured rat hepatocytes. *Toxicol. Appl. Pharmacol.* **60**, 241–252.

Chiarpotto, E., Olivero, E., Albano, E., Poli, G., Gravela, E., and Dianzani, M. U. (1981). Studies on lipid peroxidation using whole liver cells: Influence of damaged cells on the prooxidant effect of ADP-Fe^{3+} and CCl_4. *Experientia* **37,** 396–397.

Christie, N. T., and Costa, M. (1984). In vitro assessment of the toxicity of metal compounds. IV. Disposition of metals in cells: Interactions with membranes, glutathione, metallothionein, and DNA. *Biol. Trace Elem. Res.* **6,** 139–158.

Chvapil, M. (1973). New aspects in the biological role of zinc: A stabilizer of macromolecules and biological membranes. *Life Sci.* **13,** 1041–1049.

Cole, E. S., and Glass, J. (1983). Transferrin binding and iron uptake in mouse hepatocytes. *Biochim. Biophys. Acta* **762,** 102–110.

Cook, J. A., Marconi, E. A., and DiLuzio, N. R. (1974). Lead, cadmium, endotoxin interaction: Effect on mortality and hepatic function. *Toxicol. Appl. Pharmacol.* **28,** 292–302.

Craik, J. D., and Elliott, K. R. F. (1979). Kinetics of 3-O-Methyl-D-glucose transport in isolated rat hepatocytes. *Biochem. J.* **182,** 503–508.

Darwish, H. M., Hoke, J. E., and Ettinger, M. J. (1983). Kinetics of Cu(II) transport and accumulation by hepatocytes from copper-deficient mice and the brindled mouse model of Menkes Disease. *J. Biol. Chem.* **258,** 13621–13626.

Darwish, H. M., Cheney, J. C., Schmitt, R. C., and Ettinger, M. J. (1984a). Mobilization of copper(II) from plasma components and mechanism of hepatic copper transport. *Am. J. Physiol.* **246,** G72–G79.

Darwish, H. M., Schmitt, R. C., Cheney, J. C., and Ettinger, M. J. (1984b). Copper efflux kinetics from rat hepatocytes. *Am. J. Physiol.* **246,** G48–G55.

Di Monte D., Bellomo, G., Thor, H., Nicotera, P., and Orrenius, S. (1984). Menadione-induced cytotoxicity is associated with protein thiol oxidation and alteration in intracellular Ca^{2+} homeostasis. *Arch. Biochem. Biophys.* **235,** 343–350.

Din, W. S., and Frazier, J. M. (1985). Protective effect of metallothionein on cadmium toxicity in isolated rat hepatocytes. *Biochem. J.* **230,** 395–402.

Doolittle, R. L., and Richter, G. W. (1981). Isoferritins in rat Kupffer cells, hepatocytes, and extrahepatic macrophages. Biosynthesis in cell suspensions and cultures in response to iron. *Lab. Invest.* **45,** 567–574.

Dudley, R. E., Svoboda, D. J., and Klaassen, C. D. (1982). Acute exposure to cadmium causes severe liver injury in rats. *Toxicol. Appl. Pharmacol.* **65,** 302–313.

Failla, M. L., and Cousins, R. J. (1978a). Zinc uptake by isolated rat liver parenchymal cells. *Biochim. Biophys. Acta* **538,** 435–444.

Failla, M. L., and Cousins, R. J. (1978b). Zinc accumulation and metabolism in primary cultures of adult rat liver cells. Regulation by glucocorticoids. *Biochim. Biophys. Acta* **543,** 293–304.

Failla, M. L., Cousins, R. J., and Mascewik, M. J. (1979). Cadmium accumulation with metabolism by rat liver parenchymal cells in primary monolayer culture. *Biochim. Biophys. Acta* **583,** 63–72.

Farber, J. L. (1981). The role of calcium in cell death. *Life Sci.* **29,** 1289–1295.

Fariss, M. W., and Reed, D. J. (1985). Mechanism of chemical-induced toxicity. II. Role of extracellular calcium. *Toxicol. Appl. Pharmacol.* **79,** 296–306.

Fariss, M. W., Pascoe, G. A., and Reed, D. J. (1985). Vitamin E reversal of the effect of extracellular calcium on chemically induced toxicity in hepatocytes. *Science* **227,** 751–754.

Frazier, J. M., and Kingsley, B. S. (1976). Kinetics of cadmium transport in the isolated perfused liver. *Toxicol. Appl. Pharmacol.* **38,** 583–593.

Frazier, J. M., and Kingsley, B. S. (1977). Cadmium kinetics in the isolated perfused rat liver: An example of a general method for investigating metal metabolism. In ''Clinical Chemistry and Chemical Toxicology of Metals'' (S. S. Brown, ed.), pp. 33–36. Elsevier/North-Holland, Amsterdam.

Gebhardt, R., and Jung, W. (1982). Biliary secretion of sodium fluorescein in primary monolayer cultures of adult rat hepatocytes and its stimulation by nicotinamide. *J. Cell Sci.* **56,** 233–244.

Gerson, R. J., and Shaikh, Z. A. (1982). Uptake and binding of cadmium and mercury to metallothionein in rat hepatocyte primary cultures. *Biochem. J.* **208,** 465–472.

Gerson, R. A., and Shaikh, Z. A. (1984). Differences in the uptake of cadmium and mercury by rat hepatocyte primary cultures. Role of a sulfhydryl carrier. *Biochem. Pharmacol.* **33,** 199–203.

Gewirtz, D. A., Randolph, J. K., and Goldman, I. D. (1981). Efflux in isolated hepatocytes as a possible correlate of secretion *in vivo:* Induced exit of the folic acid analog methotrexate, by dibutyryl cyclic AMP or isobutyl methylxanthine. *Biochem. Biophys. Res. Commun.* **101,** 366–374.

Gewirtz, D. A., Randolph, J. K., and Goldman, I. D. (1982). Catecholamine-induced release of the folic acid analogue, methotrexate, from rat hepatocytes in suspension. An alpha-adrenergic phenomenon. *Mol. Pharmacol.* **22,** 493–499.

Grisham, J. W. (1979). Use of hepatic cell cultures to detect and evaluate the mechanisms of action of toxic chemicals. *Int. Rev. Exp. Pathol.* **20,** 123–210.

Grohlich, D., Morley, C. D. G., Miller, R. J., and Bezkorovainy, A. (1977). Iron incorporation into isolated rat hepatocytes. *Biochem. Biophys. Res. Commun.* **76,** 682–690.

Grohlich, D., Morley, C. G. D., and Bezkorovainy, A. (1979). Some aspects of iron uptake by rat hepatocytes in suspension. *Int. J. Biochem.* **10,** 797–802.

Guttas, J. J., Carter, T., and Horwitz, A. (1983). Plasma membrane protection against the acute effects of inorganic lead on the respiratory rates of intact liver cells. *J. Toxicol. Environ. Health* **12,** 731–736.

Hidalgo, H. A., Koppa, V., and Bryan, S. E. (1978). Induction of cadmium-thionein in isolated rat liver cells. *Biochem. J.* **170,** 219–225.

Hill, K. E., and Burk, R. F. (1982). Effect of selecium deficiency and vitamin E deficiency on glutathione metabolism in isolated rat hepatocytes. *J. Biol. Chem.* **257,** 10668–10672.

Hoffmann, E. O., Cook, J. A., DiLuzio, N. R., and Coover, J. A. (1975). The effects of acute cadmium administration in the liver and kidney of the rat. *Lab. Invest.* **32,** 655–664.

Hogberg, J., and Kristoferson, A. (1979). Inhibition of amino acid uptake in isolated hepatocytes by selenite. *FEBS Lett.* **107,** 77–80.

Horne, D. W., Briggs, W. T., and Wagner, C. (1979). Studies on the transport mechanism of 5-methyltetrahydrofolic acid in freshly isolated hepatocytes: Effect of ethanol. *Arch. Biochem. Biophys.* **196,** 557–565.

Hultcrantz, R., Hogberg, J., and Glaumann, H. (1983). Studies on the rat liver following iron overload: An analysis of iron and lysosomal enzymes in isolated parenchymal and non-parenchymal cells. *Virchows Arch.* **43,** 67–74.

Inmon, J., Stead, A., Waters, M. D., and Lewtas, J. (1981). Development of a toxicity test system using primary rat liver cells. *In Vitro* **17,** 1004–1010.

Jacobs, A., Hoy, T., Humphrys, J., and Perera, P. (1978). Iron overload in Chang cell cultures: Biochemical and morphological studies. *Br. J. Exp. Pathol.* **59,** 489–498.

Jewell, S. A., Bellomo, G., Thor, H., Orrenius, S., and Smith, M. T. (1982). Bleb formation in hepatocytes during drug metabolism is caused by disturbances in thiol and calcium ion homeostasis. *Science* **217,** 1257–1259.

Kingsley, B. S., and Frazier, J. M. (1979). Cadmium transport in isolated perfused rat liver: Zinc–cadmium competition. *Am. J. Physiol.* **236,** C139–C143.

Kinter, W. B., and Pritchard, J. B. (1977). Altered permeability of cell membranes. *In* "Handbook of Physiology, Sect. 9, Reactions to Environmental Agents" (D. H. K. Lee, H. L. Falk, S. D. Murphy, and S. R. Geiger, eds.), pp. 563–576. Am. Physiol. Soc., Washington, D.C.

Klaassen, C. D. (1976). Biliary excretion of metals. *Drug Metab. Rev.* **5,** 165–196.

Klaassen, C. D., and Stacey, N. H. (1982). Use of isolated hepatocytes in toxicity assessment. *In*

"Toxicology of the Liver" (G. L. Plaa and W. R. Hewitt, eds.), pp. 147–179. Raven, New York.

Klaassen, C. D., and Watkins, J. B. (1984). Mechanisms of bile formation, hepatic uptake, and biliary excretion. *Pharmacol. Rev.* **36**, 1–67.

Koster, J. F., and van Berkel, T. J. C. (1983). The effect of diethyldithiocarbamate on the lipid peroxidation of rat-liver microsomes and in rat hepatocytes. *Biochem. Pharmacol.* **32**, 3307–3310.

Le Cam, A., Rey, J. F., Fehlmann, M., Kitabgi, P., and Freychet, P. (1979). Amino acid transport
· in isolated hepatocytes after partial hepatectomy in the rat. *Am. J. Physiol.* **236**, E594–E602.

Lerman, S. A., Clarkson, T. W., and Gerson, R. J. (1983). Arsenic uptake and metabolism by liver cells is dependent on arsenic oxidation state. *Chem.-Biol. Interact.* **45**, 401–406.

Lodola, A. (1981). Effects of cobalt chloride on haem synthesis in isolated hepatocytes. *FEBS Lett.* **123**, 137–140, 1981.

Long, R. M., and Moore, L. (1986). Elevated cytosolic calcium in rat hepatocytes exposed to carbon tetrachloride. *J. Pharmacol. Exp. Ther.* **238**, 186–191.

Maridonneau-Parini, I., Mirabelli, F., Richelmi, P., and Bellomo, G. (1986). Cytotoxicity of phenazine methosulfate in isolated rat hepatocytes is associated with superoxide anion production, thiol oxidation and alterations in intracellular calcium ion homeostasis. *Toxicol. Lett.* **31**, 175–181.

Mittelstaedt, R. A., and Pounds, J. G. (1984). Subcellular distribution of lead in cultured rat hepatocytes. *Environ, Res.* **35**, 188–196.

Moore, M., Thor, H., Moore, G., Nelson, S., Moldeus, P., and Orrenius, S. (1985). The toxicity of acetaminophen and N-acetyl-p-benzoquinone imine in isolated hepatocytes is associated with thiol depletion and increased cytosolic Ca^{2+}. *J. Biol. Chem.* **260**, 13035–13040.

Morselt, A. F. W., Copius Peereboom-Stegeman, J. H. J., Jongstra-Spaapen, E. J., and James, J. (1983a). Investigation of the mechanism of cadmium toxicity at cellular level. I. A light microscopical study. *Arch. Toxicol.* **52**, 91–97.

Morselt, A. F. W., Copius Peereboom-Stegeman, J. H. J., Puvion, E., and Maarschalakerweerd, V. J. (1983b). Investigation of the mechanism of cadmium toxicity at cellular level. II. An electron microscopical study. *Arch. Toxicol.* **52**, 99–108.

Müller, L. (1983). Influence of paracetamol (acetaminophen) on cadmium-induced lipid peroxidation in hepatocytes from starved rats. *Toxicol. Lett.* **15**, 159–165.

Müller, L. (1984). Differential sensitivity of integrity criteria as indicators of cadmium-induced cell damage. *Toxicol. Lett.* **21**, 21–27.

Müller, L. (1986a). Consequences of cadmium toxicity in rat hepatocytes: Mitochondrial dysfunction and lipid peroxidation. *Toxicology* **40**, 285–295.

Müller, L. (1986b). Consequences of cadmium toxicity in rat hepatocytes: effects of cadmium on the glutathione-peroxidase system. *Toxicol. Lett.* **30**, 259–265.

Müller, L., and Ohnesorge, F. K. (1982). Different response to liver parenchymal cells from starved and fed rats to cadmium. *Toxicology* **25**, 141–150.

Müller, L., and Ohnesorge, F. K. (1984). Cadmium-induced alteration of the energy level in isolated hepatocytes. *Toxicology* **31**, 297–306.

Octave, J. N., Schneider, Y. J., Crichton, R. R., and Trouet, A. (1983). Iron mobilization from cultured hepatocytes: Effect of desferrioxamine B. *Biochem. Pharmacol.* **32**, 3413–3418.

Okuno, F., Orrego, H., and Israel, Y. (1983). Calcium requirement for anoxic liver cell injury. *Res. Commun. Chem. Pathol. Pharmacol.* **39**, 437–444.

Orrenius, S., Thor, H., and Bellomo, G. (1984). Alterations in thiol and calcium in homeostasis during hydroperoxide and drug metabolism in hepatocytes. *Biochem. Soc. Trans.* **12**, 23–28.

Page, M. A., Baker, E., and Morgan, E. H. (1984). Transferrin and iron uptake by rat hepatocytes in culture. *Am. J. Physiol.* **246**, G26–G32.

Poli, G., Gravela, E., Chiarpotto, E., Albano, E., and Dianzani, M. U. (1979). Studies on lipid peroxidation in isolated liver cells. *IRCS Med. Sci.* **7**, 558.

Potter, S. D., and Matrone, G. (1977). A tissue culture model for mercury-selenium interactions. *Toxicol. Appl. Pharmacol.* **40**, 201–215.

Pounds, J. G., Wright, R., Morrison, D., and Casciano, D. A. (1982a). Effect of lead on calcium homeostasis in the isolated rat hepatocyte. *Toxicol. Appl. Pharmacol.* **63**, 389–401.

Pounds, J. G., Wright, R., and Kodell, R. L. (1982b). Cellular metabolism of lead: A kinetic analysis in the isolated rat hepatocyte. *Toxicol. Appl. Pharmacol.* **66**, 88–101.

Puvion, E., and Lange, M. (1980). Functional significance of perichromatin granule accumulation induced by cadmium chloride in isolated rat liver cells. *Exp. Cell Res.* **128**, 47–58.

Puvion-Dutilleul, F., and Puvion, E. (1981). Relationship between chromatin and perichromatin granules in cadmium-treated isolated hepatocytes. *J. Ultrastruct. Res.* **74**, 341–350.

Rama, R., Octave, J. N., Schneider, Y. J., Sibille, J. C., Limet, J. N., Mareschal, J. C., Trouet, A., and Crichton, R. R. (1981). Iron mobilization from cultured rat fibroblasts and hepatocytes. Effect of various drugs. *FEBS Lett.* **127**, 204–206.

Richelmi, P., Baldi, C., Manzo, L., Berte, F., Martino, P. A., Mirabelli, F., and Bellomo, G. (1984). Erythromycin estolate impairs the mitochondrial and microsomal calcium homeostasis: correlation with hepatotoxicity. *Arch. Toxicol. Suppl.* **7**, 298–302.

Rothstein, A. (1959). Cell membrane as site of action of heavy metals. *Fed. Proc., Fed. Am. Soc. Exp. Biol.* **18**, 1026–1038.

Rudd, C. J., and Herschman, H. R. (1978). Metallothionein accumulation in response to cadmium in a clonal rat liver cell line. *Toxicol. Appl. Pharmacol.* **44**, 511–521.

Ryberg, D., and Alexander, J. (1984). Inhibitory action of hexavalent chromium (Cr(VI)) on the mitochondrial respiration and a possible coupling to the reduction of Cr (VI). *Biochem. Pharmacol.* **33**, 2461–2466.

Santone, K. S., Acosta, D., and Bruckner, J. V. (1982). Cadmium toxicity in primary cultures of rat hepatocytes. *J. Toxicol. Environ. Health* **10**, 169–177.

Schanne, F. A. X., Kane, A. B., Young, E. E., and Farber, J. L. (1979). Calcium dependence of toxic cell death: A final common pathway. *Science* **206**, 700–702.

Schmitt, R. C., Darwish, H. M., Cheney, J. C., and Ettinger, M. J. (1983). Copper transport kinetics by isolated rat hepatocytes. *Am. J. Physiol.* **244**, G183–G191.

Schwarz, L. R., and Greim, H. (1981). Isolated hepatocytes: An analytical tool in hepatoxicology. *In* "Frontiers in Liver Disease" (P. D. Berk and T. C. Chalmers, eds.), pp. 61–79. Neuherberg, F.R.G.

Schwarz, L. R., Summer, K. H., and Schwenk, M. (1979). Transport and metabolism of bromosulfophthalein by isolated rat liver cells. *Eur. J. Biochem.* **94**, 617–622.

Schwenk, M. (1980). Transport system of isolated hepatocytes. Studies on the transport of biliary compounds. *Arch. Toxicol.* **44**, 113–126.

Schwenk, M., Wiedmann, T., and Remmer, H. (1981). Uptake, accumulation and release of ouabain by isolated rat hepatocytes. *Naunyn-Schmiedeberg's Arch. Pharmacol.* **316**, 340–344.

Sciortino, C. V., Failla, M. L., and Bullis, D. B. (1982). Identification of metallothionein in parenchymal and non-parenchymal liver cells of the adult rat. *Biochem. J.* **204**, 509–514.

Seglen, P. O., and Gordon, P. B. (1981). Vanadate inhibits protein degradation in isolated rat hepatocytes. *J. Biol. Chem.* **256**, 7699–7701.

Sibille, J. C., Schneider, Y. J., Octave, J. N., Trouet, A., and Crichton, R. R. (1981). Uptake of iron from iron-saturated transferrin by cultured rat hepatocytes. *Arch. Int. Physiol. Biochem.* **89**, B35.

Singhal, R. L., Merali, Z., Kacew, S., and Sutherland, D. J. B. (1974). Persistence of cadmium-induced metabolic changes in liver and kidney. *Science* **183**, 1094–1096.

Skilleter, D. N., and Cain, K. (1981). Effects of thiol agents on the accumulation of cadmium by rat liver parenchymal and Kupffer cells. *Chem.-Biol. Interact.* **37**, 289–298.

Skilleter, D. N., and Paine, A. J. (1979). Relative toxicities of particulate and soluble forms of beryllium to a rat liver parenchymal cell line in culture and possible mechanisms of uptake. *Chem.-Biol. Interact.* **24**, 19–33.

Smith, A., and Morgan, W. T. (1981). Hemopexin-mediated transport of heme into isolated rat hepatocytes. *J. Biol. Chem.* **256**, 10902–10909.

Smith, M. T., and Sandy, M. S. (1985). Role of extracellular Ca^{2+} in toxic liver injury: comparative studies with the perfused rat liver and isolated hepatocytes. *Toxicol. Appl. Pharmacol.* **81**, 213–219.

Smith, M. T., Thor, H., and Orrenius, S. (1981). Toxic injury in isolated hepatocytes is not dependent on extracellular calcium. *Science* **213**, 1257–1259.

Sorensen, E. M. B., and Acosta, D. (1984). Cadmium-induced hepatotoxicity in cultured rat hepatocytes as evaluated by morphometric analysis. *In Vitro* **20**, 763–770.

Sorensen, E. M. B., Smith, N. K. R., Boecker, C. S., and Acosta, D. (1984). Calcium amelioration of cadmium-induced cytotoxicity in cultured rat hepatocytes. *In Vitro* **20**, 771–779.

Stacey, N. H. (1986a). Protective effects of dithiothreitol on cadmium-induced injury in isolated rat hepatocytes. *Toxicol. Appl. Pharmacol.* **82**, 224–232.

Stacey, N. H. (1986b). The amelioration of cadmium-induced injury in isolated hepatocytes by reduced glutathione. *Toxicology* **42**, 85–93.

Stacey, N. H., and Kappus, H. (1982a). Comparison of methods of assessment of metal-induced lipid peroxidation in isolated rat hepatocytes. *J. Toxicol. Environ. Health* **9**, 277–285.

Stacey, N. H., and Kappus, H. (1982b). Cellular toxicity and lipid peroxidation in response to mercury. *Toxicol. Appl. Pharmacol.* **63**, 29–35.

Stacey, N. H., and Klaassen, C. D. (1979). Uptake of ouabain by isolated hepatocytes from livers of developing rats. *J. Pharmacol. Exp. Ther.* **211**, 360–363.

Stacey, N. H., and Klaassen, C. D. (1980). Cadmium uptake by isolated rat hepatocytes. *Toxicol. Appl. Pharmacol.* **55**, 448–455.

Stacey, N. H., and Klaassen, C. D. (1981a). Interaction of metal ions with cadmium-induced cellular toxicity. *J. Toxicol. Environ. Health* **7**, 149–158.

Stacey, N. H., and Klaassen, C. D. (1981b). Comparison of the effects of metals on cellular injury and lipid peroxidation in isolated rat hepatocytes. *J. Toxicol. Environ. Health* **7**, 139–147.

Stacey, N. H., and Klaassen, C. D. (1981c). Uptake of zinc by isolated rat hepatocytes. *Biochim. Biophys. Acta* **640**, 693–697.

Stacey, N. H., and Klaassen, C. D. (1981d). Inhibition of lipid peroxidation without prevention of cellular injury in isolated rat hepatocytes. *Toxicol. Appl. Pharmacol.* **58**, 8–18.

Stacey, N. H., and Klaassen, C. D. (1981e). Copper toxicity in isolated rat hepatocytes. *Toxicol. Appl. Pharmacol.* **58**, 211–220.

Stacey, N. H., and Klaassen, C. D. (1982). Lack of protection against chemically induced injury to isolated hepatocytes by omission of calcium from the incubation medium. *J. Toxicol. Environ. Health* **9**, 267–276.

Stacey, N. H., and Priestly, B. G. (1978). Lipid peroxidation in isolated rat hepatocytes. Relationship to toxicity of CCl_4, ADP/Fe^{+++} and diethyl maleate. *Toxicol. Appl. Pharmacol.* **45**, 41–48.

Stacey, N. H., Cook, R., and Priestly, B. G. (1978). Comparative stability of alanine aminotransferase in rat plasma and hepatocyte suspensions. *Aust. J. Exp. Biol. Med. Sci.* **56**, 379–381.

Stacey, N. H., Cantilena, L. R., Jr., and Klaassen, C. D. (1980). Cadmium toxicity and lipid peroxidation in isolated rat hepatocytes. *Toxicol. Appl. Pharmacol.* **53**, 470–480.

Stacey, N. H., Wong, K. L., and Klaassen, C. D. (1983). Protective effects of chromium on the toxicity of cadmium *in vivo. Toxicology* **28**, 147–153.

Stahl, A., Anundi, I., and Hogberg, J. (1984). Selenite biotransformation to volatile metabolites in an isolated hepatocyte model system. *Biochem. Pharmacol.* **33**, 1111–1117.

Starke, P. E., and Farber, J. L. (1985). Ferric iron and superoxide ions are required for the killing of cultured hepatocytes by hydrogen peroxide. *J. Biol. Chem.* **260**, 10099–10104.

Starke, P. E., Hoek, J. B., and Farber, J. L. (1986). Calcium-dependent and calcium-independent mechanisms of irreversible cell injury in cultured hepatocytes. *J. Biol. Chem.* **261**, 3006–3012.

Story, D. L., Gee, S. J., and Tyson, C. A. (1983). Response of isolated hepatocytes to organic and inorganic cytotoxins. *J. Toxicol. Environ. Health* **11**, 483–501.

Stowe, H. D., Wilson, M., and Goyer, R. A. (1972). Clinical and morphological effects of oral cadmium toxicity in rabbits. *Arch. Pathol.* **94**, 389–405.

Tarao, K., Olinger, E. J., Ostrow, J. D., and Balistreri, W. F. (1982). Impaired bile acid efflux from hepatocytes isolated from the liver of rats with cholestasis. *Am. J. Physiol.* **243**, G253–G258.

Thor, H., Hartzell, P., and Orrenius, S. (1984). Potentiation of oxidative cell injury in hepatocytes which have accumulated Ca^{2+}. *J. Biol. Chem.* **259**, 6612–6615.

Tyson, C. A., Green, C. E., LeValley, S. E., and Stephens, R. J. (1982). Characterization of isolated Fe-loaded rat hepatocytes prepared by collagenase perfusion. *In Vitro* **18**, 945–951.

Tyson, C. A., LeValley, S. E., Chan, R., Hobbs, P. D., and Dawson, M. I. (1984). Biological evaluation of some ionophore-polymeric chelator combinations for reducing iron overload. *J. Pharmacol. Exp. Ther.* **228**, 676–681.

Vallee, B. L., and Ulmer, D. D. (1972). Biochemical effects of mercury, cadmium and lead. *Annu. Rev. Biochem.* **41**, 91–128.

Vonk, R. J., Jekel, P. A., Meijer, D. K. F., and Hardonk, M. J. (1978). Transport of drugs in isolated hepatocytes. The influence of bile salts. *Biochem. Pharmacol.* **27**, 397–405.

Walum, E., and Marchner, H. (1983). Effects of mercuric chloride on the membrane integrity of cultured cell lines. *Toxicol. Lett.* **18**, 89–95.

Webb, M. (1977). Metabolic targets of metal toxicity. *In* "Clinical Chemistry and Chemical Toxicology of Metals" (S. S. Brown, ed.), pp. 51–64. Elsevier/North-Holland, Amsterdam.

Weiner, A. L., and Cousins, R. J. (1980). Copper accumulation and metabolism in primary monolayer cultures of rat liver parenchymal cells. *Biochim. Biophys. Acta* **629**, 113–115.

Weiner, A. L., and Cousins, R. J. (1983a). Differential regulation of copper and zinc metabolism in rat liver parenchymal cells in primary cultures (41675). *Proc. Soc. Exp. Biol. Med.* **173**, 486–494.

Weiner, A. L., and Cousins, R. J. (1983b). Hormonally produced changes in caeruloplasmin synthesis and secretion in primary cultured rat hepatocytes. Relationship to hepatic copper metabolism. *Biochem. J.* **212**, 297–304.

Wiebkin, P., Prough, R. A., and Bridges, J. W. (1982). The metabolism and toxicity of some organotin compounds in isolated rat hepatocytes. *Toxicol. Appl. Pharmacol.* **62**, 409–420.

Willson, R. (1977). Zinc: A radical approach to disease. *New Sci.* **1**, 558–560.

Young, S. P., and Aisen, P. (1980). The interaction of transferrin with isolated hepatocytes. *Biochim. Biophys. Acta* **633**, 145–153.

Young, S. P., and Aisen, P. (1982). Transferrin receptors and the uptake and release of iron by isolated hepatocytes. *Hepatology* **1**, 114–119.

Zaman, Z., and Verwilghen, R. L. (1981). Influence of zinc on iron uptake by monolayer cultures of rat hepatocytes and the hepatocellular ferritin. *Biochim. Biophys. Acta* **675**, 77–84.

8

The Metabolism and Toxicity of Xenobiotics in a Primary Culture System of Postnatal Rat Hepatocytes

**DANIEL ACOSTA,* DAVID B. MITCHELL,*,[1]
ELSIE M. B. SORENSEN,* AND JAMES V. BRUCKNER†**

*College of Pharmacy
Department of Pharmacology and Toxicology
The University of Texas
Austin, Texas 78712

†The University of Georgia
Athens, Georgia 30602

I. INTRODUCTION

Cultured rat hepatocytes have been used extensively as *in vitro* models to evaluate potentially toxic xenobiotics, to elucidate the cellular pathways of the metabolism of chemicals, and to determine the mechanisms by which xenobiotics produce toxic or cellular injury (Grisham, 1979; Acosta *et al.*, 1980, 1985b; Acosta and Mitchell, 1981). Many different types of approaches to evaluate chemical toxicity *in vitro* have been investigated. These include the use of nonmammalian cells, established mammalian cell lines, primary cultures of fetal, postnatal, and adult mammalian cells, homogenates, freshly isolated cells, and isolated perfused organs. Although each approach has certain attributes and deficiencies, the use of primary cell cultures offers a number of distinct advan-

[1]Present address: The Procter and Gamble Company, Miami Valley Laboratories, Cincinnati, Ohio 45247.

189

tages. Despite the fact that propagated or established cell lines are easier to maintain than primary cultures, the cells found in cell lines usually retain few functional characteristics of normal parenchymal cells. Since cell lines are separated by many mitotic events from the parent cells from which they were originally derived, their toxicological responses to xenobiotics, most likely, are far removed from the *in vivo* situation. Thus, findings obtained with cells in untransformed primary culture may be regarded as more reliable for predicting anticipated effects *in vivo*.

One may argue that tissue slices or perfused organ systems also may serve as suitable experimental models for *in vitro* toxicity evaluation because the structure of individual cells and their interrelationships remain largely unchanged. However, each of these preparations has inherent drawbacks. The perfused liver has a relatively short viability period, it is expensive and technically difficult relative to other preparations, and it seems to have reproducibility problems among different laboratories (Smith *et al.*, 1977). Inadequate oxygen diffusion and substrate penetration, with resulting necrosis, are problems encountered with liver slices which lead to shortened viability of this preparation (Smith *et al.*, 1977).

The primary drawback of cellular homogenates and fractions is that structural integrity and intracellular relationships are disrupted during preparation. This introduction of artifacts is believed largely responsible for the qualitative and quantitative differences observed between *in vivo* and *in vitro* systems. Ultrastructural relationships are disrupted and physicochemical irregularities are introduced during these manipulative procedures. Schenkman *et al.* (1973), for example, reported that 50–70% of microsomal protein from the endoplasmic reticulum may be lost during standard procedures used to isolate microsomal fractions. Others have shown that mitochondria are involved in the control of hepatic cell mixed-function oxidations and that the rate of xenobiotic metabolism is slowed considerably in their absence (Hildebrandt and Estabrook, 1971; Schenkman *et al.*, 1973). It should be recognized that the actual intracellular environment cannot be duplicated in tissue homogenate–cell fraction preparations, only approximated at best.

This chapter describes our rationale and approach to the use of a primary culture system of postnatal rat hepatocytes as an experimental model to study the metabolic activation of xenobiotics and the subsequent toxicity of the formed reactive intermediates. We will describe the development of the postnatal hepatocyte culture system which retains several important liver-specific, differentiated functions, the selection of key xenobiotics which demonstrate the metabolic capability of the hepatocytes to form toxic metabolites, and the development of sensitive indices of cytotoxicity to evaluate the ability of the *in vitro* system to monitor the injury produced by the selected hepatotoxic compounds.

II. RETENTION AND MAINTENANCE OF LIVER-SPECIFIC DIFFERENTIATED FUNCTIONS IN PRIMARY CULTURE OF POSTNATAL RAT HEPATOCYTES

Initial attempts at establishing primary cultures of viable hepatic cells were disappointing. Early investigators met extreme difficulties in obtaining functional mammalian hepatocytes. Early techniques for the isolation of liver cells were based mainly on physical and/or chemical means (Evans et al., 1952). Cells isolated by these physical or chemical methods were judged to be largely nonviable using dye exclusion tests (Fry et al., 1976). Likewise, liver cells isolated by these early techniques lost their differentiated functions very rapidly when placed into culture. A major development in the isolation of rat hepatocytes was the introduction of the use of collagenase and hyaluronidase by Howard et al. (1967). Both this original method of collagenase–hyaluronidase digestion of liver and an improved recirculating perfusion technique developed by Berry and Friend (1969) produced cells which were highly viable and functional. Leffert and Paul (1972) introduced the use of a selective medium for parenchymal liver cells. Their use of an arginine-free, ornithine-enriched culture medium allowed selective growth of liver parenchymal cells which retained differentiated functions in culture.

Primary culture systems of adult rat hepatocytes would seem to be ideal models for metabolism and toxicity studies when compared to a system of postnatal rat hepatocytes. However, the major problems of adult hepatocytes in culture are the limited cellular life span of 7–10 days (Bonney, 1974; Laishes and Williams, 1976) and the rapid decline of many of the enzymatic activities of the cells with increasing time in culture. Thus, if toxicity studies are conducted for several days using adult hepatocyte cultures, interpretation of data becomes ambiguous in a system of cells which are degenerating metabolically.

Relatively few researchers have explored the use of postnatal rat liver as systems of primary hepatocyte cultures which retain differentiated functions over a reasonable period of time (Guillouzo et al., 1975). Liver cells from young rats, unlike those from fetal rats, have functions which are comparable to adults. Dallner and associates (1966) have reported that, by 8 days post partum, cytochrome P-450 levels, NADPH-cytochrome c reductase activity, and glucose-6-phosphatase activity in the postnatal rat are equivalent to adult levels. Other studies on the development of the microsomal enzyme system in the rat have indicated that demethylation and hydroxylation activities are characterized by a marked increase in activity, following a quiescent period during the first 15–20 days after birth (Gram et al., 1969; Henderson, 1971). The postnatal rat liver resembles regenerating liver in vivo and shows good plating efficiency and reliable monolayer formation without prior hepatectomy (Acosta *et al.*, 1978a). These young cells are able to recover from the trauma of isolation and subsequent adjustments to a new *in vitro* environment, coupled with a reduced chance of loss of their differentiated functions. The following discussion describes the establishment of a primary culture system of postnatal rat hepatocytes with several key liver-specific differentiated functions.

One of the most important criteria for judging the utility of hepatocyte culture systems is their relative retention of liver cell functions, notably the capacity to metabolize xenobiotics. We have demonstrated the viability of postnatal hepatocyte cultures for 80 days by comparing them morphologically to cultures which have been overgrown with fibroblasts. We have developed a selective medium (arginine-deficient, ornithine-supplemented medium) which has allowed the growth of functional hepatocytes but inhibited the growth of non-hepatocytes, such as fibroblasts or endothelial cells (Acosta *et al.*, 1978a; Leffert and Paul, 1972). In this same study, we have shown that the cultured hepatocytes take up sulfobromophthalein (BSP) over a period of 25 days (Fig. 1) and form a glutathione conjugate which is excreted into the culture medium (Acosta *et al.*, 1978a). These results compared favorably with the ability of the intact liver to selectively take up BSP and excrete its glutathione conjugate. In fact, maximum uptake of BSP was 27.4 μg per 10^7 cells which compares to 24 μg per 10^7 cells in two studies which utilized freshly isolated suspensions of adult rat hepatocytes (van Bezooijen *et al.*, 1976; Stege *et al.*, 1975). Another indicator of differentiated function that we examined in postnatal hepatocyte cultures was the retention of pyruvate kinase activity, specifically the L-type isoenzyme which is found only in parenchymal hepatocytes (Acosta *et al.*, 1978a). The M-type isoenzyme is found in nonparenchymal cells of the liver. The L-isoenzyme is an allosteric enzyme involved in the regulation of glycolysis and gluconeogenesis in the liver. In cultures which were grown in selective medium for 20 days we found only L-pyruvate kinase activity and no M-type activity. In cultures which were grown in nonselective medium, and thus contained a mixture of parenchymal and non-parenchymal hepatocytes, we found that the ratio of L/M isoenzymes was 1.70. The reported value for whole liver was 1.56 (van Berkel *et al.*, 1972).

To be truly meaningful models, primary cultures of rat hepatocytes must be rigorously compared to freshly isolated hepatocytes and whole animals in regard to xenobiotic metabolic capability. In general, it has been shown qualitatively that various biotransformation reactions can be maintained in liver cell cultures but the actual level of activity is lower than in freshly isolated hepatocytes (Fry and Bridges, 1979) and whole liver (Guzelian *et al.*, 1977). Numerous attempts have been made to prevent or diminish the dramatic loss of cytochrome *P*-450 activity when hepatocytes are grown in culture. Modifications such as culturing the cells on floating collagen membranes (Michalopoulos *et al.*, 1976), growing the cells on specially treated culture dishes (Bonney *et al.*, 1974), or supplementing the

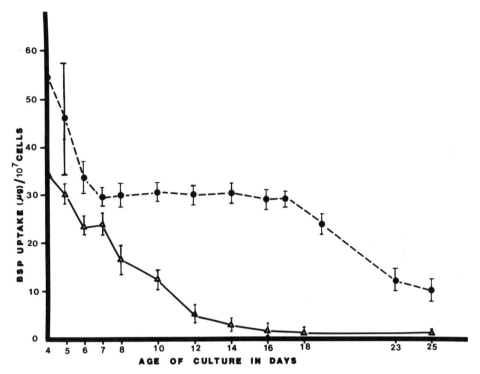

Fig. 1. Comparative uptake of bromosulfophthalein (BSP) by different-aged postnatal hepatocyte cultures grown in selective and nonselective media. Uptake of BSP was measured after a 20-min incubation period in the hepatocyte cultures. Values are reported as BSP (µg) taken up per 10^7 cells. Vertical bars indicate ±standard error of the mean. ●, Selective medium; △, nonselective medium. [Reproduced with permission from Acosta *et al.* (1978a).]

culture medium with special substrates or hormones (Bissell and Guzelian, 1979; Paine *et al.*, 1979; Sinclair *et al.*, 1979; Dickens and Peterson, 1980) have partially resolved the problem.

Thus, a major objective of our laboratory has been to develop a system of cultured postnatal hepatocytes which retains cytochrome *P*-450 levels over an extended period of time. Initial attempts were unsuccessful in maintaining cytochrome *P*-450 levels in hepatocyte cultures for several days in culture (Acosta *et al.*, 1979a). A medium enriched with several hormones (insulin, hydrocortisone, thyroxine, and testosterone) did not prevent the decline in cytochrome *P*-450 when compared to whole liver. However, more recently we have been able to maintain *P*-450 levels comparable to those found in freshly isolated hepatocytes for seven days in culture (Nelson *et al.*, 1982). By supplementing the culture medium with insulin, hydrocortisone, and nicotinamide and by removing cystine, the cultures of rat hepatocytes retained cytochrome P-450 levels which were essentially the same as freshly isolated hepatocytes—0.5 nmol per mg microsomal protein. In addition, pretreatment of the cultured hepatocytes with phenobarbital resulted in a 2-fold increase of cytochrome P-450 when compared to untreated controls. Similar results were obtained in adult hepatocyte cultures grown in cystine-free medium by Paine *et al.* (1982). Hence, the ability to maintain high levels of P-450 in culture allows one to study more confidently the metabolism of a xenobiotic and the potential toxicity of the formed metabolites in a hepatocyte culture model.

Although the importance of maintaining high levels of P-450 in cultured hepatocytes cannot be overemphasized, it is equally important that the cells have the capability to metabolize xenobiotics via a functional P-450 system. To demonstrate the metabolic capacity of postnatal hepatocyte cultures, we first examined the metabolism of p-nitroanisole, a P-450-mediated reaction (Minck *et al.*, 1973). The cultures were able to demethylate p-nitroanisole to p-nitrophenol (Fig. 2), which was further conjugated to a sulfate or glucuronide derivative (Acosta *et al.*, 1979b). p-Nitroanisole was demethylated at a rate of 0.3 nmol/10^6 cells/min and p-nitrophenol was conjugated at a rate of 0.5 nmol/10^6 cells/min, rates which represent 60 and 62%, respectively, of the reported values for isolated

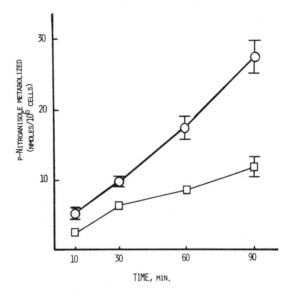

Fig. 2. Metabolism of p-nitroanisole by 4-day-old postnatal hepatocyte cultures. O-Demethylation of the substrate was measured after the indicated incubation periods by a colorimetric determination of free p-nitrophenol (□) and total p-nitrophenol (free and conjugated, ○). Vertical bars represent ±standard error of the mean.

adult rat hepatocytes (Vainio, 1973). Other examples of metabolism of xenobiotics in our primary culture system of postnatal rat hepatocytes will be presented in Section IV.

Other investigators have examined a variety of liver-specific functions in cultured postnatal hepatocytes. Their results compare favorably to the intact liver. Armato and associates (1975) have demonstrated the retention of certain differentiated functions in cultured postnatal hepatocytes, such as the ability to take up and bind bilirubin, an important function of the intact liver. Guillouzo *et al.* (1975) have reported that primary cultures of postnatal hepatocytes have the ability to synthesize and secrete albumin and fibrinogen, two key proteins normally synthesized by the liver. Chessebeuf and co-workers (1974) have shown that postnatal hepatocytes in culture retain other "*in vivo*" metabolic characteristics, including high activity of aldolase B and metabolism of various hormones such as corticosterone, testosterone, and progesterone.

III. TOXICITY EVALUATION OF XENOBIOTICS IN CULTURED HEPATOCYTES

A. Development of *in Vitro* Criteria of Cytotoxicity

The use of early, sensitive markers of cytotoxicity allows investigators to gain the maximum amount of information following toxic injury to cultured cells by drugs or chemicals. Our laboratory (Mitchell and Acosta, 1981; Santone *et al.*, 1982) as well as other laboratories (Bhuyan *et al.*, 1976; Salocks *et al.*, 1981) have shown that the commonly used tests to measure *in vitro* toxicity (e.g., viability stains, protein content, cell counts) are relatively insensitive measures of cytotoxicity. Our approach has been one of developing tests which both assess the functional capability of the cells and are clinically relevant to the *in vivo* situation. For example, it has been known for many years that damaged or diseased tissues release enzymes and that levels of specific intracellular enzymes in the blood can be measured as an index of tissue injury. Thus, in our laboratory we have used leakage of intracellular enzymes from injured cells, including glutamate–oxaloacetate transaminase (GOT), glutamate–pyruvate transaminase (GPT), lactate dehydrogenase (LDH), argininosuccinate lyase (ASAL) and creatine phosphokinase (CPK), into the culture medium as indices of cell membrane cytotoxicity (Anuforo *et al.*, 1978; Acosta *et al.*, 1978b). The obvious value of enzyme leakage tests in evaluating the cytotoxicity of xenobiotics led to our decision to publish a detailed, step-by-step description of the assay procedures in a journal which specializes in methodological detail of *in vitro* techniques (Mitchell *et al.*, 1980). Other investigators have utilized the *Journal of*

Tissue Culture Methods to describe methods for evaluating the cytotoxicity of xenobiotics. These contributions include techniques for assessing clonal growth (Wininger *et al.*, 1979), cell number and cell growth (Katz, 1975; Tolnai, 1975a; Suilovsky, 1977; Hayner *et al.*, 1982), cell-specific functions (Tolnai, 1975b; Santone and Acosta, 1982), and membrane injury (Wenzel and Acosta, 1975; Fisher, 1976; Wenzel and Reed, 1976; Sorensen *et al.*, 1984; Nealon *et al.*, 1984).

The leakage of intracellular potassium is another parameter of cell membrane integrity which can be used in addition to the enzyme leakage tests (Ramos *et al.*, 1983; Santone and Acosta, 1984). Because potassium is smaller than enzymes released from injured cells, potassium leakage may provide more subtle evidence of membrane damage by xenobiotics. Klaassen and co-workers have shown that potassium leakage is a sensitive measure of the toxicity of heavy metals to isolated rat hepatocytes (Stacey *et al.*, 1980; Stacey and Klaassen, 1981).

Organelle membrane fragility tests also can be used to assess the stability of lysosomal and mitochondrial membranes by *in situ* cytochemical enzyme measurements (Acosta and Wenzel, 1974, 1975). In essence, these fragility tests are based on the principle that uninjured organelles and their enclosed enzymes are not as accessible to substrates which react with the enzymes and thus show less formation of the products of the enzymatic reactions. Damaged organelles, on the other hand, allow entry of substrates to the sequestered enzymes via the injured organelle membranes. The end products of the enzymatic reactions, therefore, can be quantified by microscopic and spectrophotometric techniques (Acosta *et al.*, 1978c; Ramos *et al.*, 1983). In Section III,B,1, examples of drug and chemical toxicity in cultured hepatocytes as evaluated by these parameters of membrane integrity will be presented.

Whereas the cell membrane integrity tests are useful measures of the acute, direct cytotoxic effects of xenobiotics in cultured cells, these tests are less sensitive for evaluating subtle (and perhaps indirect) changes in cellular function and metabolism as the result of exposure to toxic compounds. Detection of early cell injury and altered cellular functions plays an important role in gaining a better understanding of the mechanism of toxicity of a selected agent and in determining which types of cell injury may lead to irreversible cell injury. Thus, the use of methods which assess specific liver cell functions and metabolic activities after exposure to potential hepatotoxins should help to reveal the injurious effects of the chemical agents and should be more reliable in correlating *in vitro* toxicity to liver toxicity *in vivo*. For instance, we have been able to assess the metabolic and functional capabilities of the hepatocyte cultures by their ability to both form urea and maintain a stable lactate/pyruvate (L/P) ratio. By supplementing the culture medium with insulin, hydrocortisone, and nicotinamide, the lactate/pyruvate ratio was maintained in culture comparable to the 7:1 ratio observed in the intact rat liver (Santone and Acosta, 1982), and the levels of urea formed in the postnatal hepatocyte cultures were consistent with findings observed in adult rat hepatocytes in culture (Santone and Acosta, 1982).

In addition to the metabolic indicators of hepatocyte integrity, we have assessed the effects of xenobiotics on key intracellular enzymes and substrates as potential indicators of the site of toxicity of the chemical agents. Changes in activity of glucose-6-phosphatase suggest that the endoplasmic reticulum may be a potential target of a hepatotoxin, while alterations in ATP levels and succinate dehydrogenase activity indicate that mitochondria may be the site of toxicity (Acosta and Mitchell, 1983). Because of the importance of glutathione as a protective mechanism against toxic intermediates which may accumulate in the cell after exposure to a xenobiotic, we have measured cellular glutathione levels as markers of cytotoxic interactions with xenobiotics (Mitchell and Acosta, 1982). Finally, we have utilized changes in uptake of calcium into cultured cells, mitochondria, and microsomes after exposure to xenobiotics as a sensitive measure of cytotoxicity (Ramos *et al.*, 1984; Santone and Acosta, 1985). The homeostasis of the cell is highly dependent on calcium dynamics and maintenance of strict intracellular/extracellular ratios of calcium. In Section III, B,2, we will present several examples of xenobiotics evaluated in the hepatocyte culture system by methods which assess changes in cellular functions and metabolic activities.

Another criterion which we have used to evaluate the cytotoxicity of xenobiotics at the *in vitro* level is morphological changes in the cells as evaluated by phase contrast, bright field, transmission, and scanning electron microscopy. These various microscopic methods have allowed us to assess cytotoxic morphological changes in cells, such as cell shape and size, cell swelling, surface blebbing, alterations in microvilli, cytoplasmic vacuolation, and ultrastructural changes (Acosta and Sorensen, 1983; Sorensen *et al.*, 1984). Examples of some of our morphological studies will be presented in Section III,B,3.

B. Use of Cultured Postnatal Hepatocytes to Evaluate Toxicity of Xenobiotics

1. Evaluation of Cell Membrane Damage

We have conducted several studies in which leakage of intracellular enzymes from injured cells has been used to assess the concentration and time-response effects of several hepatotoxic agents in primary cultures of postnatal rat hepatocytes (Anuforo *et al.*, 1978; Acosta *et al.*, 1980, 1982; Mitchell and Acosta, 1981; Santone *et al.*, 1982; Acosta and Sorensen, 1983). In essence, this indicator of cytotoxicity measures the stability of the plasma membrane and its integrity to retain important cytosolic enzymes. In one study, we compared and determined the sensitivity of several key intracellular enzymes found in hepatocytes [argininosuccinate lyase, lactate dehydrogenase, glutamate–oxaloacetate transaminase, glutamate–pyruvate transaminase, and acid phosphatase (AP)] in assessing the cytotoxicity of tetracycline and norethindrone (Anuforo *et al.*, 1978) and acetaminophen and papaverine (Acosta *et al.*, 1980). For all of the enzymes measured, we showed that a linear relationship existed between enzyme leakage and increased dose and duration of exposure to the toxic chemical. However, there were differences in the degree of leakage among the several enzymes. ASAL proved to be the most sensitive in evaluating cytotoxicity of the agents, and AP was the least sensitive. The other three enzymes were intermediate to ASAL and AP in assessing cytotoxicity. In clinical studies, Campanini *et al.* (1970) and Sims and Rautanen (1975) showed that ASAL was a more sensitive indicator of liver damage than the transaminase enzymes, and thus our *in vitro* study correlates quite well with *in vivo* toxicity results. By analyzing the differences in types of enzyme leakage, one may also determine the site of toxicity of the xenobiotic. For instance, treatment of the hepatocyte cultures with tetracycline resulted in a greater release of ASAL and GPT when compared to GOT and AP. The two latter enzymes are found partially or totally membrane bound to organelles; GOT is found partially bound to mitochondria, and AP is completely enclosed in lysosomes. ASAL and GPT are soluble cytoplasmic enzymes and are not associated with organelles. Thus, tetracycline is thought to exert most of its effects on the plasma membrane because of increased leakage of the two cytosolic enzymes.

We have also demonstrated that the enzyme leakage tests are useful in evaluating a series of potentially hepatotoxic compounds which have similar chemical structures: the tricyclic antidepressants (Mitchell and Acosta, 1981). In this particular study, we were able to rank four tricyclic antidepressants, imipramine, desimipramine, amitriptyline, and nortriptyline, as to their severity of cytotoxicity in the hepatocyte culture system. We found that the compounds exerted a dose- and time-dependent order of toxicity as reflected by leakage of LDH and GPT. The dibenzocycloheptadiene derivatives, amitriptyline and its demethylated pharmacologically active metabolite, nortriptyline, were more toxic than the dibenzazepine derivatives, imipramine and its respective demethylated metabolite, desimipramine. There was no significant difference between the cytotoxicity of the parent compounds and their metabolites. Our results also showed that the commonly used trypan blue viability test was inconsistent in evaluating the cytotoxicity of the compounds, while the enzyme leakage tests were far superior in assessing toxicity. Our results confirm the findings of other investigators who have demonstrated the insensitivity of viability stains in detecting early cell injury produced by toxicants (Bhuyan *et al.*, 1976; Miller *et al.*, 1979).

In order to demonstrate that the cytotoxicity produced by the hepatotoxic compounds is specific and selective for the hepatocytes, we have also evaluated the effects of compounds which are known not to be severely hepatotoxic, namely caffeine and sodium salicylate (Anuforo et al., 1978; Acosta et al., 1980). Even with concentrations as high as 5 mM and durations of exposure up to 24 hr, these two compounds produced minimal enzyme leakage in our culture system of postnatal hepatocytes. Our results suggest that the cultured hepatocytes are able to differentiate hepatotoxic compounds from agents which are non-hepatotoxic and are useful experimental models for hepatotoxicity studies.

2. Functional and Metabolic Alterations

To establish the functionality and similarity of the postnatal hepatocyte cultures to the liver in vivo, we have examined a number of parameters which are characteristic of functional hepatocytes. These have included the uptake and glutathione conjugation of sulfobromophthalein (Acosta et al., 1978a), ratio of L- and M-isoenzymes of pyruvate kinase (Acosta et al., 1978a), cytochrome P-450 levels (Nelson et al., 1982), maintenance of stable lactate/pyruvate ratios and urea levels (Santone and Acosta, 1982), marker enzymes of organelle integrity (Acosta and Mitchell, 1983), and calcium dynamics (Santone and Acosta, 1985). These activities can be monitored to not only assure continued functionality and uniformity of the hepatocyte cultures but to reveal injurious effects of toxic chemicals as well. By employing these parameters of cellular functions with indices of cell membrane integrity, the overall cytotoxicity of a particular compound can be more fully evaluated. Findings in such a toxicity testing system should be more reliable in predicting the likelihood of liver injury in vivo than findings in other systems employing propagated cell lines which retain few functional characteristics of hepatocytes in intact liver.

To illustrate the value and sensitivity of these parameters of functional and metabolic integrity in evaluating the cytotoxicity of xenobiotics, we conducted the following study on the toxicity of cadmium in primary cultures of postnatal rat hepatocytes (Santone et al., 1982). In this study, cytotoxicity was evaluated by measuring cell viability, total cellular protein, LDH leakage, intracellular levels of urea, and lactate/pyruvate ratios. Whereas all of the indices of cytotoxicity showed that cadmium was toxic to the cultured hepatocytes, there was a difference in the duration of exposure to cadmium that led to changes in the indices. After 1 hour of exposure, cadmium produced substantial increases in L/P ratios and decreases in urea content in the cultured hepatocytes, while there were only modest increases in LDH leakage and virtually no changes in cell viability or protein content. However, if the cells were allowed to recover for 24 hr after exposure to cadmium by replacing the cadmium-containing medium with fresh, untreated medium, cell viability and total cellular protein were shown to decrease by 40%. These results suggest that the two parameters of metabolic integrity are more sensitive and better indicators of early cell injury produced by cadmium than are the nonspecific parameters such as cellular protein and cell viability, which are more indicative of the later stages of cell injury or cell death. By employing a battery of cytotoxicity tests in evaluating hepatotoxic agents *in vitro,* one is able to establish a better picture of the cell injury process induced by the selected compound.

As mentioned above we have also utilized other parameters of metabolic integrity to assess the cytotoxicity of xenobiotics. In Section IV, these other indicators will be discussed in relation to the metabolism of xenobiotics and the formation of toxic metabolites.

3. Morphological Alterations

Although functional and biochemical changes produced by toxicants provide imortant information on the mechanism of toxicity, these changes should be correlated to morphological manifestations of exposure to toxicants in order to better establish the pattern of cell injury and the overall cellular responses to the toxic compound. In this regard, our laboratory has conducted several studies in which we have evaluated the morphological alterations produced by cadmium chloride in primary cultures of postnatal rat hepatocytes (Acosta and Sorensen, 1983; Sorensen and Acosta, 1984; Sorensen *et al.,* 1984). Initially, we demonstrated that cadmium exposure to cultured hepatocytes resulted in distinct ultrastructural changes, ranging from necrosis at higher doses to cisternal swelling of smooth endoplasmic reticulum, nuclear inclusions, mitochondrial cisternal swelling, peripheral blebbing, vacuolation, and loss of microvilli at intermediate and lower doses (Acosta and Sorensen, 1983). Our *in vitro* observations of cadmium toxicity paralleled the reported *in vivo* effects of cadmium on the liver, namely, degenerative changes in mitochondria and endoplasmic reticulum, vacuolation, and cell necrosis (Hoffman *et al.,* 1975). We also showed, in this same study, that the presence or absence of calcium in the culture medium when the hepatocytes were exposed to cadmium altered the severity of the toxic effects of cadmium; the presence of calcium in the medium reduced functional and morphological manifestations of cadmium cytotoxicity, while the absence of calcium exacerbated the cytotoxicity of cadmium. Thus, the presence of calcium lessened the effects of cadmium on LDH leakage, total urea levels, cell viability, and morphological alterations.

Because of the significance of these interactions of cadmium and calcium at the cellular level, we conducted further studies on the quantitation of morphological alterations produced by cadmium–calcium interaction in cultured hepatocytes, using morphometric analysis of scanning electron micrographs (Sorensen *et al.,* 1984; Sorensen and Acosta, 1984). In this particular study, differences in surface blebbing, alterations in microvilli, variations in cell swelling, and changes in cell shape were used to assess the effects of cadmium with or without calcium. Scanning electron microscopy revealed that in the absence of calcium cadmium exposure of the cultured hepatocytes resulted in more swollen and spherical cells, reduced microvillar number, and enhanced cellular blebbing. By categorizing the cells into three different groups—flattened, intermediate, and spherical—and by noting the degree of blebbing, quantitative morphometric analysis could be performed by indicating, as a percentage, the relative volume of the three different categories and blebs after exposure to different concentrations of cadmium. These quantitative morphometric measurements correlated quite well with our previous biochemical studies in which we demonstrated that calcium could ameliorate some of the cytotoxic effects of cadmium in cultured hepatocytes (Acosta and Sorensen, 1983; Sorensen and Acosta, 1984).

IV. METABOLIC ACTIVATION AND TOXICITY OF XENOBIOTICS IN CULTURED HEPATOCYTES

Although we were able to demonstrate the ability of cultured postnatal hepatocytes to metabolize simple substrates such as BSP and *p*-nitroanisole and form their glutathione, sulfate, and glucuronide conjugates (Acosta *et al.,* 1978a, 1979b) as described previously in Section II, phase I metabolism of xenobiotics via the microsomal cytochrome *P*-450 system has assumed greater importance because of its role in the activation of compounds to toxic metabolites. To demonstrate that our *in vitro* system is able to metabolically activate a compound to toxic intermediate(s), we first examined the metabolism of cyclophosphamide by the postnatal hepatocyte cultures (Acosta and Mitchell, 1981). Before the toxicity of cyclophosphamide can be expressed *in vivo,* especially to cancer cells, the compound must be activated first in the liver to toxic metabolite(s). We demonstrated that the hepatocyte cultures were able to metabolize cyclophosphamide to alkylating metabolites, primarily phosphoramide mustard and 2-chloroethylamine (Fig. 3), as quantified by the *p*-nitrobenzylpyridine assay (Friedman and Boger, 1961). In addition, we also showed that the formed metabolites were toxic to dividing fibroblasts cocultured with the hepatocytes. In essence, this study indicated that the metabolites were produced in sufficient quantities and were stable enough to be transported out of the hepatocytes and to enter the fibroblasts where their toxic effects became evident in cells which rapidly replicate and synthesize large amounts of DNA. Cyclophosphamide showed minimal toxicity to the parenchymal hepatocytes because these cells do not replicate to any significant degree in culture. If pure cultures of fibroblasts alone were exposed to cyclophosphamide, toxicity was not observed because the fibroblasts cannot metabolically activate cyclophosphamine to its toxic intermediates. The toxicity

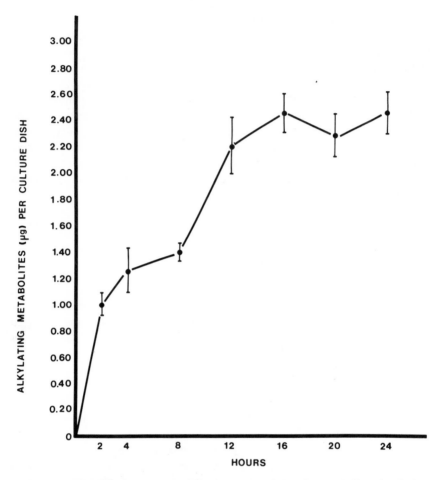

Fig. 3.   Estimation by the *p*-nitrobenzylpyridine assay of alkylating metabolites of cyclophosphamide produced by 1-day-old postnatal hepatocyte cultures as a function of time of exposure to 1 × 10^{-3} *M* cyclophosphamide. Vertical bars indicate ±standard error of the mean. [Reproduced with permission of the publisher from Acosta and Mitchell (1981).]

of cyclophosphamide to dividing fibroblasts in the coculture system with hepatocytes was significantly mitigated by administering SKF-525A, an inhibitor of the cytochrome *P*-450 monooxygenase system, at the same time the cells were exposed to cyclophosphamide. This study has provided strong evidence that postnatal hepatocytes in primary culture are metabolically competent to activate protoxic compounds via a functional cytochrome *P*-450 monooxygenase system.

Another well-known hepatotoxin which expresses its hepatotoxicity only after metabolic activation to toxic metabolite(s) is acetaminophen (Mitchell *et al.*, 1973; Potter *et al.*, 1974). Initially, in our culture system of postnatal rat hepatocytes, we showed that it took approximately 6 hr of treatment with acetaminophen before toxicity (as expressed by enzyme leakage) was observed in the cultures (Fig. 4), whereas the other hepatotoxins investigated previously showed toxicity at earlier times (Anuforo *et al.*, 1978; Acosta *et al.*, 1980; Mitchell and Acosta, 1981). We suggested that the time lag observed in the cytotoxicity of acetaminophen may represent the time required for conversion and accumulation of sufficient quantities of toxic metabolites to saturate and deplete cytoplasmic stores of glutathione (Acosta *et al.*, 1980; Mitchell and Acosta, 1982).

To substantiate the cytotoxicity study on acetaminophen, we have recently

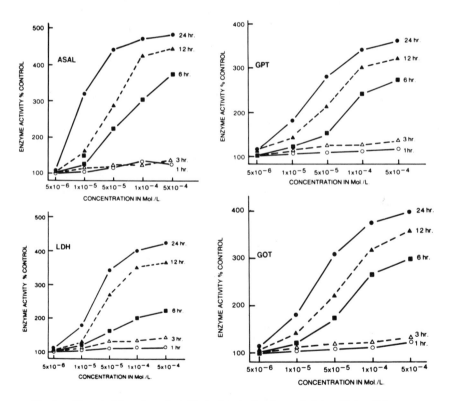

Fig. 4. Time- and dose-dependent effects of acetaminophen on leakage of intracellular enzymes from 4-day-old cultured postnatal hepatocytes. ASAL, argininosuccinate lyase; LDH, lactate dehydrogenase; GPT, glutamate pyruvate transaminase; GOT, glutamate oxaloacetate transaminase. Standard errors were less than 5%. [Reproduced with permission from Acosta *et al.* (1980).]

conducted some studies on the metabolism of the agent in our culture system. We were able to show that cultured postnatal hepatocytes are capable of metabolizing acetaminophen to conjugates in similar fashion as whole rats and as isolated rat hepatocytes (Acosta *et al.*, 1985a). The formation of the two principal conjugates, glucuronide and sulfate, was dependent on acetaminophen dose and time of exposure. The glucuronide conjugation is thought to be important in the detoxification of large doses of acetaminophen by the liver (Mitchell *et al.*, 1973). The levels of sulfate conjugate in our study were lower than the amount of glucuronide conjugate, and this correlates with the reported limited capacity of sulfation for detoxification of acetaminophen (Suolinna and Mantyla, 1980). Thus, these intial studies demonstrated that cultured postnatal hepatocytes are capable of phase II conjugation metabolism of acetaminophen.

Because the hepatotoxicity of acetaminophen is closely linked to the formation of a toxic intermediate which may covalently bind to important cellular macromolecules, we also examined the capacity of our culture system to generate reactive intermediates from the metabolism of acetaminophen. The experiments were designed both to evaluate the distribution of covalent binding of reactive metabolites of [^{3}H]acetaminophen to proteins in the various subcellular fractions isolated from treated hepatocytes and to determine if the reactive intermediates were sufficiently stable to migrate throughout the cell. A substantial amount of covalent binding of the reactive intermediates generated by the metabolism of acetaminophen occurred by 4 hr after exposure of the hepatocytes (Table I). Thus, covalent binding was evident at or before the time cytotoxicity was seen in the cultures. The radiolabel was localized primarily in the microsomal fraction (37%), where the cytochrome *P*-450 monooxygenase system responsible for the metabolic activation of acetaminophen is centered. Approximately 25% of the

TABLE I

Covalent Binding of [3H]Acetaminophen to Subcellular Fractions[a]

Time (hr)	Nuclear fraction	Mitochondrial fraction	Cytoplasmic fraction	Microsomal fraction
4	99 ± 3 (19)	99 ± 3 (21)	109 ± 4 (23)	170 ± 1 (37)
8	109 ± 2 (20)	112 ± 3 (21)	130 ± 3 (24)	185 ± 4 (35)
12	117 ± 5 (8)	133 ± 3 (21)	177 ± 12 (27)	220 ± 13 (34)
24	107 ± 3 (21)	104 ± 1 (17)	173 ± 4 (29)	229 ± 8 (37)

[a]Data are expressed in terms of dpm bound/gm protein; each value represents the mean ± standard error, $n = 4$. The cells were treated with a mixture of 1.0 μCi of [3H]acetaminophen per culture dish diluted in culture medium containing 5×10^{-3} M unlabeled acetaminophen. The number in parentheses indicates the percentage of total radiolabel bound to the individual cellular fraction. Reproduced with permission from Acosta *et al.* (1985a).

radiolabel was located in the cytoplasmic protein fraction, which indicated the ability of the reactive intermediates to bind to target proteins outside of the endoplasmic reticulum. The remaining radiolabel was distributed evenly between the nuclear fraction and the mitochondrial fraction. The amount of reactive intermediates bound to the nuclear fraction was probably somewhat less than the reported 17–20% because this fraction is partially contaminated with cellular debris and unbroken cells upon differential centrifugation. The relative percentages of radiolabel bound to proteins of the four subcellular fractions remained quite constant as a function of time, although there was a steady increase from 23 to 29% of total radioactivity found in the cytoplasmic fraction as the incubation time was increased from 4 to 24 hr. The amount of radiolabel bound in the microsomal fraction was constantly almost twice the values reported in the nuclear and mitochondrial fractions.

These results with cultured postnatal hepatocytes compared favorably with the *in vivo* studies of Jollow *et al.* (1973) who reported the same general distribution of bound radiolabel in the subcellular fractions but at only one duration of exposure with acetaminophen. The microsomal studies of Potter *et al.* (1974), however, confirmed our time-related studies in that they showed increased binding of the radiolabel as a function of time.

Although covalent binding studies of a particular compound reveal that a reactive intermediate has been formed and has interacted with macromolecules within the hepatocytes, it does not necessarily mean that there is a direct correlation between covalent binding and the subsequent hepatotoxicity, as Gillette (1981) has suggested. For instance, Devalia *et al.* (1982) have reported that the toxicity of acetaminophen to isolated rat hepatocytes can be diminished by certain agents without affecting the amount of material bound covalently to cellular protein. These investigations suggested that covalent binding per se may not be the primary cause of cell death but may initiate a series of events which lead to cell death. In this regard, in order to evaluate the cytotoxicity of acetaminophen in our culture system of postnatal rat hepatocytes and to relate it to the covalent binding of the reactive intermediate(s) to the various subcellular fractions, we examined alterations of key marker enzymes or activities associated with several subcellular fractions as indicators of toxicity produced by acetaminophen and its reactive intermediate(s).

The integrity of the microsomal or endoplasmic reticulum fraction was evaluated by measuring cytochrome P-450 levels and glucose-6-phosphatase activity. The cytochrome P-450 monooxygenase system is responsible for the biotransformation of acetaminophen to the reactive intermediate presumed to cause the resulting toxicity to liver cells (Mitchell *et al.*, 1973). The dose-dependent effects of acetaminophen on actual cytochrome P-450 levels in the microsomal fraction of parenchymal hepatocyte cultures is demonstrated in Table II. Significant decreases in cytochrome P-450 content were evident with both doses of

TABLE II

The Effects of Acetaminophen on Cytochrome *P*-450 Levels and Glucose-6-Phosphatase (G-6-Pase) Activity in Cultured Postnatal Hepatocytes[a]

Time (hr)	Control		1×10^{-4} M		5×10^{-3} M	
	P-450	G-6-Pase	*P*-450	G-6-Pase	*P*-450	G-6-Pase
0	0.67 + 0.02	140.8 + 2.7				
4	0.66 + 0.02	132.6 + 0.4	0.54 + 0.1[b] (18)	102.4 + 2.9[b] (29)	0.50 + 0.1[b] (24)	82.2 + 1.0[b] (38)
8	0.60 + 0.02	124.6 + 1.2	0.51 + 0.2[b] (15)	94.0 + 2.3[b] (25)	0.45 + 0.01[b] (25)	74.3 + 1.3[b] (40)
12	0.59 + 0.02	100.8 + 0.6	0.45 + 0.2[b] (12)	51.4 + 0.8[b] (50)	0.43 + 0.01[b] (27)	30.6 + 0.6[b] (70)
24	0.51 + 0.01	107.2 + 1.8	0.42 + 0.01[b] (18)	42.2 + 1.5[b] (61)	0.36 + 0.02[b] (30)	18.6 + 0.5[b] (82)

[a] G-6-Pase activity is expressed in terms of nmol P_i liberated/5 min incubation/mg microsomal protein, and *P*-450 levels are expressed as nmol cytochrome *P*-450/mg microsomal protein of primary cultures of 24-hr-old parenchymal hepatocytes. Each value represents the mean ± standard error, $n = 4$. The number in parentheses indicates the percentage decrease compared to control value at the same point. Reprinted with permission from Acosta *et al.* (1985a).

[b] Statistically different from controls, $p < 0.05$.

acetaminophen as early as 4 hr after treatment. The levels declined over the entire 24-hr treatment period for the $5 \times 10^{-3} M$ treatment, even though the majority of the apparent damage occurred within the first 4 hr of exposure. The results indicated that the cytochrome P-450 system had become a susceptible and early target of the reactive intermediate formed in the same microsomal fraction.

Glucose-6-phosphatase is considered a key enzyme in carbohydrate metabolism whose intracellular distribution has been localized almost exclusively in the microsomal fraction (de Duve, 1975). The effects of acetaminophen treatment on glucose-6-phosphatase activity in the microsomal fraction of parenchymal hepatocytes is described in Table II. Two specific doses of acetaminophen were chosen: previous data had shown that the $1 \times 10^{-4} M$ dose was generally not cytotoxic, whereas the $5 \times 10^{-3} M$ dose produced definite cell injury and death. However, glucose-6-phosphatase activity declined by 60% at the end of the 24-hr treatment with the presumably nontoxic dose of acetaminophen and by 83% at the end of the 24-hr exposure with the toxic dose. The results suggest that the localization of the enzyme in the microsomal membrane system had made it a susceptible target for early damage produced by the toxic intermediate because both doses caused a 23–38% decrease in enzyme activity as early as 4 hr after exposure.

In conclusion, the two marker enzymes assayed in the microsomal fraction were early primary targets of acetaminophen toxicity. The activity of each enzyme declined dramatically and the toxicity was discernible as early as 4 hr after treatment.

The integrity of the cytoplasmic subcellular fraction was assessed by evaluating alterations in urea levels in the cultured postnatal hepatocytes. Urea formation is limited almost entirely to the liver and the enzymes of the urea cycle are located in the cytoplasm of parenchymal hepatocytes (Titheradge and Haynes, 1980). The effects of acetaminophen treatment on urea levels are described in Table III. The experiments were designed to simultaneously measure the concentration of intracellular urea and the amount of urea excreted from the cells into the culture medium. In this way, the total concentration of urea could be reported. There were gradual decreases in total urea levels as the dose was increased for the 4- and 8-hr exposure periods. By 12 hr of treatment, there were significant decreases in urea levels at each of the three concentrations used in the study. Thus, acetaminophen produced dose- and time-dependent alterations of urea synthesis. We have previously shown in our laboratory that assessment of urea levels in cultured hepatocytes after exposure to cadmium chloride proved to be a sensitive indicator of cellular injury (Santone et al., 1982). In comparison to the microsomal fraction, acetaminophen caused later changes in the cytoplasmic fraction as represented by urea formation. Thus, it appears that cell injury produced by acetaminophen treatment of cultured hepatocytes occurred first in the endoplasmic reticulum, followed by changes in the cytoplasm.

TABLE III

The Effects of Acetaminophen on Total Urea Levels in Cultured Postnatal Hepatocytes[a]

Time (hr)	Controls	$1 \times 10^{-4} M$	$1 \times 10^{-3} M$	$5 \times 10^{-3} M$
4	13.40 + 0.4	12.70 + 0.15	11.70 + 0.35	10.30 + 0.2[b]
8	11.96 + 0.1	11.82 + 0.1	10.07 + 0.1[b]	8.94 + 0.1[b]
12	11.50 + 0.15	9.65 + 0.15[b]	8.73 + 0.15[b]	7.59 + 0.2[b]

[a]Total urea represents the sum of the urea found excreted in the culture medium and the urea found intracellularly in 24-hr-old hepatocytes and is expressed in terms of μ urea/mg protein per culture dish. Each value represents the mean ± standard, $n = 4$.
[b]Statistically different from controls, $p < 0.05$.

To evaluate possible toxic effects of acetaminophen on mitochondrial integrity, changes in activity of the mitochondrial marker enzyme, succinate dehydrogenase (SDH), were evaluated after exposure of the cultured hepatocytes. The data presented in Table IV suggest that succinate dehydrogenase is not a primary target of the toxic effects of acetaminophen. The $1 \times 10^{-4} M$ dose of acetaminophen produced no significant decrease in SDH activity, while the $5 \times 10^{-3} M$ dose required a 24-hr period of exposure before a significant decline of SDH activity was detectable. Succinate dehydrogenase is an extremely stable enzyme complex that is tightly embedded in the inner mitochondrial membrane (Singer, 1974). This apparent membrane protection of succinate dehydrogenase may be

TABLE IV

The Effects of Acetaminophen on Mitchondrial Succinate Dehydrogenase Activity in Cultured Postnatal Hepatocytes[a]

Time (hr)	Controls	$1 \times 10^{-4} M$	$5 \times 10^{-3} M$
0	3.29 + 0.14		
4	3.47 + 0.09	3.38 + 0.13	3.21 + 0.07
8	3.23 + 0.10	3.18 + 0.08	3.14 + 0.09
12	3.12 + 0.02	3.09 + 0.04	3.07 + 0.10
24	3.16 + 0.04	2.89 + 0.06	2.70 + 0.11[b]

[a]Succinate dehydrogenase activity is expressed as μmol ferricyanide reduced/min/mg protein of primary cultures of 24-hr-old parenchymal hepatocytes. Each value represents the mean ± standard error, $n = 4$.
[b]Statistically different from controls, $p < 0.05$.

responsible for the lack of significant toxic effects of acetaminophen on mitocondria. One may postulate that the reactive metabolites of acetaminophen had arylated other critical cellular constituents in the endoplasmic reticulum and cytoplasm prior to substantial migration of the electrophiles to the mitochondria. However, significant covalently bound radiolabel was detectable in the mitochondrial fraction as early as 4 hr after acetaminophen exposure. It must be concluded that no positive correlation can be drawn between covalent binding in the mitochondrial fraction and alterations in mitochondrial function or integrity.

One of the primary objectives of this study on acetaminophen was to evaluate the time frame of injury that could be detected by the hepatocyte culture model. Covalent binding of the reactive intermediate(s) of [^{3}H]acetaminphen increased as a function of exposure time to the parent compound. The radiolabel was distributed among the subcellular fractions but was primarily localized in the microsomal and cytoplasmic fractions. It is important to recognize that covalent binding in a subcellular fraction did not always parallel decreased functions or activities in that fraction. The results illustrate that the two markers of the microsomal fraction, cytochrome P-450 and glucose-6-phosphatase, were the primary targets of acetaminophen toxicity. The other subcellular fractions, cytoplasmic and mitochondrial, seemed to be secondary targets of acetaminophen toxicity.

V. CONCLUSIONS

The role of the liver in metabolism of drugs and chemicals has long been recognized. Its role in the metabolic activation of xenobiotics to potentially toxic metabolites further illustrates the significance of the liver in the toxicological effects of compounds. Our results support the premise that primary cultures of postnatal rat hepatocytes serve as a useful and reliable experimental model in the investigation of metabolism-mediated cytotoxicity of xenobiotics. The cultured hepatocytes have the capacity to perform integrated drug metabolism, i.e., both phase I oxidation and phase II conjugation reactions were demonstrated. Additionally, the system permits the use of early, sensitive measures of cytotoxicity based on differentiated hepatocyte function to evaluate the potential toxicity of xenobiotics.

ACKNOWLEDGMENT

This research was supported in part by EPA Cooperative Agreement CR-811215.

REFERENCES

Acosta, D., and Mitchell, D. B. (1981). Metabolic activation and cytotoxicity of cyclophosphamide in primary cultures of postnatal rat hepatocytes. *Biochem. Pharmacol.* **30,** 3225–3230.

Acosta, D., and Mitchell, D. B. (1983). Subcellular localization of hepatocyte injury due to metabolic activation of acetaminophen. *Toxicologist* **3,** 113.

Acosta, D., and Sorensen, E. M. B. (1983). Role of calcium in cytotoxic injury of cultured hepatocytes. *Ann. N.Y. Acad. Sci.* **407,** 78–92.

Acosta, D., and Wenzel, D. G. (1974). Injury produced by free fatty acids to lysosomes and mitochondria in cultured heart muscle and endothelial cells. *Atherosclerosis* **20,** 417–426.

Acosta, D., and Wenzel, D. G. (1975). A permeability test for the study of mitochondrial injury in *in vitro* cultured heart muscle and endothelioid cells. *Histochem. J.* **7,** 45–46.

Acosta, D., Anuforo, D. C., and Smith, R. V. (1978a). Primary monolayer cultures of postnatal rat liver cells with extended differentiated functions. *In Vitro* **14,** 428–436.

Acosta, D., Puckett, M., and McMillin, R. (1978b). Ischemic myocardial injury in cultured heart cells: Leakage of cytoplasmic enzymes from injured cells. *In Vitro* **14,** 728–732.

Acosta, D., Puckett, M., and McMillin, R. (1978c). Ischemic myocardial injury in cultured heart cells: *in situ* lysosomal damage. *Experientia* **34,** 1388–1389.

Acosta, D., Anuforo, D. C., McMillin, R., Soine, W. H., and Smith, R. V. (1979a). Comparison of cytochrome P-450 levels in adult rat liver, postnatal rat liver, and primary cultures of postnatal rat hepatocytes. *Life Sci.* **25,** 1413–1418.

Acosta, D., Anuforo, D. C., Smith, R. V., McMillin, R., and Soine, W. H. (1979b). Cytochrome P-450 levels and *O*-demethylation activity in cultures of rat hepatocytes. *Fed. Proc.* **38,** 366.

Acosta, D., Anuforo, D. C., and Smith, R. V. (1980). Cytotoxicity of acetaminophen and papaverine in primary cultures of rat hepatocytes. *Toxicol. Appl. Pharmacol.* **53,** 306–314.

Acosta, D., Mitchell, D. B., Santone, K. S., Bock, A., and Lewis, W. (1982). Lack of cytotoxicity of ticrynafen in primary cultures of rat liver cells. *Toxicol. Lett.* **10,** 385–388.

Acosta, D., Mitchell, D. B., and Bruckner, J. V. (1985a). Hepatotoxicity: An *in vitro* approach to the study of metabolism and toxicity of chemicals and drugs using cultured rat hepatocytes. *Saf. Eval.* **2,** 305–317.

Acosta, D., Sorensen, E. M. B., Anuforo, D. C., Mitchell, D. B., Ramos, K., Santone, K. S., and Smith, M. A. (1985b). An *in vitro* approach to the study of target organ toxicity of drugs and chemicals. *In Vitro* **21,** 495–504.

Anuforo, D. C., Acosta, D., and Smith, R. V. (1978). Hepatotoxicity studies with primary cultures of rat liver cells. *In Vitro* **14,** 981–988.

Armato, U., Draghi, E., and Andreis, P. G. (1975). Studies on the persistence of differentiated functions in rat hepatocytes set into primary tissue cultures. *Cell Differ.* **4,** 147–153.

Berry, M. N., and Friend, D. S. (1969). High-yield preparation of isolated rat liver parenchymal cells from rat liver: A biochemical and fine structural study. *J. Cell Biol.* **43,** 506–520.

Bhuyan, B. K., Loughman, B. E., Fraser, T. J., and Day, K. J. (1976). Comparison of different methods of cell viability after exposure to cytotoxic compounds. *Exp. Cell Res.* **97,** 275–280.

Bissell, D. M., and Guzelian, P. S. (1979). Ascorbic acid deficiency and cytochrome P-450 in adult rat hepatocytes in primary monolayer cultures. *Arch. Biochem. Biophys.* **192,** 569–576.

Bonney, R. J. (1974). Adult liver parenchymal cells in primary culture: Characteristics and cell recognition standards. *In Vitro* **10,** 130–142.

Bonney, R. J., Becker, J. E., Walker, P. R., and Potter, V. R. (1974). Primary monolayer cultures of adult rat liver parenchymal cells suitable for the study of the regulation of enzyme synthesis. *In Vitro* **9,** 399–413.

Campanini, R. Z., Tapia, R. A., Sarnat, W., and Natelson, S. (1970). Evaluation of serum argininosuccinate lyase (ASAL) concentrations as an index to parenchymal liver disease. *Clin. Chem.* **16**, 44–53.

Cheesebeuf, M., Olsson, A., Bournot, P., Desgres, J., Guigeuet, M., Maume, G., Maume, B. F., Perissel, J., and Padieu, P. (1974). Long term cell cultures of rat liver epithelial cells retaining some hepatic functions. *Biochemie* **56**, 1365–1379.

Dallner, G., Siekowitz, and Palade, G. E. (1966). Biogenesis of endoplasmic reticulum membranes. *J. Cell. Biol.* **30**, 73–95.

de Duve, C. (1975). Exploring cells with a centrifuge. *Science* **189**, 186–194.

Devalia, J. L., Ogilvie, R. C., and McLean, A. E. M. (1982). Dissociation of cell death from covalent binding of paracetamol by flavones in a hepatocyte system. *Biochem. Pharmacol.* **31**, 3745–3749.

Dickins, M., and Peterson, R. E. (1980). Effects of a hormone-supplemented medium on cytochrome P-450 content and monooxygenase activities of rat hepatocytes in primary culture. *Biochem. Pharmacol.* **29**, 1231–1238.

Evans, V. J., Earle, W. R., Watson, E. P., Waltz, H. K., and Mackey, C. J. (1952). The growth *in vitro* of massive cultures of liver cells. *J. Natl. Cancer Inst.* **12**, 1245–1248.

Fisher, P. B. (1976). Quantitation of membrane damage by 51Cr release. *Tissue Cult. Assoc. Man.* **2**, 363–365.

Friedman, O. M., and Boger, E. (1961). Colorimetric estimation of nitrogen mustards in aqueous media. *Anal. Chem.* **33**, 906–910.

Fry, J. R., and Bridges, J. W. (1979). Use of primary hepatocyte cultures in biochemical toxicology. *Rev. Biochem. Toxicol.* **1**, 201–247.

Fry, J. R., Jones, C. A., Wiebkin, P., Bellemann, P., and Bridges, J. W. (1976). The enzymic isolation of adult rat hepatocytes in a functional and viable state. *Anal. Biochem.* **71**, 341–350.

Gillette, J. R. (1981). An integrated approach to the study of chemically reactive metabolites of acetaminophen. *Arch. Intern. Med.* **141**, 375–379.

Gram, T. E., Guarino, A. M., Schroder, D. H., and Gillette, J. R. (1969). Changes in certain kinetic properties of hepatic microsomal aniline hydroxylase and ethylmorphine demethylase associated with postnatal development and maturation in male rats. *Biochem. J.* **113**, 681–685.

Grisham, J. W. (1979). Use of hepatic cell cultures to detect and evaluate the mechanisms of action of toxic chemicals. *Int. Rev. Exp. Pathol.* **20**, 124–210.

Guillouzo, A., Feldmann, C., Boisnard, M., Sapin, C., and Benhamou, J. P. (1975). Ultrastructural location of albumin and fibrinogen in primary cultures of new-born rat liver tissue. *Exp. Cell Res.* **96**, 239–246.

Guzelian, P. S., Bissell, D. M., and Meyer, U. A. (1977). Drug metabolism in adult rat hepatocytes in primary culture. *Gastroenterology* **172**, 1232–1239.

Hayner, N. T., Driscoll, J., Ferayouni, L., Spies-Karotkin, G., and Jauregui, H. O. (1982). A sensitive method for protein determination in freshly isolated and cultured cells. *J. Tissue Cult. Methods* **7**, 77–80.

Henderson, P. T. (1971). The metabolism of drugs in rat liver during the perinatal period. *Biochem. Pharmacol.* **20**, 1225–1232.

Hildebrandt, A., and Estabrook, R. W. (1971). Evidence for the participation of cytochrome D5 in hepatic microsomal mixed function oxidation reactions. *Arch. Biochem. Biophys.* **143**, 66–79.

Hoffman, E. O., Cook, J. A., Di Luzio, N. R., and Coover, J. A. (1975). The effects of acute cadmium administration in the liver and kidney of the rat: Light and electron microscopic studies. *Lab. Invest.* **32**, 655–664.

Howard, R. B., Christensen, A. K., Gibbs, F. A., and Pesch, L. A. (1967). The enzymatic preparation of isolated intact parenchymal cells from the rat liver. *J. Cell Biol.* **35**, 675–684.

Jollow, D. J., Mitchell, J. R., Potter, W. Z., Gillette, J. R., and Brodie, B. B. (1973). Acetaminophen-induced hepatic necrosis. II. Role of covalent binding in vivo. *J. Pharmacol. Exp. Ther.* **187,** 195–202.

Katz, I. (1975). Vital and cytochemical staining of cell cultures with a fluorescent dye to determine viability. *Tissue Cult. Assoc. Man.* **1,** 41–42.

Laishes, B. A., and Williams, G. M. (1976). Conditions affecting primary cell cultures of functional adult rat hepatocytes. II. Dexamethasone enhanced longevity and maintenance of morphology. *In Vitro* **12,** 821–832.

Leffert, H. L., and Paul, D. (1972). Studies on primary cultures of differentiated fetal liver cells. *J. Cell Biol.* **52,** 559–568.

Michalopoulos, G., Sattler, G. L., and Pitot, H. C. (1976). Maintenance of microsomal cytochrome b_5 and P-450 in primary cultures of parenchymal liver cells on collagen membranes. *Life Sci.* **18,** 1139–1144.

Miller, M. R., Castellot, J. J., and Pardee, A. B. (1979). A general method for permeabilizing monolayer and suspension cultured animal cells. *Exp. Cell Res.* **120,** 421–425.

Minck, K., Schupp, R. R., Illing, H. P. A., Kahl, G. F., and Netter, K. J. (1973). Interrelationship between demethylation of *p*-nitroanisole and conjugation of *p*-nitrophenol in rat liver. *Arch. Pharmacol.* **279,** 347–353.

Mitchell, D. B., and Acosta, D. (1981). Evaluation of the cytotoxicity of tricyclic antidepressants in primary cultures of rat hepatocytes. *J. Toxicol. Environ. Methods* **7,** 83–92.

Mitchell, D. B., and Acosta, D. (1982). In vitro hepatotoxicity of the glutathione depletors diethylmaleate, iodoacetamide, and acetaminophen. *Toxicologist* **2,** 67.

Mitchell, D. B., Santone, K. S., and Acosta, D. (1980). Evaluation of cytotoxicity in cultured cells by enzyme leakage. *J. Tissue Cult. Methods* **6,** 113–116.

Mitchell, J. R., Jollow, D. J., Patter, W. Z., Gillette, J. R., and Brodie, B. B. (1973). Acetaminophen-induced hepatic necrosis. I. Role of drug metabolism binding *in vivo*. *J. Pharmacol. Exp. Ther.* **187,** 185–194.

Nealon, D. G., Sorensen, E. M. B., and Acosta, D. (1984). A fluorescence polarization procedure for the evaluation of the effects of cadmium and calcium on plasma membrane fluidity. *J. Tissue Cult. Methods* **9,** 11–17.

Nelson, K. F., Acosta, D., and Bruckner, J. V. (1982). Long-term maintenance and induction of cytochrome P-450 in primary cultures of rat hepatocytes. *Biochem. Pharmacol.* **31,** 2211–2214.

Paine, A. J., Williams, L. J., and Legg, R. F. (1979). Apparent maintenance of cytochrome P-450 by nicotinamide in primary cultures of rat hepatocytes. *Life Sci.* **24,** 2185–2192.

Paine, A. J., Hockin, L. J., and Allen, C. M. (1982). Long term maintenance and induction of cytochrome P-450 in rat liver cell culture. *Biochem. Pharmacol.* **31,** 1175–1178.

Potter, W. Z., Thorgeirsson, S. S., Jollow, D. J., and Mitchell, J. R. (1974). Acetaminophen-induced hepatic necrosis. V. Correlation of hepatic necrosis, covalent binding and glutathione depletion in hamsters. *Pharmacology* **12,** 129–143.

Ramos, K., Combs, A. B., and Acosta, D. (1983). Cytotoxicity of isoproterenol to cultured heart cells: Effects of antioxidants on modifying membrane damage. *Toxicol. Appl. Pharmacol.* **70,** 317–323.

Ramos, K., Combs, A. B., and Acosta, D. (1984). Role of calcium in isoproterenol cytotoxicity to cultured myocardial cells. *Biochem. Pharmacol.* **33,** 1989–1992.

Salocks, C. B., Hsieh, D. P., and Byard, J. L. (1981). Butylated hydroxytoluene pretreatment protects against cytotoxicity and reduces covalent binding of aflatoxin B_1 in primary hepatocyte cultures. *Toxicol. Appl. Pharmacol.* **59,** 331–345.

Santone, K. S., and Acosta, D. (1982). Measurement of functional metabolic activity as a sensitive parameter of cytotoxicity in cultured hepatocytes. *J. Tissue Cult. Methods* **7,** 137–142.

Santone, K. S., and Acosta, D. (1984). The role of extracellular calcium in CCl₄ injury of cultured rat hepatocytes. *Toxicologist* **4**, 133.

Santone, K. S., and Acosta, D. (1985). Interference of intracellular calcium dynamics in CCl₄ injury of cultured rat hepatocytes. *Toxicologist* **5**, 74.

Santone, K. S., Acosta, D., and Bruckner, J. V. (1982). Cadmium toxicity in primary cultures of rat hepatocytes. *J. Toxicol. Environ. Methods* **10**, 169–177.

Schenkman, J. B., Cinti, D. L., and Moldeus, P. (1973). The mitochondrial role in hepatic cell mixed-function oxidations. *Ann. N.Y. Acad. Sci.* **212**, 420–427.

Sims, F. H., and Rautanen, P. (1975). Serum argininosuccinate lyase: Observations on sensitivity and specificity of test in the detection of minimal hepatocellular damage. *Clin. Biochem.* **8**, 213–221.

Sinclair, J. F., Sinclair, P. R., and Bonkowsky, H. L. (1979). Hormonal requirements for the induction of cytochrome P-450 in hepatocytes cultured in a serum-free medium. *Biochem. Biophys. Res. Commun.* **86**, 710–717.

Singer, T. P. (1974). Determination of the activity of succinate, NADH, choline, and alpha-glycerophosphate dehydrogenases. *Methods Biochem. Anal.* **22**, 123–140.

Smith, R. V., Acosta, D., and Rosazza, J. P. (1977). Cellular and microbial models in the investigation of mammalian metabolism of xenobiotics. *Adv. Biochem. Eng.* **5**, 69–100.

Sorensen, E. M. B., and Acosta, D. (1984). Cadmium-induced hepatotoxicity as evaluated by morphometric analysis. *In Vitro* **20**, 771–779.

Sorensen, E. M. B., Smith, N. K. R., Boecker, C. S., and Acosta, D. (1984). Calcium amelioration of cadmium-induced cytotoxicity in cultured rat hepatocytes. *In Vitro* **20**, 771–779.

Sorensen, E. M. B., Ramirez-Mitchell, R., and Acosta, D. (1984). Stereographic evaluation of cellular changes *in vivo* or *in vitro*. *J. Tissue Cult. Methods* **9**, 23–28.

Stacey, N. H., and Klaassen, C. D. (1982). Comparison of the effects of metals on cellular injury and lipid peroxidation in isolated rat hepatocytes, *J. Toxicol. Environ. Health* **7**, 139–147.

Stacey, N. H., Cantilena, L. R., and Klaassen, C. D. (1980). Cadmium toxicity and lipid peroxidation in isolated rat hepatocytes. *Toxicol. Appl. Pharmacol.* **53**, 470–480.

Stege, T. E., Loose, L. D., and DiLuzio, N. R. (1975). Comparative uptake of sulfobromophthalein by isolated Kupffer and parenchymal cells. *Proc. Soc. Exp. Biol. Med.* **149**, 455–461.

Sudilovsky, O. (1977). A tissue culture method for toxicity testing of biomaterials. *Tissue Cult. Assoc. Man.* **3**, 607–612.

Suolinna, E. M., and Mantyla, E. (1980). Glucuronide and sulfate conjugation in isolated liver cells from control and phenobarbital- or PCB-treated rats. *Biochem. Pharmacol.* **29**, 2963–2968.

Titheradge, M. A., and Haynes, R. C. (1980). The hormonal stimulation of ureogenesis in isolated hepatocytes through increases in mitochondrial ATP production. *Arch. Biochem. Biophys.* **201**, 44–55.

Tolnai, S. (1975a). A method for viable cell count. *Tissue Cult. Assoc. Man.* **1**, 37–38.

Tolnai, S. (1975b). Methods for the demonstration of phagocytosis by peritoneal mononuclear phagocytes. *Tissue Cult. Assoc. Man.* **1**, 39–40.

Vainio, H. (1973). Drug hydroxylation and glucuronidation in liver microsomes of phenobarbitol treated rats. *Xenobiotica* **3**, 715–721.

van Berkel, T. J. C., Koster, J. F., and Hulsmann, W. C. (1972). Distribution of L- and M-type pyruvate kinase between parenchymal and Kupffer cells of rat liver. *Biochim. Biophys. Acta* **276**, 425–429.

van Bezooijen, C. F. A., Grell, T., and Knook, D. L. (1976). Bromosulfophthalein uptake by isolated liver parenchymal cells: *Biochem. Biophys. Res. Commun.* **69**, 356–361.

Wenzel, D. G., and Acosta, D. (1975). The use of cultured cells for in situ measurement of mitochondrial fragility. *Tissue Cult. Assoc. Man.* **1**, 221–223.

Wenzel, D. G., and Reed, B. L. (1976). The use of cultured cells for in situ measurement of lysosomal fragility. *Tissue Cult. Assoc. Man.* **2,** 291–293.

Wininger, M. T., Kulik, F. A., and Ross, W. D. (1979). *In vitro* clonal cytotoxicity assay for chemicals using Chinese hamster ovary cells (CHO-KI). *Tissue Cult. Assoc. Man.* **5,** 1091–1093.

9

The Analysis of Carcinogen-Induced Pleiotropic Drug Resistance in Rat Hepatocytes

BRIAN I. CARR

Department of Medical Oncology and Therapeutics Research
City of Hope National Medical Center
Duarte, California 91010

I. INTRODUCTION

A. Human Liver Cancer Resists Cytotoxic Chemotherapy

Hepatocellular carcinoma is one of the most common cancers, worldwide, due to its high incidence rates in Southeast Asia and sub-Saharan Africa. Despite the availability of a wide variety of potent cell toxins for use in cancer chemotherapy (cytotoxic chemotherapy), hepatocellular carcinoma has been a very disappointing disease to treat. This is thought to be due to the high degree of resistance of the neoplastic liver cells to almost all antineoplastic agents that have so far been evaluated. This poses a very interesting biological problem. What is the nature of the adaptive change that hepatocytes acquire when they become transformed into cancer cells? The change(s) allows them to withstand the cytocidal and anti-proliferative action of a variety of chemicals that exert their toxicity through several different mechanisms.

B. Chemically and Virally Transformed Cells Resist Cytotoxicity

The idea that the cells of an experimental animal, which had been previously exposed to a toxin, might be altered so as to be resistant to the toxic effects of a

215

THE ISOLATED
HEPATOCYTE

subsequent exposure to the same or a different toxin originated with the experiments of MacNider (1916, 1936a,b,c, 1937) and Hunter (1928). Subsequently, Haddow (1935, 1938a,b; Haddow and Robinson, 1937; Haddow et al., 1937) showed that there was a structure–activity relationship for the polycyclic aromatic hydrocarbon carcinogens and their analogs with respect to their growth inhibitory properties: the more carcinogenic the chemical compound, the more growth inhibitory were its actions. Haddow showed that this applied both to normal growth and to the Jensen sarcoma and that carcinogens inhibited growth in the following order: normal tissues > spontaneously occurring tumors > chemically induced tumors. He showed that a chemically induced tumor was relatively resistant to the toxic and growth inhibitory actions of both the inducing and unrelated carcinogens.

This led to his hypothesis on cellular inhibition and the origin of cancer (Haddow, 1938a,b,c), that "cancer cells arise and commence a career of proliferation under conditions which impair the life of normal cells." This was based upon the parallelism between the growth inhibitory actions and carcinogenicity of both ionizing radiation and chemical carcinogens. The relative sensitivity to growth inhibition of normal cells compared to tumorous cells taken from tissues and placed in culture was confirmed by Lasnitzki (1949). There was a differential sensitivity between normal and neoplastic cells in culture to the toxic and antiproliferative effects of chemical carcinogens using a variety of chemically transformed cells (Starikova and Vasiliev, 1962; Berwald and Sachs, 1963; Huberman and Sachs, 1966) and virally transformed cells (Diamond, 1965; Berwald and Sachs, 1963; Alfred et al., 1964; Gelboin et al., 1969). The paradox that chemical carcinogens were both cell toxins and cell stimulants (and showed a relative inhibition of normal but not tumor cell growth) was reviewed by Vasiliev and Guelstein (1963), Diamond (1968), Prehn (1964); and Melzer (1980).

C. Carcinogen-Induced Adaptive Resistance
   in Rodent Hepatocytes

The acquired resistance to both toxicity and inhibition of cell proliferation has been studied most intensively in the rodent liver. Many hepatocarcinogens have been shown to inhibit DNA synthesis and mitosis of the regenerating liver after a partial hepatectomy (Laws, 1959; Hso-Kwang-Hwa, 1962; Gelstein, 1962). However, when a hepatocarcinogen is chronically administered to a rodent, despite the toxicity of the carcinogen (or because of it), foci of altered hepatocyte populations begin to grow and form hyperplastic nodules. Clearly, such nodules have developed a resistance to the antiproliferative and toxic actions of the hepatocarcinogens that induced their growth. These new cell populations also respond differently to a growth stimulant. Whereas chronic administration of a hepatocarcinogen inhibits the bulk of the liver from the regenerative response to a two-thirds partial hepatectomy, carcinogen-induced centrilobular basophilic nodules and enzyme deficient areas are able to respond with an increased proliferative activity to a two-thirds partial hepatectomy (Laws, 1959; Maini and Stich, 1962; Rabes and Szymkowiak, 1979; Kitagawa, 1971). Without the addition of a growth stimulant such as a partial hepatectomy, chronic hepatocarcinogen administration is associated with an increase in rat liver weight and liver cell number (Laird and Barton, 1959, 1960, 1961).

In addition to the antiproliferative effects of hepatocarcinogens on normal cells and the resistance of hyperplastic nodules to their antiproliferative actions, rat livers that have been altered by hepatocarcinogens also become resistant to the hepatotoxic and necrogenic actions of both inducing and different hepatocarcinogens *in vivo* (Farber *et al.*, 1976; Williams *et al.*, 1976; Sidransky and Verney, 1978). Similar results were found when hepatocytes from hepatocarcinogen-altered rodents were studied in tissue culture (Judah *et al.*, 1977; Laishes *et al.*, 1978; Sawada *et al.*, 1981). These results were extended to show that the feeding of a hepatocarcinogen *in vivo* could result in a resistance of the hepatocytes to the cytocidal actions of a variety of noncarcinogenic agents, some of which are used in cancer chemotherapy when tested *in vitro* (Carr, 1980, 1981; Carr and Laishes, 1981; Laishes *et al.*, 1980). In addition, differential growth inhibition by carcinogens was seen in normal rat liver cell lines compared to chemically and virally transformed rat liver cell lines *in vitro* (Iype *et al.*, 1979; Paraskeva and Parkinson, 1981; Niles *et al.*, 1981).

D. Types and Significance of Resistance

Two types of resistance may be discerned by these experiments. The first is resistance to toxicity, which may be assessed as resistance to necrosis (cell death) *in vivo*, resistance to cell death using cytotoxicity assays *in vitro*, or altered cloning efficiency *in vitro*. The second is resistance to the antiproliferative action of a test toxin. This may be observed *in vivo* or *in vitro* as the ability of a test cell to proliferate under conditions in which a nonresistant cell would be unable to proliferate due to the presence of a toxin. The biological significance and mechanisms of this adaptive response to the presence of chronic exposure to a carcinogen are of great interest. Most nonhormonal chemical carcinogens are also toxins as stated. The chronic exposure of an animal or cell to sublethal doses of a toxic carcinogen, therefore, results in either the death of the cell or an adaptive set of changes that allows the cell to resist chronic exposure to the toxic milieu of the carcinogen.

A set of adaptive responses to the presence of toxins has been well characterized in both bacterial and mammalian cells. In bacteria it is known as the SOS response (for review see Little and Mount, 1982) and includes the derepression of at least 12 different operons, alteration of DNA repair capability, altered sensitivity to mutagenesis, protection against cytotoxicity, and modification of restriction enzymes. Examples include the induction of DNA glycosylases. In mammalian cells, the adaptive response includes a resistance to sister chromatid exchange, cytotoxicity, alkylation (by the fast removal of O^6 mG; Montesano *et al.*, 1979), and carcinogenesis. Since neoplastic cells thus arise and develop in a chronic toxic carcinogenic milieu, the presence of an adaptive response is a prerequisite for neoplastic cell survival and is presumably an adaptation that confers a selective advantage for the ability to proliferate in a toxic environment. It thus seems reasonable to hypothesize that, when a tumor is induced by chronic exposure to a chemical carcinogen, the tumor cells are likely to have multiple adaptive mechanisms for dealing with various cell toxins. In this view, the well-known resistance of many common epithelial cancers of adult humans to the cytocidal and growth-inhibitory actions of cancer chemotherapy might be expected to be a property of cancer cells that have arisen in a chronic, toxic, carcinogenic environs. Whether the toxic environment per se is actually causal in the induction of new growth characteristics or whether it is merely an associated characteristic of the carcinogens has never been established. However, proliferating cells themselves, particularly in the liver, have been shown to be relatively resistant to a variety of toxins without a chemical inducing agent (Carr and Laishes, 1981; Roberts *et al.*, 1983; Ruch *et al.*, 1985).

E. Candidate Mechanisms Mediating Adaptive Resistance in Rodent Liver

The mechanism(s) underlying the acquired resistance of carcinogen-altered hepatocytes is not yet clear. Farber *et al.* (1976) showed a decreased uptake of radiolabeled hepatocarcinogen in hyperplastic nodules, as well as in the surrounding liver, in carcinogen-altered rat livers compared to normal rat livers. The hyperplastic nodules possess increased DNA repair mechanisms, as judged by unscheduled DNA synthesis (Kitagawa *et al.*, 1975), and several components of the mixed-function oxygenase system in hepatic microsomes and post mitochondrial supernatants have been found to be decreased in hyperplastic nodules. These include cytochrome *P*-450, aryl hydrocarbon hydroxylase, and reduced NADPH–cytochrome-*c* reductase (Cameron *et al.*, 1976; Gravela *et al.*, 1975; Okita *et al.*, 1976). These results led to the hypothesis that the basis for the resistance to cytotoxicity was an inability of the carcinogen-altered hepatocyte to metabolize the toxin to its ultimate reactive (toxic) form. In addition, other alterations in the pattern of drug metabolizing enzymes are induced by the action of the hepatocarcinogen, the net affect of which may be to detoxify toxic test compounds. These changes include an increase in glutathione (Fiala *et al.*, 1970), epoxide hydrase (Levin *et al.*, 1978), DT-diaphorase (Schor and Morris,

1977), glutathione-*S*-transferases (Smith *et al.*, 1977) and UDPglucuronyl-transferases (Bock *et al.*, 1982). Not only do the tumors have enhanced drug detoxifying capacity, but they enhance that capacity in the surrounding liver (Sultatos and Vesell, 1980).

II. STUDY OF DRUG RESISTANCE IN CARCINOGEN-ALTERED HEPATOCYTES

In order to investigate the mechanisms by which rat hepatocytes adapt to the chronic presence of hepatocarcinogens, I have used an *in vivo/in vitro* technique. In this system, a rat is exposed to an inducing hepatocarcinogen *in vivo,* and its hepatocytes are then placed in primary monolayer culture in order to investigate their newly acquired properties.

In principle, a hepatocarcinogen can alter hepatocytes in a number of ways that may be associated with increased resistance to the action of cytotoxins (Table I). These include (1) an alteration in the blood supply or oxygenation, a

TABLE I

Consequences of Hepatocarcinogenic Action That May Be Relevant to Mechanisms of Resistance to Toxic Damage

Altered cell property	Possible role in resistance
Alteration in blood supply or oxygenation	Decreased access of active form of toxin
Cell membrane (e.g., γ-GT, antigens, epoxide hydrase, lipid changes)	Altered transport of toxin (increased efflux, decreased influx) Decreased binding protein for toxin Decreased sensitivity to cell membrane damage
Altered metabolism (e.g., induction of cytochrome *P*-450 mixed-function oxidases; glutathione; superoxide dismutase; catalase)	Decreased activation of toxin Increased inactivation of toxin or toxic metabolites
Carcinogen–macromolecule adduct formation	Increased resistance of structurally altered macromolecule to toxic destruction Decreased binding by toxin to structurally altered macromolecule
Unscheduled DNA synthesis	Increased repair of toxic damage by macromolecules
Basophilia (RNA synthesis)	Competition by newly synthesized nucleic acids with DNA for binding to toxin

property that would not apply in the defined conditions of tissue culture; (2) an alteration of the cell membrane resulting in changes in transport, such as decreased influx of the toxin into the cell, increased efflux, or decreased sensitivity to membrane-mediated damage; (3) an altered metabolism with a decreased activation of a protoxin to its ultimate reactive form or an increased inactivation of a toxin and increased excretion; (4) an altered carcinogen–macromolecular reaction, resulting in decreased binding of the toxin to important cell structures; (5) an increase in repair enzymes for toxin-induced damage, such as unscheduled DNA synthesis; and (6) amplification of some important genes, such as P-glycoprotein genes (Riordan *et al.*, 1985). In the liver, carcinogen-induced foci and nodules have certain well-described changes that may be of importance in the induction of drug resistance (Table II). These include a decrease in the phase I enzymes resulting in decreased activation of prodrugs or protoxins, or alterations in the phase II enzymes resulting in increased conjugation and excretion of activated toxins, either by glutathione-S-transferase, by γ-glutamyl transpeptidase (transmembrane transport and mercapturic acid synthesis) and UDPglucuronyltransferases, as well as by a third miscellaneous group of compounds that are induced which include epoxide hydrase, DT-diaphorase and heat-shock proteins. All of these adaptive responses participate in protecting the cell from toxic damage. I have selected one toxin, the anthracycline antibiotic adriamycin (doxorubicin), as a probe to study the adaptive mechanisms in carcinogen-altered liver because of its clinical importance in the treatment of human hepatocellular carcinoma. Adriamycin is a potent cytotoxin which is thought to exert its cellular toxicity through two predominant actions. First, it can intercalate between the DNA base pairs and act as an alkylating agent. It is metabolized by the cell, by reductive glycosidases, to a series of products that include the relatively nontoxic

TABLE II

Some Phenotypic Changes in Carcinogen-Altered Liver of Relevance to the Development of Drug Resistance[a]

Phase I	Phase II	Miscellaneous groups
↓ Cytochrome P-450	↑ GSH	↑ Epoxide hydrase
↓ Cytochrome b_5	↑ GSH-S-transferase	↑ DT-diaphorase (quinone
↓ AHH	↑ γ-GT	reductase)
	↑ UDPglucuronosyltransferase	↓ Iron storage
	↓ Sulfotransferase	↓ Cholesterol, PUFA
		↑ DNA-alkylation repair
		↑ Heat-shock proteins

[a]Abbreviations: γ-GT, γ-glutamyl transpeptidase; AHH, aryl hydrocarbon hydroxylase; GSH, reduced glutathione.

TABLE III

Metabolism and Biological Activity of Adriamycin

METABOLISM

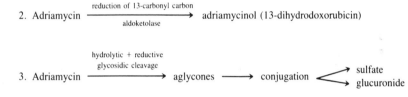

1. Adriamycin $\xrightarrow[\text{NADPH cytochrome } P\text{-450 reduction}]{\text{microsomal}}$ superoxide radicals $\longrightarrow$ hydroxyl radicals
peroxide radicals
H_2O_2
(free radical
semiquinone)

Mammalian defenses: superoxide dismutase, glutathione, catalase, peroxidase (chemical defenses include anti-oxidants such as selenium salts, vitamin C, α-tocopherol, butylated hydroxyanisole, and butylated hydroxytoluene)

2. Adriamycin $\xrightarrow[\text{aldoketolase}]{\text{reduction of 13-carbonyl carbon}}$ adriamycinol (13-dihydrodoxorubicin)

3. Adriamycin $\xrightarrow[\text{glycosidic cleavage}]{\text{hydrolytic + reductive}}$ aglycones $\longrightarrow$ conjugation $\Big\langle$ sulfate
glucuronide

BIOLOGICAL ACTIVITY

1. Base-pair intercalation by virtue of planar ring structure (apparently noncovalent)

2. Free radical reactions: (a) lipid peroxidation; (b) DNA-strand scission after covalent binding

CONSEQUENCES OF ADRIAMYCIN ACTION
Inhibiton of DNA synthesis
Inhibition of DNA-primed RNA synthesis
Inhibition of cell proliferation
Chromosomal damage
DNA-strand scission
Cancer induction in rat mammary gland
Inhibition of mitochondrial electron transport
Myocardial toxicity
Mutagenesis

---

adriamycin aglycones, as well as to adriamycinol. Second, adriamycin participates in a series of electron transfer reactions through its quinone group and can serve as an electron acceptor for microsomal and nuclear flavoproteins, resulting in the generation of a series of activated oxygen radicals which are thought to be toxic for the cell (Table III). The radicals that are produced include the superoxide radical as well as hydrogen peroxide and hydroxyl radicals through a series of secondary reactions of the superoxide anion (for review see Muggia *et al.*, 1982).

III. CYTOTOXICITY STUDIES *IN VITRO*

A. Induction of Pleiotropic Drug Resistance

In order to investigate the action of hepatocarcinogens on the induction of new hepatocyte phenotypes, rats were fed a variety of hepatocarcinogens mixed into their basal diets. These included 2-acetylaminofluorene (AAF, 0.02%), aflatoxin B_1 [nominal 1 part per million (ppm)], 3'-methyl-4-dimethylaminoazobenzene (DAB, 0.6%), diethylnitrosamine (DEN, 50 ppm in the drinking water), or ethionine (0.25%). After a variable period of dietary carcinogen administration, primary monolayer cultures of hepatocytes were made and cytotoxicity assays were performed (see legend to Fig. 1). All of these carcinogenic regimens produced a significant proportion of resistant hepatocytes using several different test toxins (Carr, 1981). The toxins used were adriamycin (9×10^{-5} M) cycloheximide (7 mM), methotrexate (5.4×10^{-3} M), and aflatoxin B_1 (1×10^{-6} M). Figure 1 illustrates the time course for the induction of resistance as measured *in vitro* using primary monolayer cultures of adult rat hepatocytes after the rat had been exposed *in vivo* to the regimen of Solt and Farber (1976). The attraction of this model is that the animal is exposed to carcinogen for only a limited time. After a single dose of DEN (200mg/kg) the rat is given dietary AAF for 2 weeks, in the middle of which a two-thirds partial hepatectomy is performed as a growth stimulant. Thereafter, no further exposure to carcinogen is made, although most of the rats still develop hepatocellular carcinomas many months later. It can be seen that despite cessation of carcinogen exposure by week 4, there is a rapid induction of resistance to all 4 test toxins that does not appear to diminish with time.

B. Stability of Resistant Phenotype

Figure 2 illustrates the stability of the resistant phenotype after limited exposure to the carcinogen AAF. Rats were exposed to AAF for 1 week (Fig. 2A), 2 weeks (Fig. 2B), 4 weeks (Figure 2C), or 12 weeks (Fig. 2D). On the horizontal axis, the zero point on all four of the charts corresponds to the time of cessation of AAF. It can be seen that on return to a normal diet after 1 or 2 weeks of AAF exposure, the resistant hepatocyte phenotype reverts to normal sensitivity. This reversion to normal is much slower after 4 weeks of AAF feeding. However, after 12 weeks of AAF feeding, a duration that corresponds with the induction of a proportion of irreversible neoplastic nodules in the liver, the resistant phenotype appears to be irreversible, and experiments conducted up to 18 months show no loss of resistance in the hepatocytes. It is of interest that the hepatocytes which are tested come from the whole liver and not just from the

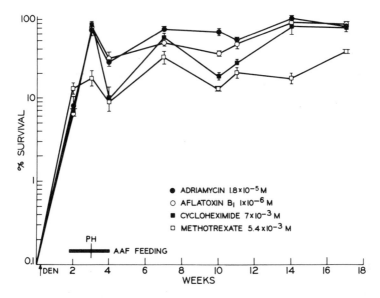

Fig. 1. Time course for the induction of drug resistance in hepatocytes *in vitro* from rats treated by the regimen of Solt and Farber (1976). Adriamycin ($1.8 \times 10^{-4} M$) (●); aflatoxin B$_1$ ($1 \times 10^{-6} M$) (○); cycloheximide ($7 \times 10^{-3} M$) (■); methotrexate ($5.4 \times 10^{-3} M$) (□). Methods: Hepatocyte cultures were prepared by the high pressure collagenase perfusion technique (Seglen, 1973). Briefly, collagenase (Cooper Biomedical, Cappel, Worthington Malvern, PA.,), 190 mg in 300 ml L-15 medium + NaHCO$_3$ 2.2g/liter, was perfused into the portal vein using a high-pressure perfusion pump at 50 ml/min, such that the temperature of the medium being infused at the portal vein was 37°C without recirculation. When the perfusing medium was used up, the hepatocytes were dissociated from the liver and placed in a tissue culture dish, and the cells were plated in Liebowitz (L-15) medium with serum (10% calf serum, Gibco) at 1×10^6 viable cells in a screw-cap tissue culture dish of 25 cm^2 surface area (Falcon Plastics, Oxnard, California) using 4 ml of L-15 medium with HEPES 3.5 mg/ml and bovine serum albumin 2 mg/ml, together with penicillin and streptomycin. After a 3-hr attachment period at 37°C in an air incubator, the cells were washed twice with L-15 medium which was then replaced with 4 ml of the same medium with 5% serum and test toxin (experiment) or no test toxin (control). After a 24-hr incubation period, 0.8 ml of trypan blue was added to the 4 ml of medium in each flask and incubated for 5 min at 37°C. The medium was then removed and the number of viable (nonstaining) cells was counted in both the experimental and the control flasks. Percentage survival was expressed as the number of viable, attached cells in experimental flasks compared to the number of viable attached cells in control flasks. Under these conditions, none of the cell types proliferates. For the regimen of Solt and Farber, male F-344 rats (160–180 g) were given DEN (200 mg/kg) in 0.9% NaCl solution intraperitoneally. After a 2-week recovery period, the rats were fed a basal diet containing AAF 0.02% (w/w). After the first week of AAF feeding, the rats were given a two-thirds partial hepatectomy. After the 2-week AAF feeding, they were returned to a basal diet.

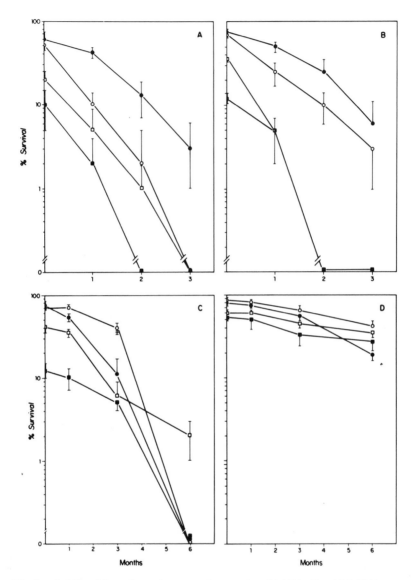

Fig. 2. Stability of the resistant phenotype after cessation of AAF feeding. Male F344 rats were fed AAF 0.02% (w/w) for 1 week (A), 2 weeks (B), 4 weeks (C), or 12 weeks (D). Thereafter, the rats were returned to a normal diet and were sacrificed at the indicated times. The 0 on the horizontal axis for each of the four graphs corresponds to the time that the rats were returned to a basal diet and AAF feeding was stopped. Adriamycin (1.8×10^{-4} M) (●); aflatoxin B_1 (1×10^{-6} M) (○); cycloheximide (7×10^{-3} M) (■); methotrexate (5.4×10^{-3} M) (□). The cells were incubated with or without a test toxin for 24 hr, and cell survival was assayed as detailed in the legend to Fig. 1.

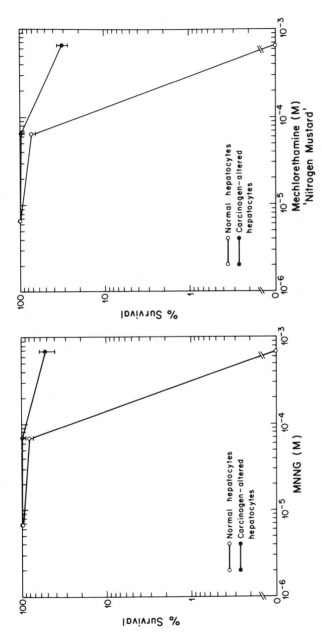

Fig. 3. Comparison of the proportion of sensitive and resistant cells in normal liver (○) and in liver from rats fed AAF 0.02 (w/w) for 3 months (●) using M-methyl-N-nitro-N-nitrosoguanidine (MNNG) or mechlorethamine (nitrogen mustard). Cell survival was assessed after a 24-hr incubation period with the test toxin exactly as described in the legend to Fig. 1.

TABLE IV

Adriamycin-Mediated Cytotoxicity in
Hepatocytes Altered by AAF Analogs[a]

Analog	Survival at 24 hr compared to controls (%)
AAF	85
Nitrofluorene	68
Aminofluorene	60
Fluorene	45
Fluorenone	7
Anthracene	4
Controls	5

[a]All analogs fed as 0.02% (w/w) diet. Adriamycin $9 \times 10^{-5} M$.

neoplastic nodules. Thus, exposure of a hepatocarcinogen appears to result in the acquisition of resistance by a large proportion of the contained hepatocytes. Since it is well known that there is a very early carcinogen-mediated decrease in the hepatic cytochrome P-450 monooxygenase enzymes which mediate the ability of hepatocytes to metabolize protoxins to their reactive forms (Cameron *et al.*, 1976; Sun *et al.*, 1985), the action of some direct-acting toxins was assessed. This was done in order to circumvent the explanation that the carcinogen-induced resistance was due to the fact that the hepatocytes were not able to activate the test toxins to their reactive toxic forms. Figure 3 shows the cytotoxic effects on normal and carcinogen-altered hepatocytes (3 months AAF feeding) of M-methyl-N-nitro-N-nitrosoguanidine (MNNG) and mechlorethamine, two carcinogens that are thought not to require metabolic activation. The AAF-altered hepatocytes are as resistant to these two compounds, which do not require metabolism for their toxicity, as they are to the four test compounds shown in Fig. 1. A structure activity study was performed (Table IV) in which the effect of feeding different AAF analogs was examined for changes in hepatocyte sensitivity to adriamycin. No induction of resistance was noted, except for the aminofluorene and nitrofluorene, which have mild carcinogenic activity. The noncarcinogenic analogs did not induce measurable resistance, except for fluorene.

IV. CELLULAR PHARMACOKINETICS OF ADRIAMYCIN IN SENSITIVE AND RESISTANT HEPATOCYTES

Normal hepatocytes and hepatocytes from rats fed AAF for 3 months were compared for their ability to accumulate, efflux, DNA bind, and metabolize adriamycin. Sensitive normal and AAF-altered resistant hepatocytes on day 1 of primary monolayer culture were incubated with various concentrations of adriamycin, and at appropriate times after addition of the drug, the cells were harvested and the adriamycin was extracted and measured. Figure 4A shows that there was little difference between the sensitive and resistant hepatocytes in the levels of adriamycin accumulation at any drug concentration. Similar experiments on the efflux showed that both cell types effluxed less than 5% of adriamycin fluorescence over a 6-hr period. There was thus little difference between the ability of sensitive and resistant hepatocytes to accumulate or to efflux adriamycin. Since adriamycin is metabolized to adriamycinol and the inactive aglycones (Table III), it seemed important to compare the ability of the sensitive and resistant cells to degrade adriamycin to its less toxic metabolites. For the determination of adriamycin metabolites, sensitive and resistant hepatocytes were incubated with adriamycin (1.8×10^{-5} M), the adriamycin was then extracted and its metabolites were examined by HPLC (Fig. 4C). Apart from adriamycin, only two metabolites contributed more than 5% of the extractable fluorescence, and these were two aglycones. The relative proportion of these two aglycones compared to parent adriamycin were similar for both sensitive and resistant cells. No adriamycinol was measurable. Thus, both the resistant and sensitive hepatocytes were capable of metabolizing adriamycin to its aglycones. In order to determine whether hepatocyte-associated adriamycin was bound to DNA to a similar extent in the sensitive and resistant hepatocytes, cells were incubated with adriamycin (1.8×10^{-5} M) and harvested by scraping, and the DNA was isolated using ultracentrifugation in cesium chloride. Figure 4B shows the amount of adriamycin fluorescence per microgram of DNA, measured across the gradient peaks of DNA. DNA content along the gradient was determined by the diphenylamine method. The amounts of adriamycin-bound DNA were identical for the sensitive and resistant hepatocytes. Thus, standard considerations of cellular pharmacokinetics, such as the amount of drug that is taken up, released, bound to the DNA, or metabolized, did not offer sufficient explanation for the differential resistance of carcinogen-altered hepatocytes to adriamycin.

V. MEMBRANE ACTIONS OF ADRIAMYCIN

A. Effects of Adriamycin

Three kinds of interaction of adriamycin with the hepatocyte cell membrane focused attention upon the possibility that adriamycin might exert its toxic action on the hepatocyte at the cell membrane. First, adriamycin inhibits hepatocyte proliferation (Tanaka et al., 1982), and since epidermal growth factor (EGF) and insulin are two potent hepatic mitogens, the effects of adriamycin on the ability of the hepatocyte membrane to bind these two mitogens was examined (Fig. 5).

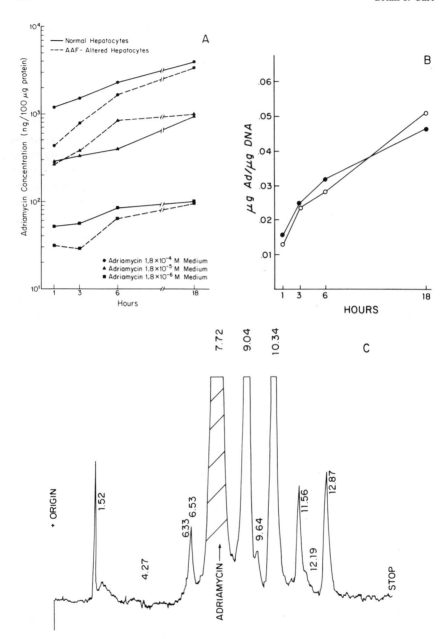

Fig. 4. (A) Comparison of the accumulation of adriamycin by normal and 3-month AAF-altered (resistant) hepatocytes. Hepatocytes were incubated with adriamycin, and at the indicated times the medium was changed; the cells were washed and then harvested by a rubber policeman and disrupted

While little inhibition of the insulin receptor action was noted, adriamycin strongly interfered with the ability of the EGF receptor to bind its ligand. This is one possible mechanism by which adriamycin might exert its antiproliferative and possibly toxic action on the hepatocyte. Second, we covalently linked adriamycin to microspheres. These microspheres were too large to enter the hepatocyte membrane, and no detectable dissociation of adriamycin from them was found. The adriamycin-linked microspheres were found to be as toxic to normal hepatocytes as was free adriamycin (Rogers et al., 1983); interestingly, the adriamycin microspheres were also toxic to the AAF (resistant) hepatocytes. This obliged a consideration of the possibility that adriamycin might exert its hepatotoxicity through a membrane effect without ever entering the hepatocyte. Third, one of the major mechanisms by which adriamycin acts on cells is through the generation of toxic free radicals that interact with the membrane lipid and cause, among other phenomena, membrane lipid peroxidation (Table III).

In order to investigate membrane lipid peroxidation in hepatocytes, adriamycin ($9 \times 10^{-5} M$) was incubated with normal and AAF-altered, resistant hepatocytes; the hepatocytes were then harvested and the amount of adriamycin-induced malonaldehyde was measured as an estimate of lipid peroxidation (Table V). The normal, sensitive hepatocytes were a far better substrate for adriamycin-induced malonaldehyde (lipid peroxidation) than the resistant hepatocytes. A similar finding was reported by others (Bartoli and Galeotti, 1979).

---

by sonication. Daunorubicin was added to control for extraction efficiency; the cells were extracted with 0.5 N HCl in 85% isopropanol, and the supernatant was further extracted with ice-cold chloroform. The organic phase was then incubated with an equal volume of a saturated aqueous solution of sodium biocarbonate and, after separation and, was dried under a stream of nitrogen gas. The residue was dissolved in chloroform, methanol, and subjected to thin-layer chromatography. Adriamycin was quantitated on the eluted and scraped spots using authentic markers in parallel lanes, and the fluorescence was measured using an excitation wavelength of 470 nm and an emission wavelength of 585 nm (Watson and Chan, 1976; Chan and Wong, 1979). (B) Measurement of adriamycin-bound DNA in normal and 3-month AAF-altered hepatocytes in vitro. Hepatocytes were incubated with adriamycin ($1.8 \times 10^{-5} M$), and after the indicated times the medium was removed and the cells were washed and then harvested by scraping with a rubber policeman. The DNA was extracted by ultracentrifugation using cesium chloride gradients (Sirica et al., 1980). The results show the relative amounts of adriamycin fluorescence per microgram of DNA, which was measured at the DNA peak of the gradient as determined by the diphenylamine method. (C) HPLC tracing of adriamycin and its metabolites in hepatocytes in vitro. Hepatocytes and adriamycin were incubated and extracted as described in previously (A). The dried residue was dissolved in tetrahydrofuran and injected into a Beckman model 332 HPLC apparatus with an automatic gradient maker. Fluorescence was determined using a Spectra/Glow filter fluorometer using a 480 nm excitation filter and a 560 nm emission filter. The column was a Bondapak-phenyl 3.9 mm $\times$ 30 cm (Waters Associates, Milford, MA), and metabolites were eluted with a linear gradient from 70% ammonium formate buffer; 30% ammonium formate buffer; tetrahydrofuran 1:1 (v/v) to 100% ammonium format buffer; tetrahydrofuran 1:1 (v/v) in 10 min (Egorin et al., 1980; Andrews et al., 1980).

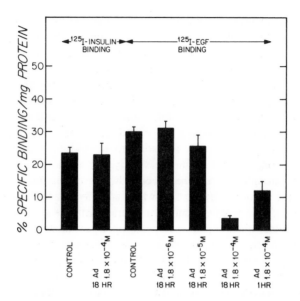

Fig. 5. The effects of adriamycin on hepatocyte membrane receptor binding. Normal hepatocytes were plated at 3.5×10^5 cells per 35-mm tissue culture dish in L-15 medium and 10% calf serum. After a 3-hr attachment period, the medium was replaced with L-15 medium containing 0.1% bovine serum albumin and 10^{-3} M phenylmethyl sulfonyl fluoride without serum. Adriamycin $(9 \times 10^5 M)$ was added to the cells and binding was performed 12 hr later in a reaction containing 0.5 ng/ml ^{125}I-labeled EGF or 0.5 ng/ml ^{125}I-labeled insulin, both prepared by the chloramine-T method. Nonspecific binding was determined in the presence of 200 ng/ml unlabeled ligand. At the end of the 18-hr reaction time at 4°C, the radioligand was removed and the cells were washed and solubilized with NaOH for gamma-counting and determination of protein content by the method of Lowry.

TABLE V

Lipid Peroxidation in Hepatocytes in Culture[a]

	OD 535 nm	
	Normal	3 month AAF
Adriamycin $(9 \times 10^{-5} M)$	0.06 ± 0.01	0.03 ± 0.01
Tert-Butylhydroperoxide $(10^{-3} M)$	0.085 ± 0.015	0.04 ± 0.01
FeSO$_4$ $(10^{-4} M)$	0.27 ± 0.03	0.07 ± 0.02
Controls (no drug)	0.02 ± 0.005	0.025 ± 0.005

[a]Freshly prepared single cell suspensions $(5 \times 10^6$ cells/ml) of normal or 3 month AAF-altered hepatocytes were incubated with no drug (controls), adriamycin $(1.8 \times 10^{-5} M)$ for 3 hr, tert-butylhydroperoxide $1.0^{-3} M$ for 0.5 hr, or FeSO$_4$ $(10^{-4} M)$ for 0.5 hr. Lipid peroxidation was measured on 2mg/ml protein solution exactly as described in the legend to Fig. 6.

Adriamycin requires metabolic activation for free radical generation, involving the microsomal NADPH cytochrome P-450 reductase. A carcinogen-induced decrease of this enzyme, resulting in decreased ability of the resistant cell to produce the reactive adriamycin semiquinone, could be responsible for the relative resistance of AAF-altered hepatocytes. Therefore, lipid peroxidation was measured in sensitive and resistant hepatocytes in culture using either an organic hydroperoxide or ferrous sulfate, neither of which are thought to require metabolic activation in order to produce free radicals. A resistance to lipid peroxidation by the AAF cells compared to the normal cells was found, similar to the differential actions of adriamycin, indicating that resistance was not due to inability to produce a toxic free radical (Table V). Extracts of normal and AAF-altered liver microsomal 105,000 g supernatant were prepared and incubated with normal hepatocytes in the presence of adriamycin. However, no difference was noted between the normal or the AAF liver supernatant in their abilities to inhibit adriamycin-mediated lipid peroxidation on normal hepatocytes. Thus, there was no gross evidence of differences in free radical scavenging activity.

Since the main substrate for lipid peroxidation is the membrane lipid, single cell suspensions of normal and AAF-altered hepatocytes were prepared, and the lipids were extracted by the Folch procedure and tested for their ability to act as a substrate for lipid peroxidation *in vitro* (Fig. 6). The separated normal hepatocyte lipid was found to be a better substrate for adriamycin-mediated lipid peroxidation than the AAF cell lipid. The principal membrane lipid substrates for peroxidation are the available double bonds of polyunsaturated fatty acids (Niehuis and Samuelsson, 1968), which are usually a reflection of the phospholipid arachidonate levels. Since hepatomas are known to have a decreased polyenoic acid content (Hartz *et al.*, 1982) compared to normal hepatocytes, the effects of altering cellular polyunsaturated fatty acid content was assessed on the cytotoxicity assay (Fig. 7). Normal hepatocytes were incubated with or without arachidonate, 20 carbons with 4 unsaturated double bonds (20:4), or with docosohexaenoic acid, 22 carbons, 6 double bonds (22:6). Both of these compounds increase the polyunsaturation of the hepatocyte membrane, and both increased the sensitivity of hepatocytes to the cytocidal action of adriamycin (Fig. 7A and 7B). Another major alteration of the hepatoma membrane lipid is a deletion of the cholesterol-negative-feedback system (Siperstein and Fagan, 1964), with a resultant increase in membrane cholesterol content. Since cholesterol increases the rigidity of membranes, the effects of cholesterol on adriamycin-mediated cytotoxicity were assessed, and were found to protect the hepatocytes (Fig. 7C).

B. Spontaneous Hepatocyte Resistance

The experiments above suggest that a major mechanism of carcinogen-induced resistance to adriamycin resides in a change in the constituents of the membrane

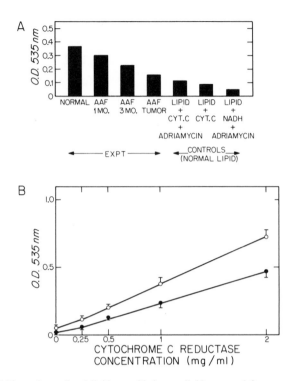

Fig. 6. Adriamycin-mediated lipid peroxidation on lipid extracted from normal and AAF 3-month-altered hepatocytes. Freshly prepared single cell suspensions (as in legend to Fig. 1) of normal and AAF-altered hepatocytes were extracted with Folch solution (0.1 M KCl-1/40 part in chloroform:methanol 2:1) (Folch *et al.*, 1976). The extracted solution was evaporated under nitrogen in an 80°C water bath for 2 hr and quantitated for lipid phosphorus (method of Bartlett, 1959). A portion of the sample (1.0 ml) and 0.5 ml of 10 N H_2SO_4 were heated in an 160°C oven for 1½ hours. Two drops of 30% H_2O_2 were added and the solution heated for a further 1½ hours. NH_4MO (0.22%, 4.6 ml) was added followed by 0.2 ml of Fiske and Subbarow reagent (Fiske and Subbarow, 1925). After heating for 7 min in a boiling water bath, the optical density was recorded at 830 nm. Assay for lipid peroxidation: liposomes were prepared from the extracted lipid by sonication in Tris buffer and were suspended in 0.1 M Tris buffer (pH 7.4) and incubated with adriamycin and enzyme at 37°C for various periods. Lipid peroxidation was determined as malonaldehyde using the thiobarbituric acid reaction (Buege and Aust, 1978). A portion of the sample (1 ml) was mixed with 50 μl of Lubrol PX solution followed by 2 ml of TBA–TCA–HCl solution. After boiling for twenty minutes, the cooled flocculent precipitate was removed by centrifugation at 2000 rpm for 10 min and absorbance of the sample was determined at 535 nm against blanks. Reagents: TBA–TCA–HCl reagent, 15% w/v trichloroacetic acid; 0.375% thiobarbituric acid; 0.25 N HCl acid. The solution is gently heated until the thiobarbituric acid solution dissolves. Lubrol PX solution: 10 mg butylate hydroxytoluene (BHT) was dissolved in a minimal amount of acetone. Lubrol PX (20 g/100 ml) was warmed and the BHT solution added.

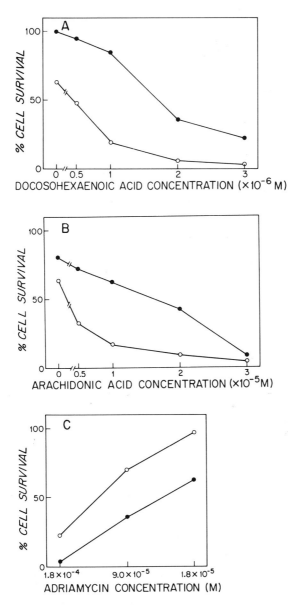

Fig. 7. Effects of fatty acids and cholesterol on adriamycin-mediated cytotoxicity. Normal hepatocytes were incubated with adriamycin with or without the indicated amount of fatty acid or cholesterol. After a 24-hr incubation period, the percentage of cell survival was assessed as indicated in the legend to Fig. 1.

phospholipids. While this may be an important contributing factor for resistance to adriamycin, it is difficult to see how this would be a sufficient explanation for the resistance to cycloheximide or methotrexate (Fig. 1). There must therefore be multiple mechanisms of resistance in hepatocytes to the various cytotoxins that are induced by chronic exposure to hepatocarcinogens. However, a word of warning is appropriate. Even for a single drug there may be multiple mechanisms by which resistance is induced. Fig. 8 shows the effects of delaying the addition of adriamycin to hepatocytes for three 24-hr periods. The hepatocytes were incubated with adriamycin for 24-hr, either at time 0, 24-hr after plating, 48-hr after plating, or 72-hr after plating. Percentage survival was determined and compared to untreated controls. Resistance to adriamycin was induced merely with the passage of time in culture, and this resistance was even more profound than resistance induced by hepatocarcinogens *in vivo*. In contrast to the AAF-altered hepatocytes (Fig. 4A), we found that there was a large decrease in the uptake of adriamycin by hepatocytes that were rendered resistant merely by being left in culture for 72-hr. Thus, normal hepatocytes change with time in culture.

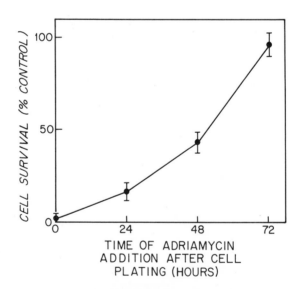

Fig. 8. Changes in the sensitivity of normal hepatocytes to the cytocidal action or adriamycin at various times after plating. Normal hepatocytes were plated as described in the legend to Fig. 1. Adriamycin ($9 \times 10^{-5} M$) was added immediately after the 3-hr attachment period, or 24, 48, or 72 hr later. After a 24-hr incubation of the cells with the adriamycin, the percentage of cell survival was measured as described in the legend to Fig. 1.

VI. CONCLUSIONS

These experiments provide evidence for the proposal that chronic exposure of murine hepatocytes to hepatocarcinogens *in vivo* leads to an adaptive resistance to the cytocidal actions of many toxins. There are many adaptive changes in the hepatocyte, and different mechanisms probably account for the resistance to different drugs and toxins. For adriamycin, a widely used cancer chemotherapeutic agent, these experiments provide evidence that the drug influences the hepatocyte membrane in several ways and that carcinogen feeding *in vivo* also induced membrane changes which probably contribute to the resistance to adriamycin cytotoxicity. Whether the same membrane changes are important in determining altered growth control, thus mechanistically linking growth and drug resistance, remains to be established. However, if the changes in rat hepatocytes induced by chemical hepatocarcinogens are a paradigm for the changes in human hepatocellular carcinoma, then pleiotropic drug resistance is likely to be a phenotype of liver cancer.

REFERENCES

Alfred, L. J., Globerson, A., Berwald, Y., and Prehn, R. T. (1964). Differential toxicity response of normal and neoplastic cells *in vitro* to 3,4-benzopyrene and 3-methylcholanthrene. *Br. J. Cancer* **18**, 159–164.

Andrews, P. A., Brenner, D. E., Chow, F.-T. E., Kubo, H., and Bachur, N. R. (1980). Facile and definitive determination of human adriamycin and daunorubicin metabolites by high-pressure liquid chromatography. *Drug Metab. Dispos.* **8**, 152–156.

Bartlett, G. R. (1959). Phosphorus assay in column chromatography. *J. Biol. Chem.* **32**, 466–468.

Bartoli, G. M., and Galeotti, T. (1979). Growth-related lipid peroxidation in tumor microsomal membranes and mitochondria. *Biochim. Biophys. Acta* **574**, 537–541.

Berwald, Y., and Sachs, S. L. (1963). *In vitro* cell transformation with chemical carcinogens. *Nature (London)* **200**, 1182–1184.

Berwald, Y., and Sachs, L. (1965). *In vitro* transformation of normal cells to tumor cells by carcinogenic hydrocarbons. *J. Natl. Cancer Inst.* **35**, 641–661.

Bock, K. W., Lilienblum, W., Pfeil, H., and Eriksson, L. C. (1982). Increase in uridine diphosphate-glucuronyltransferase activity in female liver nodules and Morris hepatumors. *Cancer Res.* **42**, 3747–3752.

Buege, J. A., and Aust, S. D. (1978). Microsomal lipid peroxidation. *Methods Enzymol.* **52**, 302–310.

Cameron, R., Sweeney, G. D., Jones, K., Lee, G., and Farber, E. (1976). A relative deficiency of cytochrome *P*-450 and aryl hydrocarbon hydroxylase in hyperplastic nodules induced by 2-acetylaminofluorene in rat liver. *Cancer Res.* **36**, 3888–3893.

Carr, B. I. (1980). Resistance of carcinogen-altered rat hepatocytes to drug-induced cytotoxicity *in vitro*: An early and stable phenotypic change. *Proc. Am. Assoc. Cancer Res.* **21**, 440.

Carr, B. I. (1981). Structurally diverse hepatocarcinogens *in vivo* induce drug resistance in rat hepatocytes when tested *in vitro*. *Br. J. Cancer* **43**, 731–732.

Carr, B. I., and Dias, C. B. (1983). Carcinogen-induced changes in hepatic interactions with free radicals. *Proc. Am. Assoc. Cancer Res.* **24**, 231.

Carr, B. I., and Laishes, B. A. (1981). Carcinogen-induced drug resistance in rat hepatocytes. *Cancer Res.* **41**, 1715–1719.

Chan, K. K., and Wong, C. D. (1979). Quantitative thin layer chromatography: Thin film fluorescence scanning analysis of adriamycin and metabolites in tissue. *J. Chromatogr.* **38**, 343–349.

Diamond, L. (1965). The effect of carcinogenic hydrocarbons on rodent and primate cells *in vitro*. *J. Cell. Comp. Physiol.* **66**, 183–198.

Diamond, L. (1968). The interaction of chemical carcinogens and cell *in vitro*. *Progr. Exp. Tumor Res.* **11**, 364–383.

Egorin, M. J., Clawson, R. E., Ross, L. A., Chow, F.-T. E., Andrews, P. A., and Bachur, N. R. (1980). Disposition and metabolism of *N,N*-dimethyldaunorubicin and *N,N*-dimethyladriamycin in rabbits and mice. *Drug Metab. Dispos.* **8**, 353–362.

Farber, E., Parker, S., and Gruenstein, M. (1976). The resistance of putative premalignant liver cell populations, hyperplastic nodules to the acute cytotoxic effect of some hepatocarcinogens. *Cancer Res.* **36**, 3879–3887.

Fiala, S., Mohindru, A., Kettering, W. G., Fiala, A. E., and Morris, P. (1976). Glutathione and gamma glutanyl transpeptidase in rat liver during chemical carcinogenesis. *J. Natl. Cancer Inst.* **57**, 591–598.

Fiske, C. H., and Subbarow, Y. (1925). The colorimetric determination of phosphorus. *J. Biol. Chem.* **66**, 375–400.

Folch, J., Lees, M., and Sloane-Stanley, G. H. (1976). A simple method for the isolation and purification of total lipides from animal tissues. *J. Biol. Chem.* **226**, 497–509.

Gelboin, H. V., Huberman, E., and Sachs, L. (1969). Enzymatic hydroxylation of benzopyrene and its relationship to cytotoxicity. *Proc. Natl. Acad. Sci. U.S.A.* **64**, 1188–1194.

Gelstein, V. I. (1962). Some biological characteristics of liver cells in the course of experimental carcinogenesis in mice. *Vopr. Onkol.* **9**, 61–68.

Gravela, E., Feo, F., Canuto, R. A., Garcia, R., and Gabriel, L. (1975). Functional and structural alterations of liver ergastoplasmic membranes during DL-ethionine hepatocarcinogenesis. *Cancer Res.* **35**, 3041–3047.

Haddow, A. (1935). Influence of certain polycyclic hydrocarbons on the growth on the Jensen rat sarcoma. *Nature (London)* **136**, 868–869.

Haddow, A. (1938a). The influence of carcinogenic compounds and related substances on the rate of growth of spontaneous tumors of the mouse. *J. Pathol. Bacteriol.* **47**, 567–579.

Haddow, A. (1938b). The influence of carcinogenic substances on sarcomata induced by the same and other compounds. *J. Pathol. Bacteriol.* **47**, 581–591.

Haddow, A. (1938c). Cellular inhibition and the origin of cancer. *Acta. Unio. Int. Cancrum* **3**, 342–352.

Haddow, A., and Robinson, A. M. (1937). The influence of various polycyclic hydrocarbons on the growth rate of transplantable tumors. *Proc. R. Soc. London* **122**, 442–476.

Haddow, A., Scott, C. M., and Scott, J. D. (1937). The influence of certain carcinogenic and other hydrocarbons on body growth in the rat. *Proc. R. Soc. London* **122**, 477–507.

Hartz, J. W., Morton, R. E., Waite, M. M., and Morris, H. P. (1982). Correlation of fatty acyl composition of mitochondrial and microsomal phospholipid with growth rate of rat hepatomas. *Lab. Invest.* **46**, 73–78.

Hso-Kwang-Hwa (1962). The effect of a carcinogenic substance (*o*-amino-azotoluene) on the reactivity of liver cells following partial hepatectomy. *Boll. Exp. Biomed.* **53**, 116–118.

Huberman, E., and Sachs, L. (1966). Cell susceptibility to transformation and cytotoxicity by the carcinogenic hydrocarbon benzo(*a*)pyrene. *Proc. Natl. Acad. Sci. U.S.A.* **56**, 1123–1129.

Hunter, W. C. (1928). Experimental study of acquired resistance of rabbit renal epithelium to uranyl nitrate. *Ann. Intern. Med.* **1**, 747–789.

Iype, P. T., Tomaszewski, J. E., and Dipple, A. (1979). Biochemical basis for cytotoxicity of 7,12-dimethylbenz(*a*)anthracene in rat liver epithelial cells. *Cancer Res.* **39**, 4925–4929.

Judah, D. J., Legg, R. F., and Neal, G. E. (1977). Development of resistance to cytotoxicity during aflatoxin carcinogenesis. *Nature (London)* **264**, 343–345.

Kitagawa, T. (1971). Responsiveness of hyperplastic lesions and hepatomes to partial hepatectomy. *Gann* **62**, 217–224.

Kitagawa, T., Michalopoulos, G., and Pitot, C. (1975). Unscheduled DNA synthesis in cells from *N*-2-fluorenylacetamide induced hyperplastic nodules of rat liver maintained in a primary culture system. *Cancer Res.* **35**, 3682–3692.

Laird, A. K., and Barton, A. D. (1959). Cell growth and the development of tumors. *Nature (London)* **183**, 1655–1657.

Laird, A. K., and Barton, A. D. (1960). Cell proliferation in precancerous liver: Linear growth curve. *Nature (London)* **188**, 417–418.

Laird, A. K., and Barton, A. D. (1961). Cell proliferation in precancerous liver: Relation to presence and dose of carcinogen. *J. Natl. Cancer Inst.* **27**, 827–839.

Laishes, B. A., Roberts, E., and Farber, E. (1978). *In vitro* measurement of carcinogen-resistant liver cells during hepatocarcinogenesis. *Int. J. Cancer* **21**, 186–193.

Laishes, B. A., Fink, L., and Carr, B. C. (1980). A liver colony assay for a new hepatocyte phenotype as a step towards purifying new cellular phenotypes that arise during hepatocarcinogenesis. *Ann. N.Y. Acad. Sci.* **349**, 373–382.

Lasnitzki, I. (1949). Some effects of urethane on the growth and mitosis of normal and malignant cells *in vitro*. *Br. J. Cancer* **3**, 501–509.

Laws, J. O. (1959). Tissue regeneration and tumor development. *Br. J. Cancer* **13**, 669–674.

Levin, K., Lu, A. Y. H., Thomas, P. E., Ryan, D., Kiver, D. E., and Griffin, M. (1978). Identification of epoxide hydrase as a neoplastic antigen in rat liver hyperplastic nodules. *Proc. Natl. Acad. Sci. U.S.A.* **75**, 3240–3243.

Little, J. W., and Mount, D. W. (1982). The SOS regulatory system of *Escherichia coli*. *Cell (Cambridge, Mass.)* **29**, 11–22.

MacNider, W. de B. (1916). A pathological study of the naturally acquired chronic nephropathy of the dog. *J. Med. Res.* **129**, 177–197.

MacNider, W. de B. (1936a). A study of the acquired resistance of fixed tissue cells morphologically altered through processes of repair. I. The liver injury induced by uranium nitrate. *J. Pharmacol. Exp. Ther.* **56**, 359–372.

MacNider, W. de B. (1936b). A study of the acquired resistance of fixed tissue cells morphologically altered through processes of repair. II. The resistance of liver epithelium altered morphologically as a result of an injury from uranium, followed by repair to the hepatotoxic action of chloroform. *J. Pharmacol. Exp. Ther.* **56**, 373–381.

MacNider, W. de B. (1936c). A study of the acquired resistance of fixed tissue cells morphologically altered through processes of repair. III. The resistance to chloroform of a naturally-acquired atypical type of liver epithelium occurring in senile animals. *J. Pharmacol. Exp. Ther.* **56**, 383–387.

MacNider, W. de B. (1937). A study of the acquired resistance of fixed tissue cells morphologically altered through processes of repair. IV. Observations on the resistance to the toxic action of chloroform. *J. Pharmacol. Exp. Ther.* **59**, 393–398.

Maini, M. M., and Stitch, H. F. (1962). Chromosomes of tumor cells. III. Unresponsiveness of precancerous hepatic tissues and hepatomas to a mitotic stimulus. *J. Natl. Cancer Inst.* **28,** 753–762.

Melzer, M. S. (1980). Paradox of carcinogens as cell destroyers and cell stimulators. A biochemical hypothesis. *Eur. J. Cancer* **16,** 15–22.

Montesano, R., Bresil, H., Margison, G. P. (1979). Increased excision of O^6-methylguanine from rat liver DNA after chronic administration of diethylnitrosomine. *Cancer Res.* **39,** 1798–1802.

Muggia, F. M., Young, C. W., and Carter, S. K., eds. (1982). "Anthracycline Antibiotics in Cancer Therapy." Nijhoff, The Hague.

Niehuis, W. G., Jr., and Samuelsson, B. (1968). Formation of malonaldehyde from phospholipid arachidonate during microsomal lipid peroxidation. *Eur. J. Biochem.* **6,** 126–130.

Niles, R. M., Loewy, B., and Krah, D. (1981). Differential sensitivity of normal and chemically-transformed epithelial cells to cholera toxin. *Cancer Res.* **41,** 4075–4079.

Okita, K., Noda, K., Fukumoto, Y., and Takemoto, G. (1976). Cytochrome P-450 in hyperplastic liver nodules during hepatocarcinogenesis with *N*-2-fluorenylacetamide in rats. *Gann* **67,** 899–902.

Paraskeva, C., and Parkinson, E. K. (1981). Sensitivity of normal and transformed rat liver epithelial cells to benzo(*a*)pyrene toxicity. *Carcinogenesis* **2,** 483–487.

Prehn, R. T. (1964). A clonal selection theory of chemical carcinogenesis. *J. Natl. Cancer Inst.* **32,** 1–17.

Rabes, H. M., and Szymkowiak, R. (1979). Cell kinetics of hepatocytes during the preneoplastic period of diethylnitrosamine-induced liver carcinogenesis. *Cancer Res.* **39,** 1298–1304.

Riordan, J. R., Deuchars, K., Kartner, N., Alon, N., Trent, J., and Ling, V. (1985). Amplification of P-glycoprotein genes in multidrug-resistant mammalian cell lines. *Nature (London)* **316,** 817–819.

Roberts, E., Ahluwalia, M. B., Lee, G., Chan, C., Sarma, D. S. R., and Farber, E. (1983). Resistance to hepatotoxins acquired by hepatocytes during liver regeneration. *Cancer Res.* **43,** 28–34.

Rogers, K. E., Carr, B. I., and Tokes, Z. A. (1983). Cell surface-mediated cytotoxicity of polymer-bound adriamycin against drug-resistant hepatocytes. *Cancer Res.* **43,** 2741–2748.

Ruch, R. J., Klaunig, J. E., and Pereira, M. A. (1985). Selective resistance to cytotoxic agents in hepatocytes isolated from partially hepatectomized and neoplastic mouse liver. *Cancer Lett.* **26,** 295–301.

Sawada, N., Furukawa, K., Gotoh, M., Mochizuki, M., and Tsukada, H. (1981). Estimation of resistance to dimethylnitrosamine of normal hepatocyte, hyperplastic nodule cell and hepatoma cells of rats in primary culture. *Gann* **72,** 318–321.

Schor, N. A., and Morris, H. P. (1977). The activity of D-T-diaphorase in experimental hepatic tumors. *Cancer Biochem. Biophys.* **2,** 5–9.

Seglen, D. O. (1973). Preparation of rat liver cells. III. Enzymatic requirements for tissue dispersion. *Exp. Cell Res.* **82,** 391–398.

Sidransky, H., and Verney, E. (1978). Acute effect of selected hepatocarcinogens on polyribosomes and protein synthesis in the livers of rats fed purified diets containing hepatocarcinogens. *Cancer Res.* **38,** 1166–1172.

Siperstein, M. D., and Fagan, V. M. (1964). Deletion of the cholesterol-negative feedback system in liver tumors. *Cancer Res.* **24,** 1108–1115.

Sirica, A. E., Hwang, C. G., Sattler, G. L., and Pitot, H. C. (1980). Use of primary cultures of adult rat hepatocytes on collagen gel-nylon mesh to evaluate carcinogen-induced unscheduled DNA synthesis. *Cancer Res.* **40,** 3259–3267.

Smith, G. J., Ohl, V. S., and Litwack, G. (1977). Ligandin, glutathione-S-transferases, and chemically-induced hepatocarcinogenesis: A review. *Cancer Res.* **37,** 8–14.

Solt, D., and Farber, E. (1976). New principle for the analysis of chemical carcinogenesis. *Nature (London)* **263**, 701–703.

Starikova, V. B., and Vasiliev, J. M. (1962). Benz(alpha)anthracene on the mitotic activity of normal and malignant rat fibroblast *in vitro. Nature (London)* **195**, 42–43.

Sultatos, L. G., and Vesell, E. S. (1980). Enhanced drug metabolizing capacity within liver adjacent to rat liver tumors. *Proc. Natl. Acad. Sci. U.S.A.* **77**, 60–603.

Sun, I., MacKellar, W. C., Crane, F. L., Barr, R., Elliott, W. L., Lem, N., Varnold, R. L., Heinstein, P. F., and Morré, D. J. (1985). Decreased NADH-oxidoreductase activities as an early response in rat liver to the carcinogen 2-acetylaminofluorene. *Cancer Res.* **45**, 157–163.

Tanaka, Y., Nagasue, N., Kanashima, R., Inokuchi, K., and Shirata, A. (1982). Effect of doxorubicin on liver regeneration and host survival after two-thirds hepatectomy in rats. *Cancer (Amsterdam)* **49**, 19–24.

Vasiliev, J. M., and Guelstein, V. I. (1963). Sensitivity of normal and neoplastic cells to the damaging action of carcinogenic substances: A review. *J. Natl. Cancer Inst.* **31**, 1123–1143.

Watson, E., and Chan, K. K. (1976). Rapid analytic method for adriamycin and metabolites in human plasma by a thin-film fluorescence scanner. *Cancer Treat. Rep.* **60**, 1611–1618.

Williams, G. M., Klaiber, M., Parker, S. E., and Farber, E. (1976). Nature of early appearing, carcinogen-induced liver lesions resistant to iron accumulation. *J. Natl. Cancer Inst.* **57**, 157–165.

10

Measurement of Chemically Induced DNA Repair in Hepatocytes *in Vivo* and *in Vitro* As an Indicator of Carcinogenic Potential

BYRON E. BUTTERWORTH

Chemical Industry Institute of Toxicology
Research Triangle Park, North Carolina 27709

I. INTRODUCTION

Carcinogenesis is a complex, multistage process which may take years to develop from the time of first exposure to an agent that can initiate the process to the appearance of tumors. For example, the lag time from the beginning of smoking cigarettes to recognizable lung cancer is of the order of 20 years (Cairns, 1975). Accordingly, identification of chemicals with carcinogenic potential is very difficult. Traditional animal cancer bioassays must be run over the lifetime of the animal. Most protocols require a preliminary acute toxicity determination, a 90-day chronic study, and a 2-year cancer bioassay with an additional year for pathology evaluation, data analysis, and report writing. Current costs to evaluate the carcinogenicity of a single agent can be in the range of five hundred thousand dollars. Thus, there is a need for assays that can provide accurate information as to the carcinogenic potential of a chemical in a shorter time and with less cost.

The information that controls the structure, function, and reproduction of a cell resides in the DNA. Both theoretical considerations and experimental evidence indicate that alteration of the genetic material is involved in the process of carcinogenesis. Chemicals that can alter the DNA are classified as genotoxic

241

agents. Workers in the laboratory of Dr. Bruce Ames of the University of California at Berkeley examined a wide variety of chemicals in a bacterial mutagenesis assay, and the correlations that they observed greatly strengthened the hypothesis that mutagens are one class of carcinogen (McCann *et al.*, 1975). The assay employed is known as the Ames test and consists of strains of *Salmonella* that have been genetically engineered to be exquisitely sensitive to chemical mutagens. Mutants are identified under selective conditions, and a rat liver homogenate is incorporated in the assay to facilitate the production of genotoxic metabolites.

The value of a short-term test for potential carcinogens was quickly recognized, and now numerous assays are available that can be applied to obtain the profile of genotoxic activity and potential carcinogenicity of a chemical in species ranging from bacteria to man. The theory supporting the use of such short-term tests has been reviewed in detail (Butterworth, 1979; Hollstein *et al.*, 1979; Hsie *et al.*, 1979; IARC, 1980, 1982; de Serres and Ashby, 1981; Bridges *et al.*, 1982; Ashby, 1983). The activity or lack thereof in these assays is scrutinized by industry, academia, and regulatory agencies alike in performing human risk assessments.

The process of correlating results in short-term tests with cancer studies has acquired the name "validation studies." Unfortunately, genetic toxicology assays are often forced to accept the unrealistic challenge of perfect correlation with cancer studies. In this exercise, if a chemical has produced tumors at any dose, in any organ, in any species, it becomes a "plus." The only "correct" result for a short-term test is then defined as a plus. Assays are often pushed to extreme limits just to get the correct response. In many cases the "right" answer is obtained, but by a mechanism unrelated to the one producing tumors in the animal. Cell culture assays cannot be expected to reflect the species, strain, sex, and organ specificities that are common in chemical carcinogenesis. If the same criterion were applied to rodent bioassays, the mouse would be shown to be a poor predictor of carcinogenicity in the rat.

Both false-negative and false-positive results may occur because cell culture tests do not reflect the complexity of interactions that take place in the whole animal. Measurements in treated animals are particularly useful because factors such as metabolism, distribution, excretion, and repair are inherently accounted for (Ashby, 1983). One must be concerned that the events measured in culture truly reflect the critical events that occur in the animal. This is best illustrated with the nitroaromatic carcinogens. The potent hepatocarcinogen technical grade dinitrotoluene (DNT), as well as the individual DNT isomers, is weakly mutagenic in the Ames test without the need for an added metabolic activation system (Couch *et al.*, 1981). No genotoxic activity is seen in cell culture assays, including metabolically competent primary hepatocytes (Bermudez *et al.*, 1979). The generation of weak activity in bacteria, however, bears no resemblance to the actual complex pattern of activation in the whole animal, which involves the enterohepatic circulation and sequential steps of metabolism by the liver, gut flora, and the liver once again (Rickert *et al.*, 1984). Only when genotoxicity, forced cell proliferation, and promotion are measured in the whole animal do the results correlate with the striking differences in carcinogenic potency of the various isomers and the sex-specific susceptibility to the carcinogenic action of DNT (Rickert *et al.*, 1984). When conducting short-term tests for potential carcinogenicity of nitroaromatic compounds, it is mandatory to examine the effects in the whole animal because gut flora are obligatory in the metabolic activation of this class of chemicals (Mirsalis and Butterworth, 1982; Mirsalis *et al.*, 1982a; Doolittle *et al.*, 1983).

Because chemicals may be acting by different mechanisms, a battery of tests is required for a realistic evaluation of the genetic toxicity of a chemical. While it is not expected or required that a chemical respond in all tests for different end points, genotoxicants are generally positive in a variety of systems and enough work must be done to demonstrate a profile of activity. Results from bacterial assays alone are not sufficient to classify a chemical as genotoxic or nongenotoxic and must be confirmed in mammalian cells. For example, vitamin C is positive in the Ames test under certain conditions (Norkus *et al.*, 1983), whereas the potent carcinogen 2,3,7,8-tetrachlorodibenzo-*p*-dioxin (TCDD) does not respond in the same assay (Geiger and Neal, 1981). Knowledge of mechanism, metabolism, and contrasting responses in various assays can be extremely useful in interpreting and giving weight to different short-term tests. Some chemicals may produce genotoxicity as a secondary result of some other biological activity and would be negative in cell culture tests for mutagenicity. Examples would be effects on the DNA polymerase, inhibition of DNA repair, induction of a physiological state that results in genetic toxicity, or forced cell proliferation resulting in possible promotional effects and an increase in the spontaneous mutational frequency. Promoters, which accelerate the multistage process of tumorigenesis, may also be negative in assays for genotoxicity. For example, studies indicate that TCDD is a potent promoter. Thus, data from promotional assays would be far more valuable than mutagenicity data in searching for a noncarcinogenic analog of TCDD.

Despite the advances in genetic toxicology, the Ames test remains the most often, and usually only, assay employed in routine testing. The system is among the simplest to perform, has the broadest data base, and is usually the first test chosen. In a voluntary testing program, a negative result in the Ames test usually brings a sigh of relief and no further demands for additional work. The more assays that are run, the greater the chance of generating a spurious positive result. Even if the weight of evidence in critically evaluating the results of several short-term tests indicates that a compound is without genotoxic activity, a single positive response in any assay often precipitates further expensive testing and a high degree of anxiety about the hazard of the chemical. In a program to develop new products, a positive Ames test usually signals the end for that chemical because of the great difficulty and massive experimentation needed to overturn a positive Ames test in the eyes of regulatory agencies. Hopefully, the field of genetic toxicology will begin to move away from the rigid "plus–minus" mentality toward a more rational consideration of all the facts concerning a chemical. If used properly, an understanding of the genotoxic potential of an agent remains key information in predicting its carcinogenic potency.

The assays discussed in this chapter are the *in vitro* (cell culture) and *in vivo* (whole animal) hepatocyte (liver) DNA repair assays. They are particularly valuable in short-term testing because the cells themselves are metabolically competent and no external activation system is required. Because hepatocytes are the most metabolically active cells in the body, and because they are often the susceptible cells for chemical carcinogenesis in the rodent, knowledge of genotoxic activity in them is often key to understanding or predicting the activity of chemical agents. There is a very low false-positive rate with these assays. Chemicals that exhibit activity in these systems almost always turn out to have carcinogenic activity. The end point is also different from mutagenesis. Thus, these assays are often chosen to complement mutagenicity testing.

II. DNA REPAIR

Measurement of chemically induced DNA repair is a means of assessing the ability of a chemical to reach and alter the DNA. DNA repair is an enzymatic process that involves recognition and excision of DNA chemical adducts, and DNA strand polymerization and ligation to restore the original primary structure of the DNA (Roberts, 1978). This process can be quantitated by measuring the amount of labeled thymidine incorporated into the nuclear DNA of cells that are not in S phase and is often called unscheduled DNA synthesis (UDS) (Rasmussen and Painter, 1966). A major advance was the demonstration that primary rat hepatocytes in culture would produce a UDS response to a wide variety of genotoxicants requiring metabolic activation (Williams, 1976). In this assay freshly isolated primary hepatocyte cultures are incubated with the chemical in question and [^{3}H]thymidine. At the time used, the cells still retain most of the metabolic competency that they had in the animal. If the compound or its metabolites bind to the DNA, they will be removed by the excision repair process and the radiolabel will be incorporated into the DNA. Quantitative autoradiography is employed to quantitate the amount of incorporation of the labeled thymidine. The primary culture rat hepatocyte DNA repair assay has proven to be particularly valuable in assessing the genotoxic activity of a wide variety of chemicals

(Williams, 1977, 1978, 1984; Williams *et al.*, 1977; Bermudez *et al.*, 1979; Williams and Laspia, 1979; Probst *et al.*, 1981; Tong *et al.*, 1981; Casciano and Gaylor, 1983; Mitchell *et al.*, 1983; Conaway *et al.*, 1984; Lawson *et al.*, 1984; Mori *et al.*, 1984a,b; Waters *et al.*, 1984; West *et al.*, 1984; Kornbrust and Dietz, 1985; Lefevre and Ashby, 1985; Maslansky and Williams, 1985). Numerous assays now have been developed for the measurement of chemically induced DNA repair in various cell lines and primary cell cultures from both rodent and human origin (Butterworth, 1983; Mitchell and Mirsalis, 1984).

A further advance was the development of an *in vivo* rat hepatocyte DNA repair assay in which the chemical is administered to the animal and the resulting DNA repair is then assessed in hepatocytes isolated from that animal (Mirsalis and Butterworth, 1980). In this system, the chemical is given by whatever route of administration is relevant. The normal processes of absorption, transportation to the liver, metabolism, detoxification, and binding to the DNA are allowed to take place. Primary hepatocyte cultures are then prepared and incubated with [^{3}H]thymidine. DNA that had been damaged *in vivo* will be repaired *in vitro* resulting in the incorporation of radioactivity. Autoradiography is employed to quantitate the amount of labeled thymidine incorporated into the DNA. Use of scintillation counting is inappropriate because the background due to cytoplasmic and replicative or synthesis phase (S phase) incorporation cannot be distinguished from DNA repair. This is easily accomplished with autoradiography. The advantage of *in vivo* assays is that they reflect the complex patterns of uptake, distribution, metabolism, detoxification, and excretion that occur in the whole animal. Further, factors such as chronic exposure, sex differences, and different routes of exposure can be studied with these systems. As was mentioned above, this can be illustrated by the potent hepatocarcinogen 2,6-dinitrotoluene (2,6-DNT). Metabolic activation of 2,6-DNT involves uptake, metabolism by the liver, excretion into the bile, reduction of the nitro group by gut flora, readsorption, and further metabolism by the liver once again to finally produce the ultimate genotoxicant (Rickert *et al.*, 1984). Thus, 2,6-DNT is negative in the *in vitro* hepatocyte DNA repair assay (Bermudez *et al.*, 1979) but is a very potent inducer of DNA repair in the *in vivo* DNA repair assay (Mirsalis and Butterworth, 1982; Mirsalis *et al.*, 1982a).

A problem with tissue-specific assays is that they may fail to detect activity of compounds that produce tumors in other target tissues. For example, no activity is seen in the *in vivo* DNA repair assay with the potent mutagen benzo(*a*)pyrene (BP), probably because an insufficient amount reaches the liver to produce a measurable response (Mirsalis *et al.*, 1982b). In contrast, BP is readily detected in the less tissue-specific *in vitro* hepatocyte DNA repair assay (Probst *et al.*, 1981; Tong *et al.*, 1981). Interestingly, BP does not produce liver tumors in feeding studies (IARC, 1973). Tumor induction by BP is related to the route of administration and predominantly affects those cells which first come in direct contact with the chemical. If BP administration to rats was preceded by a partial hepatectomy and followed by feeding a diet containing phenobarbital for a year, however, then some liver tumors were observed in addition to the tumors at the site of BP administration (Kitagawa *et al.*, 1980). An extensive literature exists on the use of the *in vivo* hepatocyte DNA repair assay (Butterworth *et al.*, 1982, 1983, 1984; Mirsalis and Butterworth, 1982; Mirsalis *et al.*, 1982a,b, 1985; Doolittle *et al.*, 1983, 1984; Helleman *et al.*, 1984; Joachim and Decad, 1984; Kornbrust *et al.*, 1984; Mitchell and Mirsalis, 1984; Ashby *et al.*, 1985; Beije and Ashby, 1985; Kornbrust and Dietz, 1985; Lowry *et al.*, 1986, 1987; Working *et al.*, 1986a,b; Butterworth and Smith-Oliver, 1987; Smith-Oliver and Butterworth, 1987). Numerous systems now exist to measure chemically induced DNA repair in specific tissues in the whole animal (Butterworth, 1983; Mitchell and Mirsalis, 1984).

The exact role that cell replication may have in carcinogenesis, as well as its predictive value for tumor formation, is still being defined. Nevertheless, many carcinogens that have apparently little or no genotoxic activity in sensitive cell culture models do possess the ability to force cell proliferation in the susceptible target organ. Continued cell turnover may result in an increase in the spontaneous mutation rate or have promotional effects. Important genes related to the regulation of cellular growth and cancer may remain stimulated during continued administration of mitogenic agents. Many such chemicals have been studied (Schulte-Hermann *et al.*, 1983), and it has been proposed that there are three useful classifications of *in vivo* hepatotoxic agents (Butterworth *et al.*, 1985; Loury *et al.*, 1987). Type 1 agents are genotoxic in the liver in the whole animal as evidenced by the induction of DNA repair, identification of DNA adducts, and other indicators of reactivity with the DNA. Examples would be dimethylnitrosamine and 2-acetylaminofluorene. These compounds have a high probability of being hepatocarcinogens in the species in which genotoxicity was measured. Type 2 agents show no indication of primary genotoxicity but cause liver hyperplasia upon prolonged administration. An S phase response is seen, with a maximum at about 24 hr after a single treatment. No cytotoxicity is produced as evidenced by unchanged liver-specific enzymes in the serum. These compounds appear to have a modest probability of being hepatocarcinogens in the species in which activity was measured. They can range from very weak agents such as di(2-ethylhexyl) phthalate (DEHP) (Butterworth, 1987; Smith-Oliver and Butterworth, 1987) to very potent hepatocarcinogens such as Wy-14,643 (Rao *et al.*, 1984). Type 3 agents, such as carbon tetrachloride, show no indication of primary genotoxicity but are cytotoxic to the liver as evidenced by liver-specific enzymes in the serum. Regenerative liver growth is produced. Compensatory cell replication is evidenced by an S phase response with a maximum at about 48 hr after a single treatment (Mirsalis *et al.*, 1982b). These compounds also would appear to have a modest probability of being weak hepatocarcinogens in the species in which activity was measured.

Cells in the process of replicative DNA synthesis (RDS) or the synthesis phase of the cell cycle are easily recognized as densely labeled nuclei in the autoradiograms. Thus, an advantage of the *in vivo* assay is that the percentage of hepatocytes in S phase may be quantitated as an indicator of cell proliferation. This ancillary information may be valuable in characterizing the biological activity and potential carcinogenicity of a chemical (Mirsalis *et al.*, 1985; Loury *et al.*, 1987).

III. PROCEDURES FOR *IN VITRO* AND *IN VIVO* RAT HEPATOCYTE DNA REPAIR ASSAYS

In many cases a particular procedure is applicable for both the *in vitro* and the *in vivo* rat hepatocyte DNA repair assays. Given below are the techniques and guidelines followed by our laboratory. In each case it is noted whether the procedure is for the *in vitro* assay, the *in vivo* assay, or both. Any reference to a particular supplier or manufacturer is meant only as an example and not as an endorsement.

A. Animal Treatment—the *in Vivo* Protocol

1. All personnel must be knowledgeable in the procedures for safe handling and proper disposal of carcinogens, hazardous chemicals, and radioisotopes. All test agents must be treated as hazardous. Disposable gloves and lab coats must be worn throughout these procedures.

2. Although any strain or sex of rat may be used, the largest data base is for male Fischer-344 rats. Young adult animals (200–275 g) are preferred.

3. Chemical administration is usually by gavage with chemicals dissolved or suspended in water or corn oil depending on solubility. An advantage of the assay is that the appropriate route of administration may be chosen. Thus, chemicals may also be administered by ip injection or inhalation or in the diet. Test chemical solution (0.2 to 1.0 ml) is administered per 100 g body weight. Controls receive the appropriate vehicle solution.

4. For DNA repair studies animals may be taken off feed for a few hours prior to sacrifice to make the process of perfusion a little easier with less food in the stomach. The period without food should never exceed 12 hr because of the possibility of altered metabolism or uptake. Water should be continually available. For measurement of cells in S phase, a more rigorous schedule of feeding and light–dark cycle must be adhered to.

5. The time from treatment to sacrifice may be varied to obtain a time course of induced repair. The routine protocol involves a time point between 1 and 3 hr and another between 12 and 16 hr. Some chemicals such as dinitrotoluene and

2-acetylaminofluorene (2-AAF) do not show a peak response until 12 hr, while dimethylnitrosamine (DMN) exhibits a maximum response 1–2 hr after treatment.

6. Dose selection will depend on the characteristics of each chemical. While high doses that do not kill the animal may be employed, one must be aware that in some instances hepatotoxicity at high doses may result in inhibition of cell attachment or DNA repair. Target doses with a new compound are usually the LD_{50} and $0.2 \times LD_{50}$. The usual range of doses is from 10 to 500 mg compound/kg body weight. One advantage of the assay is that realistic exposures can be evaluated. DMN can be detected with doses as low as 1 mg/kg.

7. Treated animals should be maintained in a vented area or other suitable location to prevent possible human exposure to expired chemicals. Contaminated cages, bedding, and carcasses should be disposed of safely.

B. Liver Perfusion—the *in Vitro* and *in Vivo* Protocols

1. The rat is anesthetized by ip injection with 0.2 ml Nembutal Sodium Solution (50 mg/ml) per 100 g body weight 10 min prior to the perfusion procedure. Make sure that the animal is completely anesthetized so that it feels no pain.

2. Secure the animal with the ventral surface up on absorbent paper on a cork board. Fold the paper in on each edge to contain perfusate overflow. The abdomen is thoroughly wetted with 70% ethanol and wiped with gauze for cleanliness to discourage loose fur from getting on the liver when the animal is opened.

3. A V-shaped incision is made through both skin and muscle from the center of the lower abdomen to the lateral aspects of the rib cage. Do not puncture the diaphragm or cut the liver. The skin and attached muscle are folded back over the chest to reveal the abdominal cavity.

4. A tube approximately 1 cm in diameter is placed under the back to make the portal vein more accessible.

5. The intestines are gently moved out to the right to reveal the portal vein. The tube under the animal is adjusted so that the portal vein is horizontal.

6. A suture is put in place (but not tightened) in the center of the portal vein, and another is put around the vena cava just above the right renal branch.

7. Perfusions are performed with a peristaltic pump, the tubing of which is sterilized by circulation of 70% ethanol followed by sterile water. A valve is placed in the line so that one may switch from the EGTA solution to the collagenase solution without disrupting the flow. Solutions are kept at 37°C with a minimal distance to the animal and shortest possible tube length to minimize cooling in the line.

8. A typical pump would be the Cole-Parmer Masterflex pump (K7562-10)

with a Masterflex pump head (K-7020-50) and Cole-Parmer silicone tubing (K-6411-43).

9. The flow of the 37°C EGTA solution is begun at 8 ml/min and run to waste.

10. The portal vein is cannulated, bevel up, with a 20 GA 1.25-in. Angiocath (Deseret Medical, Inc. #2878). The point should enter the vein about 3 mm below the suture and be inserted to about 3 mm above the suture. Extreme care must be taken at this time to keep the catheter exactly parallel to and inside of the vein. The delicate vein is easily nicked by the sharp needle. The inner needle is removed and the plastic catheter is further inserted to about one-third the length of the vein and tied in place by the suture. Blood should emerge from the catheter. The tube with the flowing EGTA is then inserted in the catheter (avoid bubbles) and taped in place.

11. Because pressure will begin to build in the circulatory system, the vena cava is immediately cut below the right renal branch and the liver is allowed to drain of blood for 1.5 min. The liver should rapidly clear of blood and turn a tan color. If all lobes do not clear uniformly, the catheter has probably been inserted too far into the portal vein.

12. The suture around the vena cava is then tightened and the flow increased to 20 ml/min for 2 min. The liver should swell at this point. In some cases gentle massaging of the liver or adjusting the orientation of the Angiocath may be necessary for complete clearing. At this point the vena cava above the suture may be clipped to release some of the pressure in the liver.

13. The flow is then switched to the 37°C collagenase solution for 12 min. During this period the liver is covered with sterile gauze wetted with sterile saline or WEI and a 40-W lamp is placed 2 in. above the liver for warming.

14. The perfusate is allowed to flow onto the paper and is collected by suction into a vessel connected via a trap to the vacuum line. If a treated animal is being used, the perfusate should be disposed of as hazardous waste.

15. After the perfusion is over the catheter and gauze are removed. The liver is carefully removed by (1) cutting away the membranes connecting it to the stomach and lower esophagus, (2) cutting away the diaphragm, and (3) cutting any remaining attachments to veins or tissues in the abdomen.

16. The liver is held by the small piece of attached diaphragm and rinsed with sterile saline or WEI.

17. The liver is placed in a sterile petri dish and taken to a sterile hood to prepare the cells.

C. Preparation of Hepatocyte Cultures—the *in Vitro* and *in Vivo* Protocols

1. The perfused liver is placed in a 60 mm petri dish and rinsed with 37°C WEI. Extraneous tissues (fat, muscle, etc.) are removed.

2. The liver is placed in a clean petri dish and 30 ml of fresh collagenase solution at 37°C is added.

3. Carefully make several incisions in the capsule of each lobe of the liver. Large rips in the capsule lead to large, unusable clumps of hepatocytes.

4. Gently comb out the cells, constantly swirling the liver while combing. A sterile, metal, dog-grooming comb with teeth spaced from 1 to 3 mm works well.

5. When only fiberous and connective tissue remain, discard the same and add 20 ml cold WEI to the cell suspension. Transfer the cell suspension to a sterile 50-ml centrifuge tube (Corning #25331) using a wide-bore sterile 10 ml pipet (Falcon #7536).

6. Allow the cells to settle on ice for 5–10 min until a distinct interface is seen. Carefully remove and discard the supernatant by suction. The supernatant will contain debris, broken cells, and nonparenchymal cells.

7. Bring the cells to 50 ml with cold WEI. Resuspend the cells by pipeting with a wide bore pipet. Gently pipet the suspension through a 4-ply layer of sterile gauze (Johnson & Johnson #8-2317, 4×4) into a sterile 50-ml centrifuge tube. This will remove any large debris remaining in the suspension.

8. Centrifuge the cells at 50 g for 5 min and discard the supernatant. Gently resuspend the pellet in ice-cold WEI with a wide-bore pipet.

9. Keep the cells on ice until ready for use. When following the *in vivo* protocol the cells should be used immediately. When following the *in vitro* protocol the cells should be used within 1 hr.

10. Determine viability and cell concentration by the method of trypan blue exclusion. The preparation should be primarily a single cell suspension with a viability of over 60%. With the proper technique, viabilities of about 90% can routinely be obtained for cells derived from control animals.

11. A thermanox #5415 25-mm round plastic coverslip No. 1½ (Lux Scientific Corp.) is placed into each well of Linbro 6-well culture dishes (Flow Labs). Be sure to keep the proper side up as marked on the package. WEC (4 ml) is added to each well. Hepatocytes will not attach to glass unless the coverslips have been boiled.

12. Approximately 400,000 viable cells are seeded into each well and distributed over the coverslip by shaking or stirring gently with a plastic 1-ml pipet. Glass pipets can scratch the coverslips.

13. The cultures are incubated for 90–120 min in a 37°C incubator with 5% CO_2, 95% relative humidity, to allow the cells to attach.

D. Treating the Cultures and Labeling the Cells— the *in Vitro* Protocol

1. Attachment is usually for 90–120 min. This period can be extended for several hours to obtain greater adhesion of the cells. After the attachment period,

cultures are washed once with 4 ml WEI per well to remove unattached cells and debris. This is done by tilting the culture slightly, aspirating the media, and adding the fresh media at 37°C. Be careful not to direct the stream from the pipet directly onto the cells.

2. Chemical solutions are prepared in [^{3}H]WEI (10 μCi/ml [^{3}H]thymidine). Serial dilutions 10-fold are generally employed. Any solvent such as dimethyl sulfoxide (DMSO) or ethanol should not exceed a 1% final concentration. Concentrations should be chosen that go just beyond cytotoxicity to about 1000-fold below the cytotoxic concentration. Typical concentration ranges are from 1 to 0.001 mM. Two to six duplicate coverslips may be treated for a backup or to compare culture to culture variability. Dimethylnitrosamine produces a strong response at the noncytotoxic dose of 10 mM. In contrast, 1,6-dinitropyrene induces DNA repair at concentrations as low as 0.00005 mM.

3. The WEI is removed and replaced with 2 ml of [^{3}H]thymidine solution containing the dissolved test chemical. The cultures are then placed in the incubator for 16–24 hr. During this period the compound will be metabolized. If DNA damage results it will be repaired, resulting in incorporation of the [^{3}H]thymidine.

4. Cultures are washed twice with 4 ml WEI per well.

5. Media is replaced with 4 ml of a 1% sodium citrate solution, and the cultures are allowed to stand for 10 min to swell the nuclei.

6. The sodium citrate solution is replaced with 3 ml of a 1:3 acetic acid:absolute ethanol solution and allowed to stand for 10 min to fix the cells. This is repeated twice more for a total fixing time of at least 30 min.

7. Wells are washed two times each with deionized distilled water.

8. Coverslips are removed from the wells and placed cell-side up on the edge of the dish covers to dry in a dust-free location.

9. When dry, coverslips are mounted cell-side up on microscope slides with Krystalon. The coverslip should be mounted about 1 cm from the unfrosted end of the slide. Each slide is given a unique identifying number.

10. At this point the cultures can be examined for gross cytotoxicity. In pilot experiments, if chemical treatment at the higher doses has resulted in there being no cells on the slides, they need not be subjected to autoradiography.

E. Labeling the Cells—the *in Vivo* Protocol

1. After the attachment period, cultures are washed once with 4 ml WEI per well to remove unattached cells and debris. This is done by tilting the culture slightly, aspirating the media, and adding the fresh media at 37°C. Be careful not to direct the stream from the pipet directly onto the cells.

2. The WEI is then removed and replaced with 2 ml of [^{3}H]thymidine solution

(10 μCi/ml). The cultures are then placed in the incubator for 4 hr. During this period some of the DNA damage that occurred in the animal will be repaired, resulting in the incorporation of [³H]thymidine.

3. The cultures are washed once with 4 ml WEI/well, then 3 ml of unlabeled thymidine solution (0.25 mM) is added to each well. Cultures are incubated overnight (14–16 hr).

4. Cultures are washed twice with 4 ml WEI per well.

5. Media is replaced with 4 ml of a 1% sodium citrate solution, and the cultures are allowed to stand for 10 min to swell the nuclei.

6. The sodium citrate solution is replaced with 3 ml of a 1:3 acetic acid:absolute ethanol solution and allowed to stand for 30 min to fix the cells. This is repeated twice more for a total fixing time of at least 30 min.

7. Wells are washed two times each with deionized distilled water.

8. Coverslips are removed from the wells and placed cell-side up on the edge of the dish covers to dry in a dust-free location.

9. When dry, coverslips are mounted cell-side up on microscope slides with Krystalon. The coverslip should be mounted about 1 cm from the unfrosted end of the slide. Each slide is given a unique identifying number.

F. Autoradiography—the *in Vitro* and *in Vivo* Protocols

1. All steps involving photographic emulsion should be done in total darkness. If absolutely necessary, a #1 red (Kodak) safelight filter may be used sparingly.

2. One to three of the slides for each dose for the *in vitro* protocol, or three of the six slides for each animal for the *in vivo* protocol, are mounted in plastic slide grips (Lipshaw). The other slides are held in reserve.

3. A 50-ml disposable plastic beaker is mounted with tape into a slightly larger jar full of water to hold the plastic beaker down in the water bath. This assembly is then placed into a 42°C water bath and allowed to reach the bath temperature.

4. Kodak NTB-2 emulsion is most commonly used. The emulsion is used undiluted or can be used diluted 1:1 with distilled water. If the emulsion is diluted, care should be taken to use ultrapure water; thoroughly mix the solution and avoid the formation of air bubbles. Undiluted emulsion is more expensive but saves a step and provides slightly higher grain counts. Emulsion is melted in a 37°C incubator for at least 3 hr. In the dark, gently pour 40–50 ml of the emulsion into the dipping cup. The unused portion can be resealed and stored under refrigeration.

5. Dip a test slide. Briefly turn on the #1 Red (Kodak) safelight and hold the slide up to it to make sure that there is enough emulsion in the cup to cover the cells and that there are no bubbles in the emulsion. Air bubbles can be removed from the surface of the emulsion by skimming the surface with a glass slide. Turn off the safelight.

6. Dip each group of slides by lowering them into the cup until they touch the bottom. Tilt the slides so that their edges touch the edge of the cup and pull them out of the emulsion with a smooth action to a 5-second count. Touch the bottom ends of the slides to a pad of paper towels to remove the bead of emulsion on the bottom. Remember, all of these steps are taking place in total darkness. Do not reuse the used emulsion.

7. Hang the slide holders in a vertical position in a rack in a light-tight box for 3–12 hr to let the emulsion dry. Pack the slides into light-tight slide boxes that contain a false bottom packed with desiccant. Seal the boxes with black electrical tape and wrap them in aluminum foil for good measure.

8. Store the slides at +4°C to −20°C for a set amount of time. Most common is 7–14 days. Shorter times yield lower backgrounds. Longer times produce higher counts.

9. After the exposure period, allow the slide boxes to thaw at room temperature for at least 3 hr. In the dark, place the slides into a rack suitable for developing and staining the slides.

10. Develop the slides at 15°C (56°F) for 3 min in Kodak D-19 developer. Tap the rack gently to the bottom of the developing dish several times to dislodge any air bubbles on the slides.

11. Rinse slides 30 sec in 15°C water, then fix in Kodak Fixer (not Rapid-Fix) for 5 min with agitation every 60 sec. Wash the slides in a bath with gently running water for 25 min. Do not hit the slides directly with the water stream.

12. Slides can be stained while still wet from development. Dip into methyl green pyronin Y solution for 10–20 sec. Follow this immediately with repeated washings in water and a final rinse in distilled water. Do not overstain the cells. Cells should have faint blue nuclei and pink cytoplasms. Overstained cells make automatic grain counting difficult. Remember, the cells are still exposed at this point. Take care not to touch the slide surface.

13. Allow the slides to air dry for a few hours. Mount a 25-mm square coverslip (Corning) over the round coverslip using a thin layer of Krystalon. Keep the slides flat overnight to dry. They are now ready for grain counting.

G. Grain Counting—the *in Vitro* and *in Vivo* Protocols

1. Grain counting can be done by hand. This, however, is a tedious process. If the assay will be used in any routine way, an automated counting system will be necessary.

2. Grain counting is best accomplished with an automated system such as the ARTEK model 982 Colony Counter interfaced to a Zeiss Universal microscope with an ARTEK high resolution TV camera. Data can be fed directly into a computer such as the VAX 11/780 via an ARTEK BCD-RS232 Omni-interface. Any proved system that accurately counts the grains, however, is acceptable.

3. For the *in vitro* protocol, 20–50 cells are counted per slide, 1–3 slides per treatment, 2–3 experiments per data point. Normally, a total of 100 cells is counted per experiment. In an initial screening experiment in which multiple doses are being examined, only noncytotoxic doses need be counted and repeats only need cover the active, noncytotoxic concentrations.

4. For the *in vivo* protocol, normally 25 cells are counted per slide, 3 slides per animal, 3 animals per data point. In an initial screening experiment in which multiple doses and time points are examined, three animals per data point are not necessary.

5. Counting usually requires a 100× objective under oil immersion. An optivar can be employed to further increase magnification.

6. Each slide is examined to make sure that the culture as a whole is viable. Signs of toxicity are the absence of cells or pyknotic (small, darkly stained) cells.

7. A patch of cells is selected as a starting point, and cells are scored in a regular fashion by bringing new cells into the field of view moving only the x-axis. If the desired number of cells has not been scored before coming to the edge of the slide, the stage is moved 1–2 fields on the y-axis and counting resumes in the opposite x-direction, parallel to the first line. If, upon visual scanning of the slide, there appears to be any difference in response in different areas of the slide, then the counting should be done selecting patches of cells from several areas of the slide.

8. The following criteria are used to determine if a cell should or should not be counted:

    a. Cells with abnormal morphology, such as those with pyknotic or lysed nuclei, should not be counted.
    b. Isolated nuclei not surrounded by cytoplasm should not be counted.
    c. Cells with unusual staining artifacts or in the presence of debris should not be counted.
    d. Heavily labeled cells in S phase should not be counted.
    e. All other normal cells encountered while moving the stage must be counted without regard as to their apparent response.

9. Counts are generally made in the mode that counts the area of the grains. This allows patches of grains that are touching to be counted without being mistaken by the machine as a single grain. When using the area mode, a correction factor to convert to grain counts must be used. This conversion factor must be determined for the particular counting setup and configuration being used. To do so, count a number of areas (10–30 discreet grains/aperture) on both the count mode and manually to determine the actual number of silver grains. Then perform a machine area count on the same aperture area. After counting 20–30 areas from at least two different slides, add all the actual counts and all the area counts. The conversion factor is calculated as

$$C = \frac{\text{actual number of grains (total)}}{\text{measured area of grains (total)}}$$

Then, machine counts can be converted to actual grains by multiplying by C.
10. For each cell the following procedure is used:

    a. The sensitivity of the machine is adjusted so that only grains are being counted and so that the configuration is the same as when the conversion factor was calculated.

    b. Place the aperture directly over the nucleus and adjust to the same size as the nucleus.

    c. Press the count button to record the nuclear counts.

    d. Keeping the aperture the same size, count an area over the cytoplasm that is adjacent to the nucleus and appears to have the highest grain counts. Press the count button to record the nuclear counts.

    e. Subtract the cytoplasmic count from the nuclear count to give the net grains/nucleus (NG) for that cell. In the case of control cells, there will usually be more grains per unit area in the cytoplasm than in the nucleus so that the NG will be a negative number. This must be reported as such. Continue counting the desired number of cells in the same manner.

    f. At all times remember that actual grains over the nucleus are being determined. If a spurious count is observed, it should be corrected before the data are calculated. If Good Laboratory Practice is being followed, any change in the data set should be noted for the record.

H. Calculations and Data Analysis

1. The in Vitro Protocol

1. Alterations in the following parameters will affect the NG observed: (1) the concentration and specific activity of [³H]thymidine used; (2) the length of time that the cultures are exposed to the chemical and the [³H]thymidine; (3) the type and dilution of photographic emulsion; and (4) The autoradiographic exposure time. Thus, in data analysis one must either use a published procedure where the expected values for negative and positive responses are known, or this must be determined for the conditions being used.

2. For the conditions described here, ≥5 NG has been chosen as a conservative estimate as to whether an individual cell is responding or is "in repair."

3. The following should be calculated for each slide: (1) the population average NG ± SD (cell to cell); (2) the percent of cells responding or in repair; and (3) the population average NG ± SD (cell to cell) for the subpopulation of cells that are in repair (optional).

4. The following should be calculated for each cell preparation: (1) the population average NG ± SE (slide to slide); (2) the percent of cells responding or in repair ± SD (slide to slide); and (3) the population average NG ± SE (slide to slide) for the subpopulation of cells that are in repair (optional).

5. Data from one cell preparation represents an n of one. Experiments must be repeated at least once and preferably twice with different cell preparations.

6. Repeat values should be presented side by side. Alternatively, the following

can be calculated for each data point: (1) the population average NG ± SE (preparation to preparation); (2) the percentage of cells responding or in repair ± SE (preparation to preparation); and (3) the population average NG ± SE (preparation to preparation) for the subpopulation of cells that are in repair (optional).

7. Using conditions similar to those described here with a cutoff of 5 NG for a cell in repair, control values in the range from −10 to −3 NG with 0–10% in repair should be expected. One of the most troublesome aspects of this assay is that high cytoplasmic backgrounds that obscure any realistic evaluation of nuclear counts are often observed. If cytoplasmic background counts exceed 30 grains per nuclear-sized area, the experiment should not be considered valid. Procedures to lower the background include lowering the concentration of [^{3}H]thymidine, incubating the cells for a shorter period of time, using fresh [^{3}H]thymidine, and decreasing the autoradiographic exposure time.

8. For the positive controls 10 mM dimethylnitrosamine or 0.001 mM 2-acetylaminofluorene, one might expect values of 15–30 NG with 70–100% of the cells with ≥5 NG.

Historical data from this laboratory indicate that, for those conditions described here, a value of ≥5 NG (population average) and ≥20% of cells responding should be considered a positive response. Counts between 0 and 5 NG would be considered a marginal response. A positive dose–response relationship and an increase in the percentage of cells in repair are required additional information to confirm a positive response for counts below 5 NG. It is important to run several control cultures over a period of time so that the historical control baseline can be determined for those conditions in any particular laboratory.

Casciano and Gaylor (1983) have described statistical criteria for evaluating chemicals as positive or negative in the *in vitro* hepatocyte DNA repair assay for the conditions that they employ in their laboratory. The criteria involve both the NG count and the percentage of cells in repair. In the *in vitro* assay, the control cells come from the same preparation as the treated cells and thus represent a true concurrent control. The unpaired *t* test for the equality of two means (slide to slide) is an appropriate statistical test for these data.

The probable reason that control NG values tend to be less than zero is that the cytoplasm (and the components therein producing the cytoplasmic background) is slightly thinner over the nucleus compared to the rest of the cell as it sits on the substrate. NG counts may vary as the result of compound-related effects on cytoplasmic grain counts. Consequently, no result may be considered positive unless the compound actually produces more grains over the nucleus than over the cytoplasm, i.e., a NG value greater than zero. Knowledge of the biology of this assay dictates that in order to have any confidence in a positive DNA repair response, the treatment must produce nuclear counts beyond the cytoplasmic background. Thus, for any statistical test employed, a lower limit of at least 0 NG is required for a positive response.

2. The in Vivo Protocol

1. Alterations in the following parameters will affect the NG observed: (1) the conditions of exposure of the animal to the chemical; (2) the concentration and specific activity of [^{3}H]thymidine used; (3) the length of time that the cells are exposed to the [^{3}H]thymidine; (4) the type and extent of dilution of the emulsion; and (5) the autoradiographic exposure time. Thus, in data analysis one must either use a published procedure where the expected values for negative and positive responses are known, or this must be determined for the conditions being used.

2. For the conditions described here, ≥ 5 NG has been chosen as a conservative estimate as to whether an individual cell is responding or is "in repair."

3. The following should be calculated for each slide: (1) the population average NG $\pm$ SD (cell to cell); (2) the percentage of cells responding or in repair; and (3) the population average NG $\pm$ SD (cell to cell) for the subpopulation of cells that are in repair (optional).

4. The following should be calculated for each animal: (1) the population average NG $\pm$ SE (slide to slide); (2) the percentage of cells responding or in repair $\pm$ SD (slide to slide); and (3) the population average NG $\pm$ SE (slide to slide) for the subpopulation of cells that are in repair (optional).

5. The following should be calculated for each data point and are the numbers presented in reports and publications: (1) the population average NG $\pm$ SE (animal to animal); (2) the percent of cells responding or in repair $\pm$ SE (animal to animal); and (3) the population average NG $\pm$ SE (animal to animal) for the subpopulation of cells that are in repair (optional).

6. For the conditions described here, historical data from 93 control rats (untreated, corn oil, water, or saline; by gavage or ip; 2–48 hr post treatment) from this laboratory yielded -4.2 ± 1.5 NG with $2 \pm 3\%$ of cells responding. No individual animal NG value greater than zero was observed. No value greater than 14% was observed for the percentage of cells responding.

7. For the positive controls 10 mg/kg dimethylnitrosamine (administered in water by gavage 2 hr before sacrifice) or 50 mg/kg 2-acetylaminofluorene (administered in corn oil by gavage 12 hr before sacrifice), one might expect values of 30–60 NG with 80–100% of the cells with ≥ 5 NG.

Historical data from this laboratory indicate that for those conditions described here, a value of ≥ 5 NG (population average) and $\geq 20\%$ of cells responding should be considered a positive response. Counts between 0 and 5 NG would be considered a marginal response. A positive dose–response relationship and an increase in the percentage of cells in repair are required additional information to confirm a positive response for counts < 5 NG. It is important to run several control animals over a period of time so that the historical control baseline can be determined for those conditions in any particular laboratory.

Different situations may be encountered when running the *in vivo* hepatocyte DNA repair assay. There may be times when several true concurrent controls will be run with the treated group. Since one laboratory cannot be expected to run more than six animals in one day, numerous concurrent controls for every experiment are prohibitive and not required. Nevertheless, if concurrent controls have been run, then the unpaired *t* test for the equality of two means using the individual animal NG as the unit of measure is a reasonable test for statistical significance.

The usual case is that over the course of a few weeks several batches of animals will be run in building a data base for a chemical. On each experimental day a control animal will be employed. If conditions and procedures in the laboratory remain constant, it is valid to pool those controls produced over the course of the study. The unpaired *t* test or a multicomparison test such as the Dunnett's multiple range test may be used to compare the mean of the control group with the means for the treated groups for statistical significance. One may even choose to pool the entire historical control data base for comparison to the treated group by the unpaired *t* test. However, with a very large *N*, relatively small increases in the treated samples (i.e., NG values less than zero), that may have no biological significance, may appear as statistically significant.

The probable reason that control NG values tend to be less than zero is that the cytoplasm (and the components therein producing the cytoplasmic background) is slightly thinner over the nucleus compared to the rest of the cell as it sits on the substrate. NG counts may vary as the result of compound related effects on cytoplasmic grain counts. Consequently, no result may be considered positive unless the compound actually produces more grains over the nucleus than over the cytoplasm, i.e., a NG value greater than zero. Knowledge of the biology of this assay dictates that, in order to have any confidence in a positive DNA repair response, the treatment must produce nuclear counts beyond the cytoplasmic background. Thus, for any statistical test employed, a lower limit of at least zero NG is required for a positive response.

I. Cells in S Phase—the *in Vivo* Protocol

1. The rationale and strategies for measuring chemical induction of hepatocyte replication have been discussed (Loury *et al.*, 1987). Because replicative DNA synthesis cannot be distinguished from repair synthesis by scintillation counting techniques, only autoradiographic methods are acceptable in assessing DNA repair and induction of S phase in the whole animal. Cells in S phase are easily recognized by a very dense pattern of silver grains over the nucleus.

2. Measurement of cells in S phase is optional. For a single dose, cells in S phase should be evaluated at 24 and 48 hr after chemical treatment. Food and

water are given *ad libitum*. Rats should be on a defined light–dark cycle that remains unchanged throughout the experiment (Loury *et al.*, 1987).

3. Cells are prepared exactly as for the DNA repair experiments. Cells in S phase are scored visually under low power (100–400×). Experience from this laboratory shows that there will be a high degree of animal to animal variability in the measurement of the percentage of hepatocytes in S phase. Consequently, concurrent controls and treated groups of at least five animals each are recommended. One thousand cells are scored per slide, for each of three slides per animal, for each of five animals per data point. Data are expressed as the percentage of cells in S phase ± SE (animal to animal). Expected control values range from 0–0.5% of cells in S phase. Values between 0.5 and 5% would be considered a modest response. Values over 5% would be considered a strong response. For the positive control 400 mg/kg carbon tetrachloride administered in corn oil 48 hr before sacrifice, one might expect values in the range of 4–8% of the hepatocytes in S phase. For the positive control 200 mg/kg 4-acetylaminofluorene administered in corn oil 24 hr before sacrifice, one might expect values in the range of 10–30% of the hepatocytes in S phase.

For true concurrent controls the unpaired t test may be used to compare the equality of the treated and control means. A one-way analysis of variance can be performed on S phase data with multiple treatment groups. S phase data are adjusted by a square root $(x + 0.5)$ transformation to increase the homogeneity of variances among treatment groups as recommended for percentage data containing zeros (Steel and Torrie, 1980). Treatment means can then be compared to control means by Dunnett's multiple range test.

J. Solutions and Media

Presented here are solutions that work using the procedures given. Other suppliers of these media, reagents, or equipment are acceptable. Any specific reference to a particular supplier or manufacturer is meant only as an example and not as an endorsement.

WEI (Williams medium E–incomplete): 500 ml Williams medium E (Flow Labs 12-502-54), 5 ml sterile 200 mM L-glutamine (Flow Labs 16-801-49), and 0.5 ml of 50 mg/ml gentamycin sulfate (Sigma G-7507). Approximately 250 ml is needed per animal.

WEC (Williams medium E–complete): 180 ml WEI and 20 ml heat-inactivated fetal bovine serum (Reheis or Gibco 200-6140) (heat inactivation 30 min at 56°C, freeze aliquots). Approximately 25 ml is needed per animal.

*EGTA perfusion solution (0.5 m*M*):* 100 ml Hanks balanced salt solution without Ca^{2+} or Mg^{2+} (Flow 18-104-54), and 19 mg EGTA [ethylene glycol bis(β-aminoethyl ether) N,N'-tetraacetic acid (Sigma E-4378)]. Dissolve EGTA in 0.1 ml 2 N NaOH, 0.5 ml 2 M HEPES (Calbiochem 391338), and 0.1 ml of 50 mg/ml gentamicin sulfate (Sigma G-7507). Filter sterilize. Approximately 100 ml is needed per animal.

Collagenase perfusion solution (100 units/ml): 500 ml WEI, 2.5 ml 2 M HEPES, 0.1 ml 2 N NaOH, and 50,000 units Type 1 collagenase (Worthington or Sigma C-0130). Leave in 37°C bath until dissolved (20–30 min). Filter sterilize. Approximately 350 ml is needed for each animal. This solution should be made up no more than 24 hr prior to use.

[³H]Thymidine solution (10 μCi/ml): 100 ml WEI, 1000 μCi (1 ml) [³H]thymidine (methyl [³H]thymidine, Amersham TRK.418, 40–60 Ci/m-mole), and 0.5 ml sterile 2 M HEPES. Make up just prior to use. [³H]Thymidine should be stored refrigerated and be no older than 2 months. Approximately 13 ml needed per animal.

Unlabeled thymidine solution (0.25 mM): 100 ml WEI and 6.1 mg thymidine (Sigma T-9250). Filter sterilize. Approximately 20 ml is needed per animal.

Methyl green pyronin Y solution: Add 7.45 g Na_2HPO_4 to 66 ml methanol. Add distilled water to bring volume to 263 ml. Stir until dissolved. Add 5 g citric acid to 59 ml methanol. Add distilled water to bring volume to 240 ml. When both solutions are dissolved, mix and add 12.5 μl phenol, 125 mg resorcinol, and 5 g methyl green pyronin Y (Roboz Surgical Co.). Allow to sit for 2 weeks before use. Keep in a dark bottle and filter before each use. Discard after approximately 6 months or when an obvious decrease in staining intensity is observed.

REFERENCES

Ashby, J. (1983). The unique role of rodents in the detection of possible human carcinogens and mutagens. International Commission for Protection Against Environmental Mutagens and Carcinogens (ICPEMC). ICPEMC Working Paper 1/1. *Mutat. Res.* **115,** 177–213.

Ashby, J., Lefevre, P. A., Burlinson, B., and Penman, M. G. (1985). An assessment of the *in vivo* rat hepatocyte DNA repair assay. *Mutat. Res.* **156,** 1–18.

Beije, B., and Ashby, J. (1985). Use of an *in vivo/in vitro* rat liver DNA repair assay to predict the relative rodent hepatocarcinogenic potency of 3 new azo mutagens. *Carcinogenesis* **6,** 611–615.

Bermudez, E., Tillery, D., and Butterworth, B. E. (1979). The effect of 2,4-diaminotoluene and isomers of dinitrotoluene on unscheduled DNA synthesis in primary rat hepatocytes. *Environ. Mutagen.* **1,** 391–398.

Bridges, B. A., Butterworth, B. E., and Weinstein, I. B., eds. (1982). "Indicators of Genotoxic Exposure," Banbury Rep. Vol. 13. Cold Spring Harbor Lab., Cold Spring Harbor, New York.

Butterworth, B. E. (1979). "Strategies for Short-Term Testing for Mutagens/Carcinogens." CRC Press, West Palm Beach, Florida.

Butterworth, B. E. (1983). Measurement of chemically induced DNA repair in rodent and human cells. *In* "New Approaches in Toxicity Testing and their Application to Human Risk Assessment" (A. P. Li, ed.), pp. 33–40. Raven Press, New York, NY.

Butterworth, B. E. (1987). The genetic toxicology of di(2-ethylhexyl)phthalate. *In* "Nongenotoxic Mechanisms in Carcinogenesis" (B. E. Butterworth and T. Slaga, eds.), Banbury Rep. Vol. 25. Cold Spring Harbor Lab., Cold Spring Harbor, New York. In press.

Butterworth, B. E., and Smith-Oliver, T. (1987). A comparison of the response of carcinogen/noncarcinogen pairs in the rat and mouse hepatocyte DNA repair assays in vivo and in vitro. *In* "Evaluation of Short-Term Tests for Carcinogens, Report of the International Programme on Chemical Safety's Collaborative Study on *In Vivo* Assays" (J. Ashby, F. J. de Serres, M. Draper, M. Ishidate, Jr., B. H. Margolin, and M. D. Shelby, eds.), Progress in Mutation Research, Vol. 6. Elsevier, Amsterdam. In press.

Butterworth, B. E., Doolittle, D. J., Working, P. K., Strom, S. C., Jirtle, R. L., and Michalopoulos, G. (1982). Chemically-induced DNA repair in rodent and human cells. *In* "Indicators of Genotoxic Exposure" (B. A. Bridges, B. E. Butterworth, and I. B. Weinstein, eds.), Banbury Rep. Vol. 13, pp. 101–114. Cold Spring Harbor Lab., Cold Spring Harbor, New York.

Butterworth, B. E., Earle, L. L., Strom, S., Jirtle, R., and Michalopoulos, G. (1983). Induction of DNA repair in human and rat hepatocytes by 1,6-dinitropyrene. *Mutat. Res.* **122**, 73–80.

Butterworth, B. E., Bermudez, E., Smith-Oliver, T., Earle, L., Cattley, R., Martin, J., Popp, J. A., Strom, S., Jirtle, R., and Michalopoulos, G. (1984). Lack of genotoxic activity of di(2-ethylhexyl)phthalate (DEHP) in rat and human hepatocytes. *Carcinogenesis* **5**, 1329–1335.

Butterworth, B. E., Loury, D., and Smith-Oliver, T. (1985). The value of measurement of both genotoxicity and forced cell proliferation in assessing the potential carcinogenicity of chemicals. *Proc. Int. Conf. Environ. Mutagens, 4th* p. 214.

Cairns, J. (1975). The cancer problem. *Sci. Am.* **233**, 64–78.

Casciano, D. A., and Gaylor, D. W. (1983). Statistical criteria for evaluating chemicals as positive or negative in the hepatocyte/DNA repair assay. *Mutat. Res.* **122**, 81–86.

Conaway, C. C., Tong, C., and Williams, G. M. (1984). Evaluation of morpholine, 3-morpholinone, and N-substituted morpholines in the rat hepatocyte primary culture/DNA repair test. *Mutat. Res.* **136**, 153–157.

Couch, D. B., Allen, P. F., and Abernethy, D. J. (1981). The mutagenicity of dinitrotoluenes in *Salmonella typhimurium. Mutat. Res.* **90**, 373–383.

de Serres, F. J., and Ashby, J., eds. (1981). "Evaluation of Short-Term Tests for Carcinogenesis." Elsevier/North-Holland, New York.

Doolittle, D. J., Sherrill, J. M., and Butterworth, B. E. (1983). The influence of intestinal bacteria, sex of the animal, and position of the nitrogroup on the hepatic genotoxicity of nitrotoluene isomers *in vivo. Cancer Res.* **93**, 2836–2842.

Doolittle, D. J., Bermudez, E., Working, P. K., and Butterworth, B. E. (1984). Measurement of genotoxic activity in multiple tissues following inhalation exposure to dimethylnitrosamine. *Mutat. Res.* **141**, 123–127.

Geiger, L. E., and Neal, R. A. (1981). Mutagenicity testing of 2,3,7,8-tetrachlorodibenzo-*p*-dioxin in histidine auxotrophs of *Salmonella typhimurium. Toxicol. Appl. Pharmacol.* **59**, 125–129.

Hellemann, A. L., Maslansky, C. J., Bosland, M., and Williams, G. M. (1984). Rat liver DNA damage by the nonhepatocarcinogen 3,2'-dimethyl-4-aminobiphenyl. *Cancer Lett.* **22**, 211–218.

Hollstein, M., McCann, J., Angelasanto, F. A., and Nichols, W. W. (1979). Short-term tests for carcinogens and mutagens. *Mutat. Res.* **65**, 133.

Hsie, A. W., O'Neill, J. P., and McElheny, V. K., eds. (1979). "Mammalian Cell Mutagenesis: The Maturation of Test Systems," Banbury Rep. 2. Cold Spring Harbor Lab., Cold Spring Harbor, New York.

IARC (1973). "Monographs on the Evaluation of the Carcinogenic Risk of Chemicals to Humans," Vol. 3, pp. 91–136. IARC, Lyon.

IARC (1980). "Monographs on the Evaluation of the Carcinogenic Risk of Chemicals to Humans, Suppl. 2, Long-Term and Short-Term Screening Assays for Carcinogens: A Critical Appraisal." IARC, Lyon.

IARC (1982). "Monographs on the Evaluation of the Carcinogenic Risk of Chemicals to Humans, Suppl. 4, Chemicals, Industrial Processes and Industries Associated with Cancer in Humans." IARC, Lyon.

Joachim, F., and Decad, G. M. (1984). Induction of unscheduled DNA synthesis in primary rat hepatocytes by benzidine-congener-derived azo dyes in the *in vitro* and *in vivo/in vitro* assays. *Mutat. Res.* **136,** 147–152.

Kitagawa, T., Hirakawa, T., Ishikawa, T., Nemoto, N., and Takayama, S. (1980). Induction of hepatocellular carcinoma in rat liver by initial treatment with benzo(a)pyrene after partial hepatectomy and promotion by phenobarbital. *Toxicol. Lett.* **6,** 167–171.

Kornbrust, D., and Dietz, D. (1985). Aroclor 1254 pretreatment effects on DNA repair in rat hepatocytes elicited by *in vivo* or *in vitro* exposures to various chemicals. *Environ. Mutagen.* **7,** 857–870.

Kornbrust, D. J., Barfknecht, T. R., and Ingram, P. (1984). Effect of di(2-ethylhexyl)phthalate on DNA repair and lipid peroxidation in rat hepatocytes and on metabolic cooperation in Chinese hamster V-79 cells. *J. Toxicol. Environ. Health* **13,** 99–116.

Lawson, T. A., Mirvish, S. S., Pour, P., and Williams, G. (1984). Persistence of DNA single-strand breaks and other tests as indicators of the liver carcinogenicity of 1-nitroso-5,6-dihydrouracil and the noncarcinogenicity of 1-nitroso-5,6-dihydrothymine. *JNCI, J. Natl. Cancer Inst.* **73,** 515–519.

Lefevre, P. A., and Ashby, J. (1985). Investigations into the reported ability of cimetidine to initiate UDS in rat hepatocyte primary cultures. *Environ. Mutagen.* **7,** 833–837.

Loury, D. J., Smith-Oliver, T., Strom, S., Jirtle, R., Michalopoulos, G., and Butterworth, B. E. (1986). Assessment of unscheduled and replicative DNA synthesis in hepatocytes treated *in vivo* and *in vitro* with unleaded gasoline or 2,2,4-trimethylpentane. *Toxicol. Appl. Pharmacol.* **85,** 11–23.

Loury, D. J., Goldsworthy, T. L., and Butterworth, B. E. (1987). The value of measuring cell replication as a predictive index of tissue-specific tumorigenic potential. *In* "Nongenotoxic Mechanisms in Carcinogenesis" (B. E. Butterworth and T. Slaga, eds.), Banbury Rep. Vol. 25. Cold Spring Harbor Lab., Cold Spring Harbor, New York. In press.

McCann, J., Choi, E., Yamasaki, E., and Ames, B. N. (1975). Detection of carcinogens as mutagens in the *Salmonella*/microsome test: Assay of 300 chemicals. *Proc. Natl. Acad. Sci. U.S.A.* **72,** 5135–5139.

Maslansky, C. J., and Williams, G. M. (1985). Methods for the initiation and use of hepatocyte primary cultures from various rodent species to detect metabolic activation of carcinogens. In "*In Vitro* Models for Cancer Research" (M. Webber and L. Sekely, eds.), pp. 43–60. CRC Press, Boca Raton, Florida.

Mirsalis, J., and Butterworth, B. (1980). Detection of unscheduled DNA synthesis in hepatocytes isolated from rats treated with genotoxic agents: An *in vivo-in vitro* assay for potential mutagens and carcinogens. *Carcinogenesis* **1,** 621–625.

Mirsalis, J. C., and Butterworth, B. E. (1982). Induction of unscheduled DNA synthesis in rat hepatocytes following *in vivo* treatment with dinitrotoluene. *Carcinogenesis* **3,** 241–245.

Mirsalis, J. C., Hamm, T. E., Sherrill, J. M., and Butterworth, B. E. (1982a). Role of gut flora in the genotoxicity of dinitrotoluene. *Nature (London)* **295,** 322–323.

Mirsalis, J. C., Tyson, K. C., and Butterworth, B. E. (1982b). The detection of genotoxic carcinogens in the *in vivo-in vitro* hepatocyte DNA repair assay. *Environ. Mutagen.* **4,** 553–562.

Mirsalis, J. C., Tyson, C. K., Loh, E. N., Steinmetz, K. L., Bakke, J. P., Hamilton, C. M., Spak, D. K., and Spalding, J. W. (1985). Induction of hepatic cell proliferation and unscheduled DNA synthesis in mouse hepatocytes following *in vivo* treatment. *Carcinogenesis* 6, 1521–1524.

Mitchell, A. D., and Mirsalis, J. C. (1984). Unscheduled DNA synthesis as an indicator of genotoxic exposure. *In* "Single-Cell Mutation Monitoring Systems" (A. A. Ansari and F. J. de Serres, eds.), pp. 165–216. Plenum, New York.

Mitchell, A. D., Casciano, D. A., Meltz, M. L., Robinson, D. E., San, R. H. C., Williams, G. M., and von Halle, E. S. (1983). Unscheduled DNA synthesis test: A report of the "Gene-Tox" Program. *Mutat. Res.* 123, 363–410.

Mori, H., Kawai, K., Ohbayashi, F., Kuniyasu, T., Yamazaki, M., Hamasaki, T., and Williams, G. M. (1984a). Genotoxicity of a variety of mycotoxins in the hepatocyte primary culture/DNA repair test using rat and mouse hepatocytes. *Cancer Res.* 44, 2918–2923.

Mori, H., Ohbayashi, F., Hirono, I., Shimada, T., and Williams, G. M. (1984b). Absence of genotoxicity of the carcinogenic sulfated polysaccharides carrageenan and dextran sulfate in mammalian DNA repair and bacterial mutagenicity assays. *Nutr. Cancer* 6, 92–97.

Norkus, E. P., Kuenzis, W., and Conney, A. H. (1983). Studies on the mutagenic activity of ascorbic acid *in vitro* and *in vivo*. *Mutat. Res.* 117, 183–191.

Probst, G. S., McMahon, R. E., Hill, L. E., Thompson, C. Z., Epp, J. K., and Neal, S. B. (1981). Chemically-induced unscheduled DNA synthesis in primary rat hepatocyte cultures: A comparison with bacterial mutagenicity using 218 compounds. *Environ. Mutagen.* 3, 33–43.

Rao, M. S., Lalwani, N. D., and Reddy, J. K. (1984). Sequential histologic study of rat liver during peroxisome proliferator [4-chloro-6-(2,3-xylidino)-2-pyrimnidinylthio]-acetic acid (Wy-14,643)-induced carcinogenesis. *JNCI, J. Natl. Cancer Inst.* 73, 983.

Rasmussen, R. E., and Painter, R. B. (1966). Radiation-stimulated DNA synthesis in cultured mammalian cells. *J. Cell Biol.* 9, 11–19.

Rickert, D. E., Butterworth, B. E., and Popp, J. A. (1984). Dinitrotoluene: Acute toxicity, oncogenicity, genotoxicity, and metabolism. *CRC Crit. Rev. Toxicol.* 13, 217–234.

Roberts, J. J. (1978). The repair of DNA modified by cytotoxic, mutagenic and carcinogenic chemicals. *Adv. Radiat. Biol.* 7, 211–436.

Schulte-Hermann, R., Schuppler, J., Timmermann-Trosiener, I., Ohde, G., Bursch, W., and Berger, H. (1983). The role of growth of normal and preneoplastic cell populations for tumor promotion in rat liver. *Environ. Health Perspect.* 50, 185–194.

Smith-Oliver, T., and Butterworth, B. E. (1987). Correlation of the carcinogenic potential of di(2-ethylhexyl)phthalate (DEHP) with induced hyperplasia rather than with genotoxic activity. *Mutat. Res.* (in press).

Steel, R. G. D., and Torrie, J. H. (1980). "Principles and Procedures of Statistics," p. 235. McGraw-Hill, New York.

Tong, C., Laspia, M. F., Telang, S., and Williams, G. M. (1981). The use of adult rat liver cultures in the detection of the genotoxicity of various polycyclic aromatic hydrocarbons. *Environ. Mutagen.* 3, 477–487.

Waters, R., Ashby, J., Burlinson, B., Lefevre, P., Barrett, R., and Martin, C. (1984). Unscheduled DNA synthesis, *In* "UKEMS (The United Kingdom Environmental Mutagen Society), Report of the UKEMS Sub-committee on Guidelines for Mutagenicity Testing," pp. 63–87. UKEMS, Swansea, England.

West, W. R., Smith, P. A., Stokes, P. W., Booth, G. M., Smith-Oliver, T., Butterworth, B. E., and Lee, M. L. (1984). Analysis and genotoxicity of a PAC-polluted river sediment. *Proc. Int. Symp. Polynucl. Aromat. Hydrocarbons, 8th* (M. Cook and A. J. Dennis, eds.), pp. 1395–1411. Battelle, Columbus, Ohio.

Williams, G. M. (1976). Carcinogen induced DNA repair in primary rat liver cell cultures; a possible screen for chemical carcinogens. *Cancer Lett.* 1, 231–236.

Williams, G. M. (1977). Detection of chemical carcinogens by unscheduled DNA synthesis in rat liver primary cell cultures. *Cancer Res.* **37,** 1845–1851.

Williams, G. M. (1978). Further improvements in the hepatocyte primary culture DNA repair test for carcinogens: Detection of carcinogenic biphenyl derivatives. *Cancer Lett.* **4,** 69–75.

Williams, G. M. (1984). DNA damage and repair tests for the detection of genotoxic agents. Food Additives and Contaminants **1,** 173–178.

Williams, G. M., and Laspia, M. F. (1979). The detection of various nitrosamines in the hepatocyte primary culture/DNA repair test. *Cancer Lett.* **66,** 199–206.

Williams, G. M., Bermudez, E., and Scaramuzzino, D. (1977). Rat hepatocyte primary cell cultures—III. Improved association and attachment techniques and the enhancement of survival by culture medium. *In Vitro* **13,** 809–817.

Working, P. K., Doolittle, D. J., Smith-Oliver, T., White, R. D., and Butterworth, B. E. (1986a). Unscheduled DNA synthesis in rat tracheal epithelial cells, hepatocytes, and spermatocytes following exposure to methyl chloride *in vitro* and *in vivo. Mutat. Res.* **162,** 219–224.

Working, P. K., Smith-Oliver, T., White, R., and Butterworth, B. E. (1986b). Induction of DNA repair in rat spermatocytes and hepatocytes by 1,2-dibromoethane: The role of glutathione conjugation. *Carcinogenesis* **7,** 467–472.

11

Genotoxicity Studies with Human Hepatocytes

STEPHEN C. STROM, DAVID K. MONTEITH,
K. MANOHARAN, AND ALAN NOVOTNY

Department of Radiology and Pharmacology
Duke University Medical Center
Durham, North Carolina 27710

I. INTRODUCTION

Two goals are basic to carcinogenesis research. One is to understand at the molecular level the process by which "normal" cells escape their growth control mechanisms and become "transformed" cells. Another is to make risk assessment estimates from *in vivo* or *in vitro* models in carcinogenic research to predict the genotoxic potential of specific chemicals for humans. Information pertinent to both of these goals may be obtained from investigations on the ability of chemical or physical agents to covalently bind to and damage DNA, induce DNA repair responses, mutate specific genetic loci, induce phenotypically altered cells, or induce chromosome- or chromatid-level aberrations. For chemical carcinogens, the cellular metabolism of procarcinogens to their ultimately carcinogenic forms is frequently a necessary prerequisite for the expression of the genotoxic potential of these chemicals. Differences in the metabolism of xenobiotics (including chemical carcinogens), attributable to species and sex differences as well as to different tissues within a single animal, may account for much of the difference in susceptibility to the carcinogenic effect (Langenbach *et al.*, 1981). In addition to these factors differences may also be encountered in the level and accuracy of DNA repair pathways, susceptibility and exposure to tumor-promoting compounds, and cellular turnover rates within a given experimental animal.

265

In view of these considerations there are no assurances that investigations of carcinogenesis done with a particular animal will yield data which are directly relevant to human carcinogenesis. It is more likely that there is *no* single animal species which is an adequate model for humans in *all* investigations of carcinogenesis.

An approach which has been taken to investigate the genotoxic potency of chemical carcinogens in systems directly relevant to the human situation is through the use of human tissues or cells in culture (Stoner *et al.*, 1982; Harris *et al.*, 1976, 1977; Autrup *et al.*, 1981; Autrup and Harris, 1983).

One organ of particular importance to chemical carcinogenesis research is the liver. Although the liver is not one of the more common sites of tumor development in inhabitants of the United States, it is a common site of tumor induction among people in many parts of the world. The liver is also most frequently involved in tumor formation in animals in the 2-year carcinogenesis bioassays. The reason for the frequent induction of liver tumors in subjects exposed to chemical carcinogens is thought to be related to the proportionally high levels of enzymes in the liver, compared to most other tissues, which are necessary for the metabolism of carcinogens. Within the liver, the parenchymal hepatocyte is the cell with the highest levels of xenobiotic metabolizing enzymes. It is also the cell type most frequently transformed by exposure to chemical carcinogens.

Rodent parenchymal hepatocytes have been used extensively in carcinogenesis research. Intact cells are preferable to cell-free activation systems as a source of carcinogen-metabolizing enzymes because, although cell-free systems such as postmitochondrial supernatant fractions (S-9) or microsomal fractions, when supplied with NADPH-generating systems, can provide monooxygenase activity to metabolize chemicals to their genotoxic forms, the critical balance between activation and detoxification of chemical carcinogens is disrupted in cell-free systems. It has been shown that the profile of metabolites generated (Selkirk, 1977) and the nature of the DNA adducts (Bigger *et al.*, 1978) that are formed following the metabolism of carcinogens differ between intact cellular systems and cell-free homogenates. Thus, most investigators favor the use of intact cells for carcinogenic investigations. The higher levels of carcinogen-metabolizing enzymes, and the capacity to metabolize a wide range of chemical classes of compounds, make the liver a logical choice as an activation system for short-term genotoxicity test systems. Although no single tissue, including the liver, would be expected to metabolize every possible chemical carcinogen to its genotoxic metabolites, the wide range of chemical classes that are metabolized by the liver make the hepatocyte a more universally useful activation system for carcinogenic investigations.

This chapter will focus on the use of the isolated human hepatocyte as a model system for genotoxicity investigations.

II. ISOLATION AND CULTURE OF HUMAN HEPATOCYTES

The near simultaneous publication of successful methods for the isolation of viable single hepatocytes from pieces of human liver by three groups (Reese and Byard, 1981; Strom *et al.*, 1982; Guguen-Guillouzo *et al.*, 1982) established the techniques which made investigations with human hepatocytes possible. Earlier work by Miyazaki *et al.* (1981) reported that viable hepatocytes could be isolated from biopsy-sized pieces of human liver; however, the small cell yield has limited the use of this technique. Bojar *et al.* (1976) reported a perfusion technique for the whole liver. Whole-organ perfusion on a human liver is costly and technically difficult. Currently, the most widely used techniques for the isolation of human hepatocytes are based on the collagenase perfusion technique described by Berry and Friend (1969). Entire lobes (Guguen-Guillouzo *et al.*, 1982) or sublobular pieces of liver tissue (Reese and Byard, 1981; Strom *et al.*, 1982) may be adequately perfused through catheters inserted into the vascular channels in the liver tissue. A calcium-free buffer (Strom *et al.*, 1982) is introduced into the liver tissue at a flow rate of between 12 and 20 ml/min/catheter. The number of catheters used for a particular piece of liver will depend on the number of vessels available for perfusion. Calcium-free buffer is perfused into the liver until the areas which are being perfused warm up to between 35 and 37°C. As with rat liver perfusion techniques, the second stage of human liver perfusion is conducted with calcium-supplemented buffer containing collagenase (0.5 mg/ml; Strom *et al.*, 1982). Buffer with collagenase is perfused for at least 20 min or until the areas of the liver being perfused swell and soften due to the breakdown of the intercellular connections.

Following perfusion, the areas of the liver which have clearly been perfused are cut out and minced in cold buffer and filtered through a nylon mesh (150 μm pore size). Parenchymal hepatocytes are separated from the nonparenchymal elements of the liver by either a 2-min centrifugation at 100 *g* or a 4-min centrifugation at 50 *g*. This centrifugation step differs slightly from the techniques with rat hepatocytes because the human hepatocytes are, in general, not completely sedimented by the usual 2-min centrifugation at 50 *g* used for rat hepatocytes.

We have identified two factors which are critical to a successful perfusion of human liver samples. Both factors involve the temperature of the liver pieces. First, it is of prime importance to chill the pieces of liver in ice-cold buffer as soon as possible after their removal. We gently squeeze the liver tissue in cold buffer to expel blood from the tissue and to prevent infarcts in the vessels which will have to be perfused. This step also introduces cold buffer into the tissue to slow cellular respiration. A suitable buffer for transporting human liver is Euro-

Collins (Collins *et al.*, 1969), which is the buffer used during the transport of liver and kidney tissue for transplantation purposes. We have obtained >90% cellular viability from the perfusion of pieces of liver tissue stored in Euro-Collins buffer for up to 12 hr. A single attempt at the perfusion of a piece of liver stored in Euro-Collins for 24 hr was unsuccessful.

The second critical factor is the temperature of the liver during collagenase perfusion. As stated above, we perfuse the tissue with calcium-free buffer until the temperature in the perfused areas is brought up to 35–37°C. Temperatures in this range must be attained to allow the collagenase to work most effectively. Warming the liver tissue with calcium-free buffer perfusion for periods in excess of 20 min may be needed to warm tissue samples which have been equilibrated in ice-cold Euro-Collins buffer. Temperatures in the liver tissue should be monitored by inserting suitable thermistor probes into areas of the liver being perfused.

III. HEPATOCYTE CELL CULTURE

Culture of human hepatocytes is, in general, the same as that used with rat hepatocytes. In our experience, following an initial short culture time (1–4 hr) required for attachment of the hepatocytes to the substrate in media containing 5 fetal bovine serum (v/v), human hepatocytes are optimally maintained in serum-free media containing 0.1 μM dexamethazone and insulin. As with rat hepatocytes, human hepatocytes in primary culture are strongly influenced by the culture conditions. Rat hepatocytes rapidly dedifferentiate in culture and experience significant losses in cytochrome P-450 (Cyt P-450) [see Guguen-Guillouzo and Guillouzo (1983) for review]. Approximately 20–30% of the initial values of Cyt P-450 remain in rat hepatocytes cultured for 48 hr (Guzelian *et al.*, 1977; Michalopoulos *et al.*, 1979). Since much of the rationale for the use of hepatocytes for genotoxicity investigations is based on the assumption that hepatocytes contain high levels of Cyt P-450, the decline in Cyt P-450 values in cultured hepatocytes is of concern. This potential problem is overcome in several ways.

First, cultures are generally used for experimental purposes on the first day of culture when the Cyt P-450 levels are highest. Culture media may also be supplemented with "hormonal cocktails" which help to prevent the loss of Cyt P-450 over the first 24 hr (Decad *et al.*, 1977; Hockin and Paine, 1983). Another consideration is that even with significant losses of Cyt P-450 content in hepatocyte cultures, the residual levels of Cyt P-450 in hepatocytes are still much higher than the constitutive levels of Cyt P-450 in most other cell types.

Cultured human hepatocytes differ from rat hepatocytes in that the human hepatocytes seem to dedifferentiate much more slowly in culture than rat hepatocytes. Human hepatocytes maintain much higher levels of Cyt *P*-450 in culture, relative to the amounts initially present in the cells immediately after isolation, than do rat hepatocytes. The Cyt *P*-450 levels in human hepatocytes over the first 4 days of culture, as determined by a spectrophotometric technique, were maintained at approximately 60% of their initial values (Strom *et al.*, 1982). In rat hepatocyte cultures, by day 4 the Cyt *P*-450 levels would have declined to approximately 10–15% of their initial values. In a recent investigation, Blaaboer *et al.* (1985) reported that in human hepatocyte cultures prepared from 5 of 7 donors there was no significant loss of Cyt *P*-450 content in cells over the first 24 hr. In the one case of hepatocytes where there was a significant loss of Cyt *P*-450, the decline was completely inhibited by the addition of 0.5 m*M* metyrapone to the culture media. The authors concluded that, since the human hepatocytes in primary culture do not lose significant amounts of Cyt *P*-450 in the majority of cases, the hepatocyte culture system should be useful for investigations of xenobiotic metabolism (Blaaboer *et al.*, 1985).

Wang *et al.* (1983) have purified and characterized six Cyt *P*-450 isozymes from human liver microsomes. *P*-450 isozymes, P-450 1, 2, 4, 5, 6, and 8, have been characterized for substrate specificity and were found to differ in their catalytic activity toward drugs and carcinogens, although considerable overlapping metabolizing capacity was seen among the different Cyt *P*-450 isozymes for *d*-benzphetamine, benzo(*a*)pyrene (BP), and acetanilide as well as for other chemicals. Antibodies to some of the human isozymes were cross reactive with rat *P*-450 isozymes, and many of the isozymes isolated from human liver were found to be immunologically cross reactive among themselves. The human *P*-450 isozymes may fall into two families: one family of immunologically related isozymes contains Cyt *P*-450 2, 4, 5, and 6. Monoclonal antibodies against forms 1 or 8 do not cross react with isozymes 2, 4, or 5, but do react with each other. Using monoclonal antibodies against specific human Cyt *P*-450 isozymes, it is now possible to quantitate the relative amounts of each of the isozymes present in microsomal samples. The results of these types of experiments indicate that Cyt $P\text{-}450_8$ accounts for a considerable proportion of the total Cyt *P*-450 present. Wang *et al.* (1983) reported that the level of Cyt $P\text{-}450_8$ varied 33-fold among 22 liver samples examined. Beaune *et al.* (1985) reported approximately a 4-fold variation in the levels of Cyt $P\text{-}450_5$ in six human liver samples examined. We (Michalopoulos *et al.*, 1985) have reported less than 3-fold variation in the levels of Cyt *P*-450 as determined spectrophotometrically from a total of eight human liver samples. The data indicate that there are significant differences in Cyt *P*-450 content in different human liver samples. Quantitative polymorphisms of specific *P*-450 isozymes may be detected by immunological methods, but this procedure is complicated somewhat by the cross reactivity of the different isozymes. The polymorphisms observed between individuals as determined by immunological methods directed at specific isozymic forms of Cyt P-450 more accurately reflect the polymorphisms observed in P-450 catalytic activity than do simple measurements of total microsomal Cyt P-450.

Specific isozymes of Cyt P-450 have been investigated in cultured human hepatocytes. At 8 days in culture, at a time when rat hepatocytes would have lost approximately 90% of their Cyt P-450 content, human hepatocytes retained approximately 50% of their original levels (Guillouzo et al., 1985). When P-450 isozymes 5 and 8 were quantitated with monoclonal antibodies raised against these two forms, there was approximately a 30–40% decline in P-450$_8$ but no apparent change in the amounts of isozyme 5. Addition of 3.2 mM phenobarbital to the culture media induced the levels of Cyt P-450 approximately 2-fold; however, there were no changes in the amounts of P-450 isozymes 5 or 8 in the induced cells.

When human hepatocytes were cocultured with rat liver epithelial cells, the decline in Cyt P-450 levels observed in the conventional cultures (maintained without rat liver epithelial cells) was completely prevented (Guillouzo et al., 1985). Thus, as was observed with rat hepatocytes in culture, survival and long-term maintenance of differentiated functions were the best in mixed cultures of hepatocytes and liver epithelial cells (Guillouzo et al., 1985; Gugen-Guillouzo et al., 1984; Begue et al., 1984; Novicki et al., 1982). Although the mechanism by which coculture of the two cell types helps to maintain differentiation in hepatocytes is not known, cell–cell interactions seem to be necessary to cause the secretion and deposition of an insoluble extracellular material containing fibronectin and type III collagen between the two cell types and around the hepatocytes. Published reports (Guillouzo et al., 1985; Gugen-Guillouzo et al., 1984; Begue et al., 1984; Novicki et al., 1982) and our own observations indicate that human hepatocytes may be maintained in conventional culture conditions for 2–3 weeks and in mixed cultures with rat liver epithelial cells for at least 3 months.

Coculture of human hepatocytes with liver epithelial cells stimulates the hepatocytes to secrete extracellular matrix to create and condition their environment in a manner which is compatible with long-term maintenance of differentiated functions. Even in conventional culture in the absence of liver epithelial cells, human hepatocytes retain most of their microsomal metabolic capacity for more than a week. Since the utility of the hepatocytes as a model system for genotoxicity research depends largely on the metabolic capacity maintained by the cells in culture, the cultured human hepatocyte is extremely useful.

IV. CRYOPRESERVATION OF HUMAN HEPATOCYTES

Because the perfusion of normal human liver samples is an infrequent procedure and it is difficult to utilize all of the viable hepatocytes obtained from a single perfusion (10^9 cells) for experimental purposes immediately after isolation, we have sought to develop effective techniques for the storage of viable hepatocytes. Cryopreservation of human hepatocytes would be helpful if hepatocytes would retain, upon thawing, levels of viability and metabolic capacity present at the time of freezing. Successfully cryopreserved hepatocytes would be useful for distribution to other laboratories with limited access to human material. A "bank" consisting of a large number of cryopreserved hepatocyte samples taken from a number of individuals would be extremely useful for many short-term genotoxicity assays. Finally, frozen hepatocytes may be ultimately useful for cellular transplantation procedures in cases of liver injury, especially if hepatocytes from a particular patient could be isolated, expanded *in vitro*, and used to "seed" a damaged liver.

To date there are no detailed investigations of techniques for the cryopreservation of human hepatocytes. We (Novicki *et al.*, 1982, and Jackson *et al.*, 1985) have recently investigated different freezing protocols with rat hepatocytes to optimize the techniques which may be useful for cryopreservation of human hepatocytes.

Rat hepatocytes may be frozen in media containing 10% fetal bovine serum and 10% DMSO at a freezing rate of 1–2°C per minute. As a cryoprotectant, DMSO is better than glycerol, and 10% DMSO (v/v) is better than 20% DMSO. After a rapid thaw to 37°C, the viability of the hepatocytes as determined by trypan blue exclusion is approximately 80% of the viability of the cells initially frozen. If attachment of the cells to a culture substrate is used as the criteria for viability, approximately 50% of the frozen and thawed hepatocytes appear to be viable. When the plating efficiency of the freshly isolated hepatocytes is taken into account, the overall plating efficiency of hepatocytes after freezing is approximately 25–40%.

Frozen–thawed rat hepatocytes contain Cyt *P*-450 levels which are slightly higher than the levels in the cells initially frozen, possibly through a selection of cells with high Cyt *P*-450 levels. Frozen–thawed hepatocytes lose Cyt *P*-450 in culture at a rate which is approximately twice that observed in cells which were never frozen (Novicki *et al.*, 1982; Jackson *et al.*, 1985).

Although techniques have been presented for freezing and storing hepatocytes, the optimal freezing, thawing protocol or subsequent culture conditions following cryopreservation have yet to be determined even for rat hepatocytes. Further experimentation, based on protocols developed for rat hepatocytes, will have to be done to determine if human hepatocytes can be frozen. We have attempted to freeze/thaw human hepatocytes and find that, in brief, the success of the cryopreservation depends markedly on the initial viability of the hepatocytes which are frozen. With initial viabilities of >80%, the frozen–thawed human hepatocytes had a culture plating efficiency approximately equal to that of rat hepatocytes. Figure 1 shows cells obtained from a recent human sample. Figure 1A shows the morphology of the hepatocytes after 48 hr of culture, and Fig. 1B

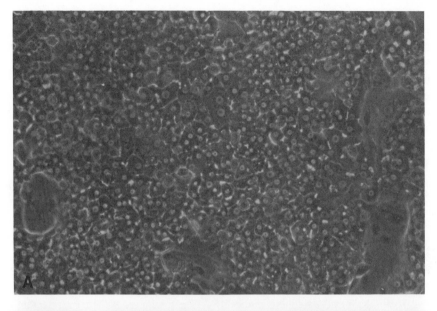

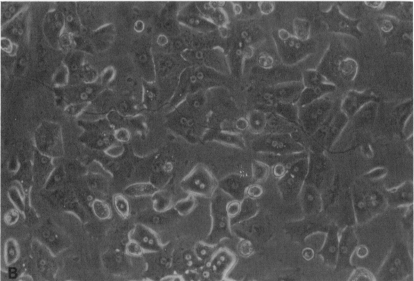

Fig. 1.   Phase contrast micrographs (×100) of human hepatocytes in primary culture. Both photographs are from the same human hepatocyte case. (A) Hepatocytes cultured for 48 hr in serum-free media. (B) Previously frozen hepatocytes cultured for 24 hr.

shows the morphology of hepatocytes frozen, thawed, and cultured for 24 hr. Hepatocytes were frozen at a rate of 1°C/min in media supplemented with 10% DMSO and 10% fetal bovine serum. The overall plating efficiency of the hepatocytes after freezing was approximately 22%. While we do not believe that an optimal freezing protocol has been developed for human hepatocytes, these early experiments are encouraging and will serve as a basis for comparison for new freezing protocols as they are being developed.

V. GENOTOXICITY STUDIES WITH HUMAN HEPATOCYTES

Human hepatocytes in primary culture are particularly well suited for investigations of xenobiotic metabolism and quantitation of DNA repair elicited by exposure of cells to genotoxic compounds. However, since human hepatocytes, like rat hepatocytes, do not replicate under normal culture conditions, direct mutagenesis assays are not possible. Mutagenic activation of chemicals by human hepatocytes may be quantitated if clonable target cells are cocultured with the hepatocytes. Other endpoints, such as chromosome level genotoxicity, may also be investigated in the target cell. We will briefly review the genotoxicity investigations which have been conducted with human hepatocytes in primary culture.

A. Investigations of DNA Repair

Because of the relatively small numbers of cells required for DNA repair studies and the relative ease with which large numbers of chemicals may be examined for genotoxicity with this assay, unscheduled DNA synthesis (UDS) has been investigated in human hepatocytes exposed to a rather long list of compounds (Table I). UDS in human hepatocytes is frequently compared to the response measured in rat hepatocytes. As Table I indicates, the UDS response measured in rat hepatocytes frequently is in qualitative agreement with the response measured in human hepatocytes. This data would suggest that, for the majority of compounds examined, the rat may be a good model for the human in investigations of carcinogenesis.

Steinmetz et al. (1985) reported that qualitative and quantitative differences were observed in the induction of UDS in hepatocytes isolated from human, rat, mouse, and hamster. J. C. Mirsalis (personal communication) has also investigated UDS in the squirrel monkey. Whereas the rat and human hepatocytes frequently respond to chemical exposures with a similar UDS response (Strom et al., 1983a,b), there was little agreement between the UDS response measured in

TABLE I

Qualitative Comparison of the UDS Responses Observed in Culture in Rat or Human
Hepatocytes Exposed the Indicated Compounds

Compounds in which rat and human hepatocytes show quantitatively the same UDS response, *in vitro*

Chemical	Response	Reference
Aminofluorene	+	Hsu *et al.* (1985)
2-Acetylaminofluorene	+	Strom *et al.* (1982; 1983)
Aflatoxin B_1	+	Butterworth *et al.* (1984)
4-Aminobiphenyl	+	Steinmetz *et al.* (1985)
Benzidine	+	Steinmetz *et al.* (1985)
Benzo (*a*) pyrene	+	Strom *et al.* (1982)
Diethylnitrosamine	+	Strom *et al.* (1982)
Dimethylnitrosamine	+	Strom *et al.* (1982)
Di-(2-ethylhexyl)phthalate	−	Butterworth *et al.* (1984)
Aniline	−	Butterworth *et al.* (1982)
Nitrobenzene	−	Butterworth *et al.* (1982)
2,6-Dinitrotoluene	−	Butterworth *et al.* (1982)
2-Nitrotoluene	−	Butterworth *et al.* (1982)
1,6-Dinitropyrene	+	Butterworth *et al.* (1983)
1,3-Dinotropyrene	+	unpub. observation
1,8-Dinitropyrene	+	unpub. observation
NNK (4-methylnitrosamino)- 1-(3 pyridyl)-1-butanone	+	Strom *et al.*, (1986)
TCDD	−	unpub. observation

Compounds where the UDS response was qualitatively different between rat and human hepatocytes *in vitro*

Chemical	Rat	Human
Unleaded gasoline	+	—[a]
2-Naphthylamine	+	—

[a]Data from one human case (Loury *et al.*, 1986). +, positive; −, negative.

human hepatocytes and the response measured in monkey, mouse, or hamster hepatocytes.

Some significant observations have already been made with suspected or known human carcinogens which have been examined in the UDS assay. The potent hepatocarcinogen aflatoxin B_1, a suspected human hepatocarcinogen, is positive for the induction of UDS in human hepatocytes at or below 1 μM (Table I; Butterworth *et al.*, 1982). Benzidine and 4-aminobiphenyl, both human bladder carcinogens, are potent inducers of UDS in human hepatocytes (Steinmetz *et al.*, 1985), suggesting that the liver may contribute to the genotoxicity observed *in vivo* in man. Another human bladder carcinogen, 2-naphthylamine, was found to be negative for the induction of UDS in human hepatocytes but slightly positive in rat hepatocytes. Both 1- and 2-naphthylamine were found to be substrates for purified human Cyt P-450s (Wang *et al.*, 1983), so the naphthylamines may simply fail to induce a significant repair response in the liver. The negative UDS response may not necessarily be indicative of failure of the liver to metabolize the naphthylamines to their genotoxic species. Further investigations of hepatocyte-mediated mutagenesis or covalent binding of naphthylamine metabolites to hepatocyte DNA may help to resolve the role of the liver in naphthylamine carcinogenesis *in vivo*. The tobacco-specific nitrosamine NNK [4-methylnitrosamino)-1-(3-pyridyl)-1-butanone] is positive for induction of UDS in human hepatocytes and positive in human hepatocyte-mediated mutagenesis of human fibroblasts (Strom *et al.*, 1986). The demonstrated genotoxicity of NNK towards human cells support the hypothesis that these compounds may be human carcinogens. Environmentally important dinitropyrene compounds, combustion products typically found in diesel exhaust, were found to be some of the most potent inducers of UDS in human hepatocytes (Butterworth *et al.*, 1983; Strom and Kornbrust, 1987). Dinitropyrenes rival aflatoxin for potency in human hepatocyte UDS assay, with both being active in the micromolar range.

B. Hepatocyte-Mediated Genotoxicity Assays

Since no one has yet successfully cloned human or rat hepatocytes in culture, direct mutagenesis assays with hepatocytes are not possible. Hepatocytes may be added to cultures with clonable target cells to quantitate the efflux of mutagenic metabolites from hepatocytes exposed to carcinogens. Human hepatocytes have been cocultured with human fibroblasts (Strom *et al.*, 1983b), or V-79 Chinese hamster fibroblasts (Moore and Gould, 1984) in hepatocyte-mediated mutagenesis assays. Strom *et al.* (1983b) found a great deal of similarity between human and Sprague–Dawley rat hepatocytes in the activation of diethylnitrosamine to mutagenic metabolites for human fibroblasts in coculture with the hepatocytes. In unpublished experiments, we have found dimethylnitrosamine (DMN), NNK, 2-acetylaminofluorene (AAF), aflatoxin B_1, and benzo(*a*)pyrene to be positive for mutagenicity in human hepatocyte-mediated assays. Moore and Gould (1984) have reported the mutagenic activation of BP by human hepatocytes in coculture with V-79 cells. These authors use the cell-mediated assays to examine the systemic (liver) vs the target cell activation of carcinogens by specific target cells, such as mammary epithelial cells which have the capacity to activate carcinogens.

Green *et al.* (1985) have combined human hepatocytes and CHO or V-79 cells in culture to investigate the hepatocyte-mediated induction of sister chromatid exchange (SCE) in the respective target cells (Green *et al.*, 1985). Both V-79 and CHO responded with a significant induction of SCE when DMN was added to the cultures in the presence of hepatocytes, but not in the absence of hepatocytes.

The mutagenicity of 5-arylamines, benzidine (BZ), *N*-acetylbenzidine (MABZ), *N,N'*-diacetylbenzidine (DABZ), 4-aminobiphenyl (4-AB), and 2-aminoanthracene (2-AA), was investigated by Neis *et al.* (1985) in human or dog hepatocyte-mediated mutagenesis of *Salmonella typhimurium* strain TA1538. The arylamines were generally more potent mutagens when activated by human hepatocytes compared to activation by dog hepatocytes. This study typifies the problems encountered with the extrapolation of genotoxicity data from one species to another. Carcinogenic arylamines are acetylated by most animal species, but wide variations in the capacity to acetylate have been noted between species, and polymorphisms in acetylation rates exist in humans. Since the acetylated arylamines may undergo further metabolism to yield genotoxic species, the acetylation rates of the subjects being studied strongly influence the genotoxic potency of the arylamines (McQueen *et al.*, 1982). Human hepatocytes, in general, are good acetylators while dogs are regarded as poor acetylators. Consequently, the ratio of acetylation to deacetylation helps to determine the tissue and species specificity of carcinogenic arylamines.

VI. XENOBIOTIC METABOLISM BY HUMAN HEPATOCYTES

The liver is the principal organ of xenobiotic metabolism *in vivo*. The considerable differences which exist in the metabolism of xenobiotics among different species result mainly from differences in hepatic metabolism of the chemical. Because of ethical considerations, human xenobiotic metabolism of many chemicals, especially potentially genotoxic compounds, could only be assessed with *in vitro* techniques. Comparative analysis of the metabolism of toxic and/or genotoxic compounds by specific tissues, such as liver, from laboratory animals and humans would provide valuable information for toxicologists, pharmaceutical and chemical manufacturers, and regulatory authorities.

Begue *et al.* (1983) investigated the metabolism of ketotifen, an orally active antianaphylactic agent, in hepatocytes isolated from two male kidney donors. Clinical pharmacological studies had already identified several metabolic pathways *in vivo* for ketotifen, including *N*-demethylation, *N*-oxidation, *N*-glucuronidation, and ketoreduction followed by *O*-glucuronidation. All of the metabolic pathways known *in vivo* were also found after *in vitro* metabolism of ketotifen by human hepatocytes in primary culture. Long-term (>7 days) metabolism of ketotifen was maintained in cocultures of human hepatocytes and rat liver epithelial cells until the experiment was terminated on day 21. Another compound of pharmacological interest, amphetamine, was investigated by Green *et al.* (1986) in hepatocyte cultures isolated from rat, rabbit, dog, squirrel monkey, and human liver. In general, the metabolic profile produced after *in vitro* metabolism was similar to the urinary metabolites of amphetamine previously identified from the respective species. Since the metabolic disposition of amphetamine was unique to each species, the recapitulation of the specific profile of metabolites by hepatocytes isolated from each of the species demonstrated the applicability of the cultured hepatocyte for interspecies comparisons of xenobiotic metabolism.

The metabolism of genotoxic compounds has also been investigated. Hsu *et al.* (1985) reported the production of phenolic and conjugated metabolites of BP by human hepatocytes isolated from immediate autopsy human subjects. Moore and Gould (1985) conducted a more extensive investigation of BP metabolism by hepatocytes from six human donors. All cases metabolized significant amounts of BP (24–35 nmol/10^7 cells plated/24 hr). These authors found considerable variability between donors in the relative proportion of each BP metabolite produced, but the predominant metabolites were the 9,10- and 7,8-dihydrodiol, 9-hydroxy-BP, and various tetrols. Less variability was noted among the different donors in the ratios of the BP metabolites conjugated to glucuronic acid. BP metabolism in a single case of previously cryopreserved hepatocytes was also investigated and was found to be significantly different from the metabolism of BP by the freshly isolated hepatocytes from the same donor. More 7,8- and 4,5-dihydrodiol and relatively less tetrol metabolites of BP were produced by previously frozen cells. Booth *et al.* (1981) examined levels of binding of AFB_1 to DNA of slices of fresh human liver. Considerable variation was noted among the six cases examined, but binding levels ranged for 0.7–8.5 ng AFB_1/mg DNA, which was somewhere between the values noted by the authors for binding to hamster and mouse liver DNA.

VII. CONCLUSIONS

We have outlined methods useful for the isolation and culture of hepatocytes from pieces of human liver. In addition, we have tried to briefly review the available literature concerning the use of human hepatocytes in pharmacological or toxicological investigations. Human hepatocytes may be isolated in large numbers and high viability with relative ease by currently available techniques. The isolated human hepatocytes have been shown to be very useful for investigations of DNA damage and repair, hepatocyte-mediated mutagenesis, cytogenetic toxicology, and xenobiotic metabolism. Just as one sees large differences among species in their susceptibility to carcinogenesis by chemicals, one may expect to see large differences in the metabolism and genotoxicity of specific chemicals toward cells isolated from different human donors. Qualitative similarities between the metabolism and genotoxicity of chemicals by human hepatocytes and hepatocytes isolated from various laboratory animals suggest that proper metabolic and genotoxic surrogates for humans may be identified through comparative experimentation with tissues or cells isolated from humans and various animal species. The use of human hepatocytes and other human cells in genetic toxicology investigations should enable a more critical evaluation of *in vitro* results from other species with respect to potential human genotoxicity with specific chemicals and should put the extrapolation of genetic toxicology data from animal species to humans on a more scientific basis.

ACKNOWLEDGMENTS

Sincere thanks to Carol Green, Jon Mirsalis, Karen Steinmetz, Charles Tyson, and Byron Butterworth for supplying unpublished data. Special thanks to Alice Mitchell for technical help and S. S. for typing the manuscript. Supported by Grant ES-023471 and contract ES-55091, and in part by CIIT, Chemical Industry Institute of Toxicology, Research Triangle Park, North Carolina.

REFERENCES

Autrup, H., and Harris, C. C., ed., (1983). "Human Carcinogenesis." Academic Press, New York.

Autrup, H., Grafstrom, R. C., Christensen, B., and Kieler, J. (1981). Metabolism of chemical carcinogens by cultured human and rat bladder epithelial cells. *Carcinogenesis* **2**, 763–768.

Beaune, P., Kremers, P., Letawe-Goujon, F., and Gielen, J. E. (1985). Monoclonal antibodies against human liver cytochrome P-450. *Biochem. Pharmacol.* **34**, 3547–3552.

Begue, J. M., Le Bigot, J. F., Guguen-Guillouzo, C., Kiechel, J. R., and Guillouzo, A. (1983). Cultured human hepatocytes: A new model for drug metabolism studies. *Biochem. Pharmacol.* **32**, 1643–1646.

Begue, J. M., Guguen-Guillouzo, C., Pasdeloup, N., and Guillouzo, A. (1984). Prolonged maintenance of active cytochrome P-450 in adult rat hepatocytes co-cultured with another liver cell type. *Hepatology* **4**, 839–842.

Berry, M. N., and Friend, D. S. (1969). High-yield preparation of isolated rat liver parenchymal cells. A biochemical and fine structural study. *J. Cell Biol.* **43**, 506–520.

Bigger, C. A. H., Tomaszewski, J., and Dipple, A. (1978). Differences between the products of binding of 7,12-dimethylbenzanthracene to DNA in mouse skin and rat liver microsomes. *Biochem. Biophys. Res. Commun.* **80**, 229–235.

Blaaboer, B. J., Van Holsteijn, I., Van Graft, M., and Paine, A. J. (1985). The concentration of cytochrome P-450 in human hepatocyte culture. *Biochem. Pharmacol.* **34**, 2405–2408.

Bojar, H., Basler, M., Fuchs, F., Dreyfurst, R., and Staib, W. (1976). Preparation of parenchymal and nonparenchymal cells from adult human liver: Morphological and biochemical characteristics. *J. Clin. Chem. Clin. Biochem.* **14**, 527–532.

Booth, S. C., Bosenberg, H., Garner, R. C., Hertzog, P. J., and Norpoth, K. (1981). The activation of aflatoxin B-1 in liver slices and in bacterial mutagenicity assays using livers from different species including man. *Carcinogenesis* **2**, 1063–1068.

Butterworth, B. E., Doolittle, D. J., Working, P. K., Strom, S. C., Jirtle, R. L., and Michalopoulos, G. (1982). Chemically-induced DNA repair in rodent and human cells. In "Indicators of Genotoxic Exposure," Banbury Rep., Vol. 13. Cold Spring Harbor Lab., Cold Spring Harbor, New York.

Butterworth, B. E., Earle, L., Strom, S. C., Jirtle, R. L., and Michalopoulos, G. (1983). Induction of DNA repair in human and rat hepatocytes by 1,6-dinitropyrene. Mutat. Res. 122, 73–80.

Butterworth, B. E., Bermudez, E., Smith-Oliver, T., Earle, L., Cattley, R., Martin, J., Popp, J. A., Strom, S. C., Jirtle, R. L., and Michalopoulos, G. (1984). Lack of genotoxic activity of di(2-ethylhexyl)phthalate (DEHP) in rat and human hepatocytes. Carcinogenesis 5, 1329–1335.

Collins, G. M., Bravo-Sugarman, M., and Terasaki, P. I. (1969). Kidney preservation for transportation. Lancet ii, 1219–1224.

Decad, G. M., Hsieh, D. P. H., and Byard, J. L. (1977). Maintenance of cytochrome P-450 and metabolism of aflatoxin B_1 in primary hepatocyte cultures. Biochem. Biophys. Res. Commun. 78, 279–287.

Green, C. E., LeValley, S. E., and Tyson, C. A. (1986). Comparison of amphetamine metabolism using isolated hepatocytes from five species including human. J. Pharmacol. Exp. Ther. (in press).

Green, C. E., Tyson, C. A., Mirsalis, J. C., and Blazak, W. F. (1985). Comparative metabolism and toxicity assessments using human hepatocytes. Toxicologist 5, 224.

Guguen-Guillouzo, C., and Guillouzo, A. (1983). Modulation of functional activities in cultured rat hepatocytes. Mol. Cell. Biochem. 53/54, 35–56.

Guguen-Guillouzo, C., Campion, J. P., Brissot, P., Glaise, D., Launois, B., Bourel, M., and Guillouzo, A. (1982). High yield preparation of isolated human adult hepatocytes by enzymatic perfusion of the isolated human adult hepatocytes by enzymatic perfusion of the liver. Cell Biol. Int. Rep. 6, 625–628.

Guguen-Guillouzo, C., Clement, B., Lescoat, G., Glaise, D., and Guillouzo, A. (1984). Modulation of human fetal hepatocyte survival and differentiation by interactions with a rat liver epithelial cell line. Dev. Biol. 105, 211–220.

Guillouzo, A., Beaune, P., Gascoin, M. N., Begue, J. M., Campion, J. P., Guengerich, F. P., and Guguen-Guillouzo, C. (1985). Maintenance of cytochrome P-450 in cultured adult human hepatocytes. Biochem. Pharmacol. 34, 2991–2995.

Guzelian, P. S., Bissell, D. M., and Meyer, U. A. (1977). Drug metabolism in adult rat hepatocytes in primary culture. Gastroenterology 72, 1232–1239.

Harris, C. C., Autrup, H., Conner, R., Barrett, L. A., McDowell, E. M., and Trump, B. F. (1976). Interindividual variation in binding of benzo(a)pyrene to DNA in cultured human bronchi. Science 194, 1067–1069.

Harris, C. C., Autrup, H., Stoner, G. D., McDowell, E. M., Trump, B. F., and Schafer, P. (1977). Metabolism of dimethylnitrosamine and 1,2-dimethylhydrazine in cultured human bronchi. Cancer Res. 37, 2309–2311.

Hockin, J. L., and Paine, A. J. (1983). The role of 5-amino-laevulinate synthetase, haem oxygenase, and ligand formation in the mechanism of maintenance of cytochrome P-450 concentration in hepatocyte culture. Biochem. J. 210, 855–857.

Hsu, I. C., Lipsky, M. M., Cole, K. E., Su, C. H., and Trump, B. F. (1985). Isolation and culture of hepatocytes from human liver of immediate autopsy. In Vitro Cell. Dev. Biol. 21, 154–159.

Jackson, B. A., Davies, J. E., and Chipman, J. K. (1985). Cytochrome P-450 activity in hepatocytes following cryopreservation and monolayer culture. Biochem. Pharmacol. 34, 3389–3391.

Langenbach, R., Rice, J., and Nesnow, S. (1981). "Organ and Species Specificity in Chemical Carcinogenesis." Plenum, New York.

Loury, D. J., Smith-Oliver, T., Strom, S. C., Jirtle, R. L., Michalopoulos, G., and Butterworth, B. E. (1986). Assessment of unscheduled and replicative DNA synthesis in hepatocytes treated in *vivo* and *in vitro* with unleaded gasoline or 2,2,4-trimethylpentane. *Toxicol. Appl. Pharmacol.* **85,** 11–23.

McQueen, C. A., Maslansky, C. J., Glowinski, I. B., Crescenzi, S. B., Weber, W., and Williams, G. M. (1982). Relationship between the genetically determined acetylator phenotype and DNA damage induced by hydralazine and 2-aminofluorene in cultured rabbit hepatocytes. *Proc. Natl. Acad. Sci. U.S.A.* **79,** 1269–1272.

Michalopoulos, G., Russell, F., and Biles, C. (1979). Primary culture of hepatocytes on human fibroblasts. *In Vitro* **15,** 796–806.

Michalopoulos, G., Strom, S. C., Novotny, A. R., Novicki, D. L., and Jirtle, R. L. (1985). Normal human hepatocytes in primary culture applications in studies of human carcinogenesis. *In* "In Vitro Models for Cancer Research" (M. M. Weber, ed.), Vol. 2, pp. 9–21. CRC Press, Boca Raton, Florida.

Miyazaki, K., Takaki, R., Nakayama, F., Yamauchi, S., Koga, A., and Todo, S. (1981). Isolation and primary culture of adult human hepatocytes. *Cell Tissue Res.* **218,** 13–21.

Moore, C. J., and Gould, M. N. (1984). Human hepatocyte and breast cell activation of suspected mammary carcinogens: Mutagenesis and metabolism. *Proc. Am. Assoc. Cancer Res.* **25,** 95.

Moore, C. J., and Gould, M. N. (1985). Metabolism of benzo(a)pyrene by cultured hepatocytes from multiple donors. *Carcinogenesis* **5,** 1577–1582.

Neis, J. M., Yap, S. H., VanGemert, P. J. L., Roelofs, H. M. J., and Handerson, P. T. H. (1985). Mutagenicity of five arylamines after metabolic activation with isolated dog and human hepatocytes. *Cancer Lett.* **27,** 53–60.

Novicki, D. L., Irons, G. P., Strom, S. C., Jirtle, R. L., and Michalopoulos, G. (1982). Cryopreservation of isolated rat hepatocytes. *In Vitro* **18,** 393–399.

Reese, J. A., and Byard, J. L. (1981). Isolation and culture of adult hepatocytes from liver biopsies. *In Vitro* **17,** 935–940.

Selkirk, J. K. (1977). Divergence of metabolic activation systems for short-term mutagenesis assays. *Nature (London)* **270,** 604–607.

Steinmetz, K. L., Spak, D. K., Green, C. E., and Mirsalis, J. C. (1985). Induction of unscheduled DNA synthesis in primary cultures of hamster, rat, mouse and human hepatocytes. *Toxicologist* **5,** 20.

Stoner, G. D., Daniel, F. B., Schenck, K. M., Schut, H. A. J., Goldblatt, P. J., and Sandwisch, D. W. (1982). Metabolism and DNA binding of benzo(a)pyrene in cultured human bladder and bronchus. *Carcinogenesis* **3,** 195–201.

Strom, S. C., and Kornbrust, D. (1987). Induction of DNA repair in rat and human hepatocytes by dinitropyrenes. In preparation.

Strom, S. C., Jirtle, R. L., Jones, R. S., Novicki, D. L., Rosenberg, M. R., Novotny, A., Irons, G., McLain, J. R., and Michalopoulos, G. (1982). Isolation, culture and transplantation of human hepatocytes. *JNCI, J. Natl. Cancer Inst.* **68,** 771–778.

Strom, S. C., Jirtle, R. L., and Michalopoulos, G. (1983a). Genotoxic effects of 2-acetylaminofluorene on rat and human hepatocytes. *Environ. Health Perspect.* **49,** 165–170.

Strom, S. C., Novicki, D. L., Novotny, A., Jirtle, R. L., and Michalopoulos, G. (1983b). Human hepatocyte-mediated mutagenesis and DNA repair activity. *Carcinogenesis* **4,** 683–686.

Strom, S. C., Novotny, A., Michalopoulos, G., Shank, R., Castonguay, A. (1986). Metabolism, mutation induction and DNA repair induced by rat or human hepatocytes exposed to dimethylnitrosamine or 4-methylnitrosamino)-1-(pyridyl)-1-butanone (NNK). *Proc. Am. Assoc. Cancer Res.* **27,** 380.

Wang, P. P., Beaune, P., Kaminski, L. S., Dannan, G. A., Kadlubar, F. F., Larrey, D., and Guengerich, F. P. (1983). Purification and characterization of six cytochrome P-450 isozymes from human liver microsomes. *Biochemistry* **22,** 5375–5383.

Index